HEAT PIPES: CONSTRUCTION AND APPLICATION

A Study of Patents and Patent Applications

A literature study compiled for the Commission of the European Communities, Technological Information and Patents Division of the Directorate-General Telecommunications, Information Industries and Innovation, Luxembourg.

HEAT PIPES: CONSTRUCTION AND APPLICATION

A Study of Patents and Patent Applications

Edited by

MARTEN TERPSTRA

and

JOHAN G. VAN VEEN

The Hague, The Netherlands

ELSEVIER APPLIED SCIENCE
LONDON and NEW YORK

ELSEVIER APPLIED SCIENCE PUBLISHERS LTD
Crown House, Linton Road, Barking, Essex IG11 8JU, England

Sole Distributor in the USA and Canada
ELSEVIER SCIENCE PUBLISHING CO., INC.
52 Vanderbilt Avenue, New York, NY 10017, USA

WITH 382 ILLUSTRATIONS

Softcover reprint of the hardcover 1st edition 1987

British Library Cataloguing in Publication Data

Terpstra, Marten
Heat pipes: construction and application:
a study of patents and patent applications.
1. Heat pipes—Design and construction—
Research
I. Title II. Veen, Johan G. van
621.402'5 TJ264

ISBN-13: 978-94-010-8043-9 e-ISBN-13: 978-94-009-3447-4
DOI: 10.1007/978-94-009-3447-4

Library of Congress CIP data applied for

Publication arrangements by Commission of the European Communities, Directorate-General Telecommunications, Information Industries and Innovation, Luxembourg

EUR 10925 EN

CONTENTS

INTRODUCTION

This analysis of invention is the result of a study of patents, patent applications and some other technical and scientific literature published in the United States, Japan, Great Britain, Western Germany and France, as well as European (Munich) and PCT (Patent Cooperation Treaty) patent applications, in the period from about January 1975 to about January 1985. It provides a systematic review of research activities on the structural and operational development of heat pipes in the period covered, and on their applications.

The study is in two parts: Part I relating to aspects of individual heat pipes and Part II concerning the application of heat pipes to heat exchangers and mechanical and electrical components, in which the construction and operation of the individual heat pipes are of secondary importance.

The material studied during the search has been derived from the search files of the European Patent Office, Rijswijk, The Netherlands, classified in accordance with the following international patent classes (IPC):

F 28 D 15/02	relating to heat exchange apparatus with the intermediate heat transfer medium in closed tubes passing into or through the conduit walls in which the medium condenses and evaporates, e.g. heat pipes.
F 24 J 2/32	relating to heat pipes of solar heat collectors.

G 21 D 5/02 relating to arrangements of nuclear reactor and engine, in which reactor produced heat is converted into mechanical energy and in which reactor and engine are structurally combined.

G 21 D 7/04 relating to arrangements for direct production of electric energy from fusion or fission reactions using thermo-electric elements,

comprising altogether about 700 basic patents and patent applications. The preselection of the literature studied and the delivery of copies of the respective documents was done by TNO-NIDER, Rijswijk, The Netherlands, while an orientating study was carried out by the authors in the library of the Technical University of Delft, The Netherlands. The authors thank both institutions for their cordial contribution to this work.

From the study, it became apparent that since 1975 the development of heat pipes has been directed to the increase of the heat transferring capacity of these apparatus utilizing latent evaporating and condensing heat. In this respect, dependant upon the use of the respective heat pipes, new materials for casings, liquids and wicks have been proposed. The use of ceramics, sintered metal powder and specific nettings having increasing or decreasing capillarity along their lengths are striking proposals.

Regarding the heat transferring liquids, it appears that water is still preferred, but it has drawbacks in its restricted application and in the fact that together with a steel casing it may form hydrogen.

As a consequence some inventions deal with the so-called hydrogen getters and with measures to avoid the entrance of air or other impurities during evacuation of the pipes.
Furthermore, many measures have been developed to obviate the formation of vapour bubbles which substantially affect the heat transfer.
The second part (Chapter 5) deals with applications of heat pipes to various heat exchanging devices and apparatus. It is obvious that a major part of such appliances is directed to direct heat exchangers such as radiators for heating purposes, solar heat exchangers, air conditioning, etc.
It will be observed that many applications of heat pipes to solar energy converting systems were published in the period from 1976 to 1980. Obviously the interest in solar heat exchangers has somewhat decreased during the last few years.
On the other hand, however, the interest in so-called indirect heat exchanging by heat pipes is increasing, which will appear from the increasing number of patent applications on dissipating heat from electrical equipment, particularly from electronic components such as semiconductors and integrated circuits.
The authors trust that the present report may be of some help to all who are concerned with research on and the construction of heat pipes, irrespective of whether their work is to do with the development of new pipes or with the application of heat pipes to thermodynamic or thermostatic apparatus.

Marten Terpstra

Johan G. van Veen

CHAPTER 1

CONSTRUCTION, OPERATION, COATINGS, HEAT TRANSFERRING LIQUIDS, SEALINGS AND JOINTS

This chapter deals with inventions which specifically are directed to some embodiments of heat pipe constructions and auxiliary means thereof developed the last ten years.
Apart from a few surprising new ideas, little changes have been applied to the well-known principles and construction of the heat pipe. The attention of the inventors appeared to be directed to improving heat tranferring efficiency and capillarity, the latter either by special wicks or by roughening or threading the inner walls of the pipes.

1.1 Construction and operation of heat pipes

To extract heat from e.g. waste water, Stephan, E. and Stiller, N., (Buderus AG), 25.3, designed a heat pipe surrounding two superposed vessels 1 and 2, vessel 1 containing cold or fresh water and vessel 2 hot or waste water. To heat fresh water in vessel 1 to a desired temperature, an additional heat source 15 in the form of a heating coil is mounted in the upper region of vessel 1.
See Fig. 1.
To transfer heat from an upper to a lower location, Kosson, R. of Grumman Aerospace Corporation,(82.11), designed a down pumping heat pipe combined with a mechanical pump and motor to provide fluid

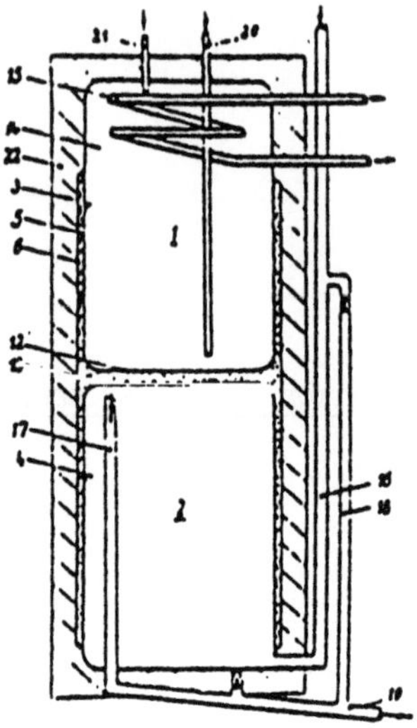

Fig. 1

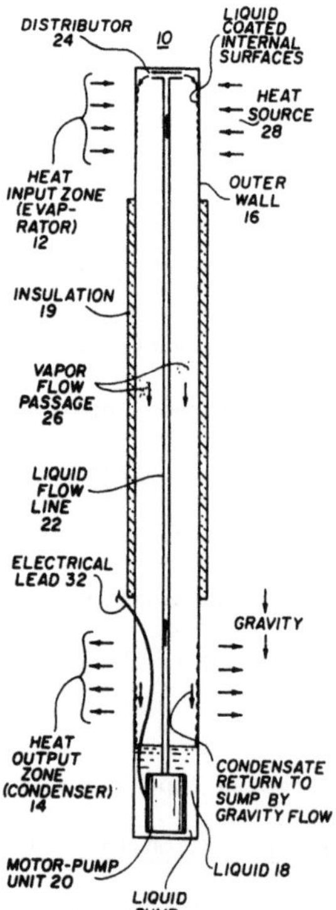
DISTRIBUTOR 24
10
LIQUID COATED INTERNAL SURFACES
HEAT SOURCE 28
HEAT INPUT ZONE (EVAP-RATOR) 12
OUTER WALL 16
INSULATION 19
VAPOR FLOW PASSAGE 26
LIQUID FLOW LINE 22
ELECTRICAL LEAD 32
GRAVITY
HEAT OUTPUT ZONE (CONDENSER) 14
CONDENSATE RETURN TO SUMP BY GRAVITY FLOW
MOTOR-PUMP UNIT 20
LIQUID SUMP
LIQUID 18

Fig. 2

transport. The device is entirely encased within a single vertically disposed casing and is shown in Fig. 2.

A cryogenic heat pipe that may be stored at normal room temperatures and pressures is developed by Basilius, A. (Hughes Aircraft Co.), 90.1.

The cryogenic working fluid is caused to expand into a storage reservoir. To activate the heat pipe from storage conditions, the condenser region must be cooled down to cryogenic temperatures. The cryogenic vapour in the system forms a condensate in the wick and only superheated gas remains in the reservoir. The adiabatic region of the heat pipe is covered with a multi-foil-vacuum super-insulation section to minimise ambient temperature effects when the device is activated and operating.

Groke, U.,(Kernforschungsanlage Julich GmbH), 103, tried to improve the efficiency of a heat pipe by mounting a collecting reservoir 3 (figure 3) into the heat pipe, which reservoir is in communication with the outer wall 2 by a tube 4 which is connected to the wall by a connecting element 7. See Fig. 3.

Fig. 3

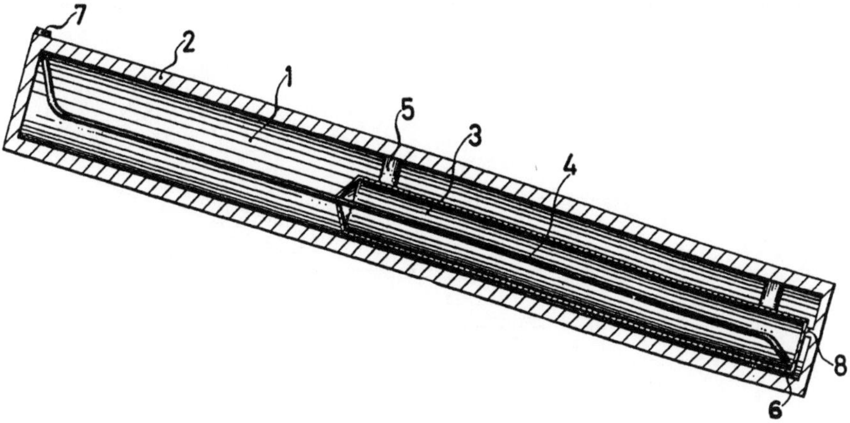

The efficiency of a closed evaporation and condensation system of a heat pump can be improved by using a heat pipe interconnecting the condenser and the evaporator, so that both are located apart from each other. The thermal efficiency of such a pump, according to Laude, F., (109), can still be improved by disposing heat exchanging balls into both the evaporator and the condenser.

The starting ability of a heat pipe constructed by White, W.J. of National Nuclear Corporation Limited, (126.1 and 126.2), having an inner tube 12 separating the vapour-phase path (inside the inner tube) from the liquid-phase return path (annularly outside it), under "negative slope" conditions,i.e. with the heat pipe extending (slightly) upwardly towards it heat-absorbing and 10A (Fig. 4), is improved by providing not only a bend 10C so that a part of the heat pipe also extends upwardly towards the heat-releasing end 10B but also by providing a reservoir 15 communicating, in that part, with the vapour-phase path (instead of the liquid-phase path). A further improvement is then obtained by providing (i)at the heat-absorbing end a weir 18 tending to separate liquid surface levels in the liquid-phase and (flooded) vapour-phase paths and (ii) a saddle-like blocking element 19,blocking the upper part of the annular liquid-phase path near the said bend.

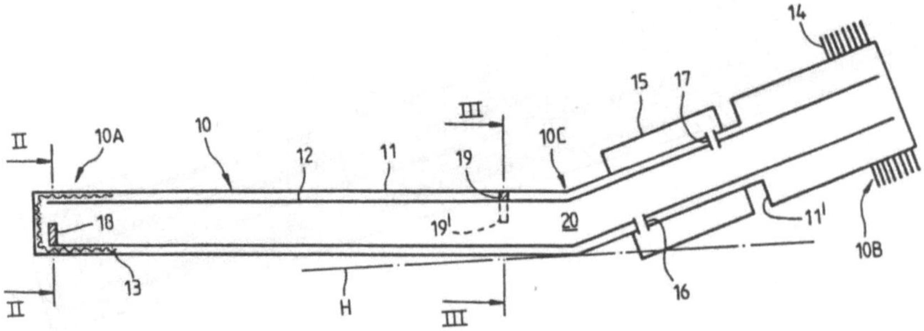

Fig. 4

The 126.1 embodiment can be improved by replacing the reservoir 15 by a pair of weirs.

A heat-transporting device of Kiseev, V.M. et al (Otdel Fiziko-Teknicheskikh Problem Energetiki Uralskogo Nauchnogo Tscentra Akademii Nauk SSSR, (132.1), includes an evaporating chamber containing a coaxially-arranged evaporator in a capillary material soaked with a heat-transfer agent. The evaporator is in thermal contact with a source of heat and has an axial bore with a transverse partition in the capillary material, a vapour-jet pump serving to transform the dynamic pressure of the heat-transfer agent in the vapour phase into the static pressure of the heat-transfer agent in the liquid phase, and a heat-exchanging chamber. The evaporating chamber is provided with two end face cavities each bounded by the corresponding end face of the evaporator and walls of the chamber. The partition in the capillary material is located contiguously with that end face of the evaporator which faces the heat-exchanging chamber and is provided with through-holes placing the end face cavities in communication with one another and is further provided with a diametrical passage, which is connected to a nozzle of the vapour-jet pump. The nozzle is located in the partition and also to a vapour header which, in its turn, is connected to vapour outlets. These outlets are provided in the form of longitudinal grooves cut in the outside surface of the evaporator between smooth annular collars,which are present on the outside surface of the evaporator at the end faces thereof to prevent vapour leaks into the end face cavities. An outlet of a first conduit is located in the axial bore of the evaporator. A zone of the heat-exchanging chamber containing the heat-transfer agent with a lower heat content is connected to a suction side of the vapour-jet pump by way of the first conduit and

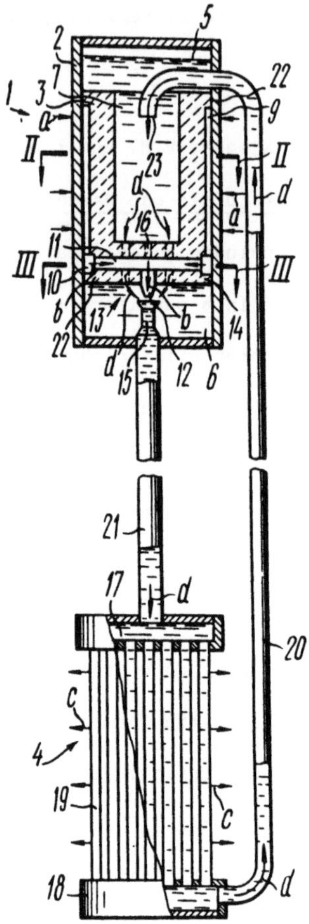

Fig. 5

another zone of said chamber containing the heat-transfer agent with a higher heat content is connected to a discharge side of the pump through a second conduit.

Fig. 5 shows a longitudinal cross section of the Kiseev device.

A somewhat similar heat transferring device is subject of (132.2), a longitudinal cross section of which being shown in Fig. 6.

Nolan, J.P.C., (Nolan Design Limited), (130), has tried to modify the Perkins tube by using a special enclosure which defines at least one vaporiser and condenser and which is designed so that condensate is free to return to the vaporiser unimpeded by wicks or other capillary

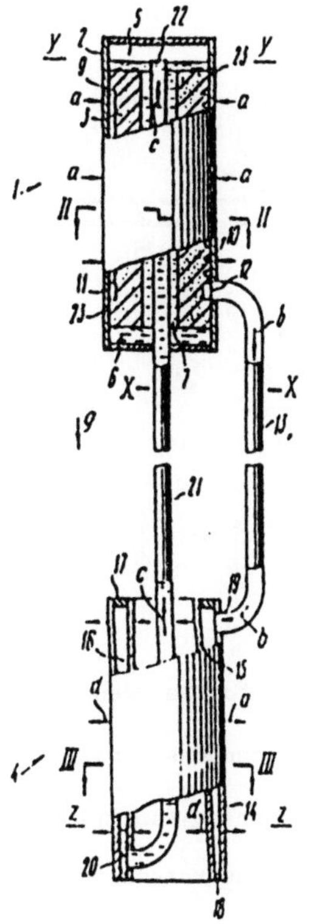

Fig. 6

means. The enclosure communicates with an openable sealing which has at least a joint or a removable cap or plug, or a valve, such as a ball valve, and a liquid trap containing a liquid compatable with the fluid in the enclosure. The trap is arranged such that the liquid therein will enter the enclosure in the event of a leak in the sealing. The adavantage of Nolan's heat pipe resides in that the device can be disassembled by the user to facilitate and to reduce the cost of maintenance.

A heat pipe provided by Fitton, G.L. of Redpoint Limited, (143.4), has a flow path for the operating fluid which includes at least one duct 15 or 22 (Fig. 7) which, within the evaporator region of the heat pipe, has an internal cross-sectional area which increases along the duct in the direction of fluid flow. The duct may be either a liquid return

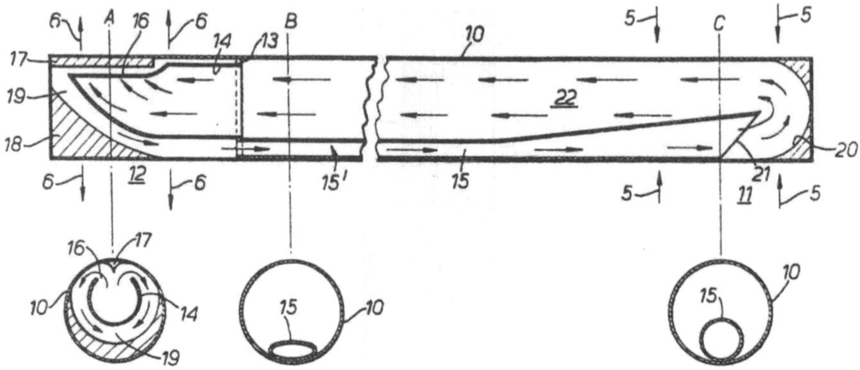

Fig. 7

pipe 15 for conveying liquid to the evaporator region or a duct 22 for conveying vapour to the condenser region of the heat pipe. Within the condenser region, a condenser tube 14 may lead to a duct 19 having a cross-sectional area which decreases in the direction of fluid flow. Preferably, the liquid return tube 15 is impermeable to the operating fluid.

A thermosyphon according to Chisholm, D.,(The Secretary of State for Industry),(159.3), may be used to transfer heat from two heat sources, one each adjacent the top and bottom of the thermosiphon vessel, to a heat sink adjacent an intermediate point along the vessel, heat at the heat sources being absorbed by evaporation of liquid in respective evaporators at the top and bottom of the vessel and heat being given up by condensation of vapour in a condenser at the intermediate point along the vessel. The liquid from the condenser is transferred to the evaporator at the bottom of the vessel by gravitation, and liquid from the condenser is transferred to the evaporator at the top of the vessel by preferably a multistage vapour lift pump.

Fig. 8 shows a vertical cross section of Chisholm's thermosiphon.

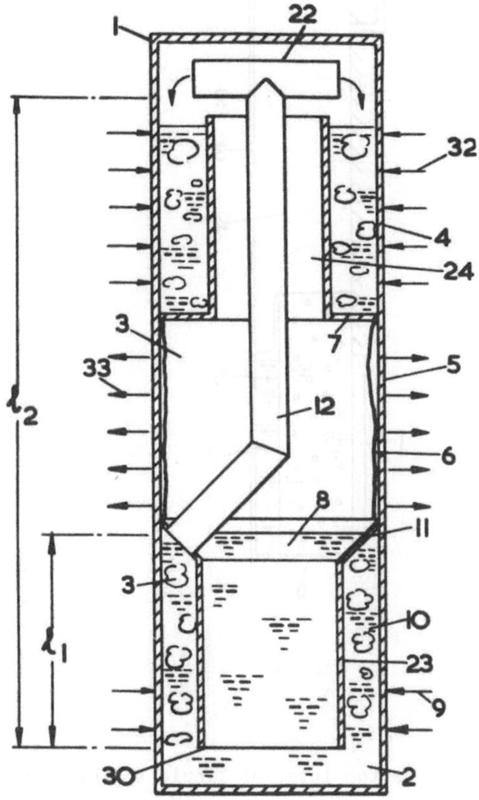

Fig. 8

One stage vapour lift pump embodiment.

Chisholm together with Chirnside, I., (159.4), also developed a thermosiphon which is shown in Fig. 9.

It comprises a cavity and a working surface 2 which is in thermal contact with a heat transmitting surface 3 of the cavity. The other walls of the cavity are thermally insulated. Vertical baffle 2 extends from one side wall of the cavity to the other, its bottom edge being a distance above the floor 6 of the cavity and its top edge at a higher level than the top of the heater.

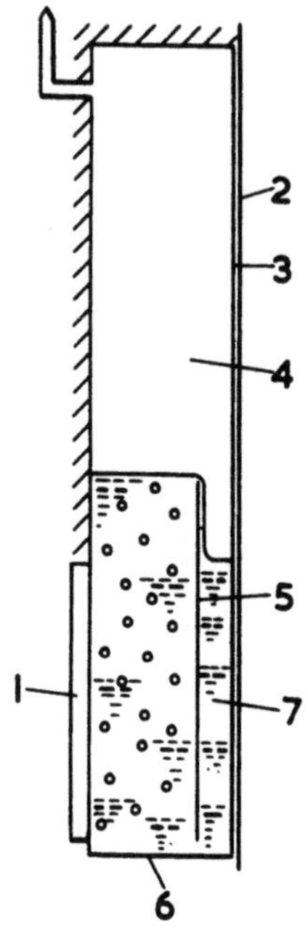

Fig. 9

During boiling, as soon as the level of liquid in the heated region rises to the top of the baffle, the liquid passes into the region adjacent to the wall 3 to set up a steady circulation of liquid downwards in that region and back under the lower edge of the baffle to the heated region. All the liquid present in the cavity therefore rapidly reaches the boiling point, and temperature variations on the working surface 2, which might be produced by turbulent flow and lack of mixing in the absence of a baffle, are kept to a minimum. Consequently, the working surface 2 is, in use, substantially isothermal.

A heat pipe whose fluid can be repeatedly frozen and thawed without damage to the casing has been constructed by D.M. Ernst, and J.L. Sanzi of Thermacore, Inc., (181.6).

An additional part is added to a conventional heat pipe.

This addition is a simple porous structure, such as a cylinder, self-supporting and free standing, which is dimensioned with its diameter not spanning the inside transverse dimension of the casing, and with its length surpassing the depth of maximum liquid.

A catalytic cartridge surrounding a heat pipe driven by a heat source is utilized as a SO_3 decomposer for thermochemical hydrogen production and has been designed by Galloway, T.R. of The United States of America, represented by the US Department of Energy, (193.3).

The cartridge has two embodiments, a cross-flow cartridge and an axial flow cartridge. In the cross-flow cartridge, SO_3 gas is flowed through a chamber and incident normally to a catalyst coated tube extending through the chamber, the catalyst coated tube surrounding the heat pipe. In the axial-flow cartridge, SO_3 gas is flowed through the annular space between concentric inner and outer cylindrical walls, the inner cylindrical wall being coated by a catalyst and surrounding the heat pipe.

Fig. 10 shows the cross flow embodiment.

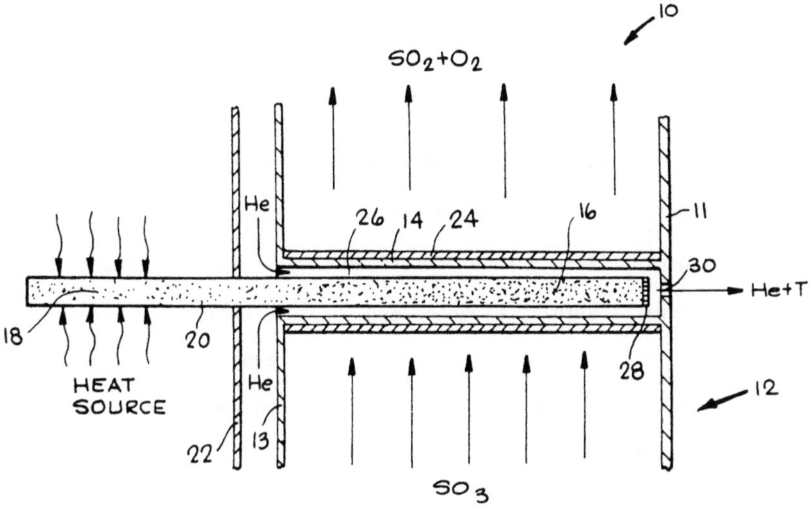

Fig. 10

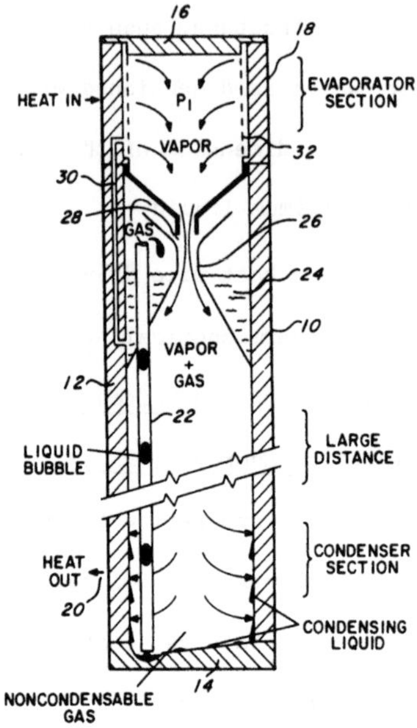

Fig. 11

An entirely passive heat pipe capable of operating against gravity and developed by Koenig, D.R. of the US Department of Energy, Patent (193.4), is shown in Fig. 11.

A return conduit into which an inert gas is introduced is used to lower the specific density of the working fluid so that it may be returned a greater vertical distance from condenser to evaporator.

Ions producing electrodes

According to Borgoyn, M.J. of the US Secretary of the Navy, Patent (196.1), the performance of a heat pipe can be improved by mounting electrodes either in the working fluid vapour or its condensate. The ion flow resulting therefrom is axially directed and, through collision phenomena, picks-up the surrounding low velocity stream, increases its momentum and generates additional pumping pressure for the condensate. Borgoyn states that the performance can be improved even when low surface tension working fluids are used.

Fig. 12 shows a longitudinal section of Borgoyn's heat pipe.

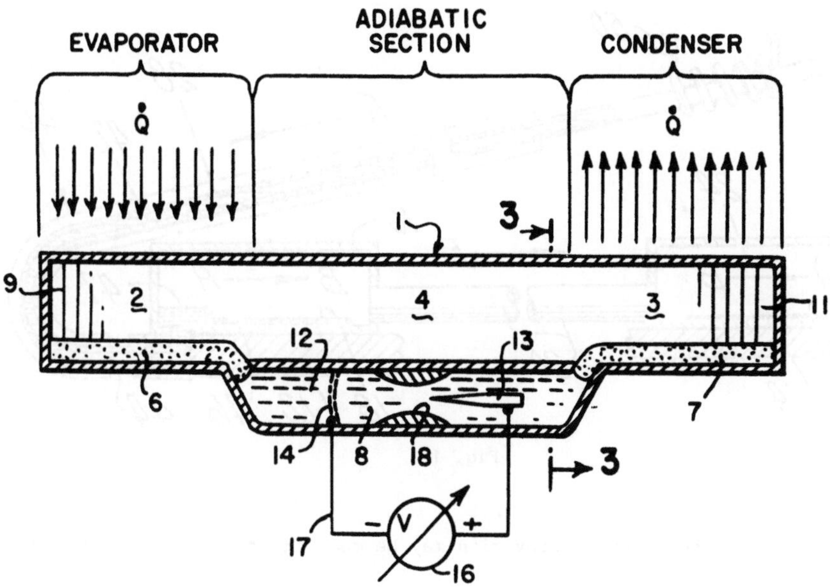

Fig. 12

Chemical heat pipe

In Figure 13 a self-driven chemical heat pipe is shown, which has been constructed by Kesten, A.C. and Haught, A.F. of United Technologies Corporation, (198.3).

The circuit includes an endothermic reaction chamber 16 and an exothermic chamber 24 connected by a pair of arms 20, 28 extending therebetween. Reactant A is endothermically reacted near the heat source 12 to create reaction product B at a pressure greater than exists at the exothermic reaction chamber 24 to promote flow thereof through one of the arms 20. Reaction product B is exothermically reacted in the exothermic reaction chamber 24 near the heat sink 14 to liberate heat and form recombined gaseous reactant A. The gaseous reactant A is converted to liquid form 42, which liquid-form reactant occludes part of the other arm 28 and flows therethrough toward the endothermic chamber 16 to complete the cycle. The liquid occlusion of the arm 28 ensures adequate unidirectional

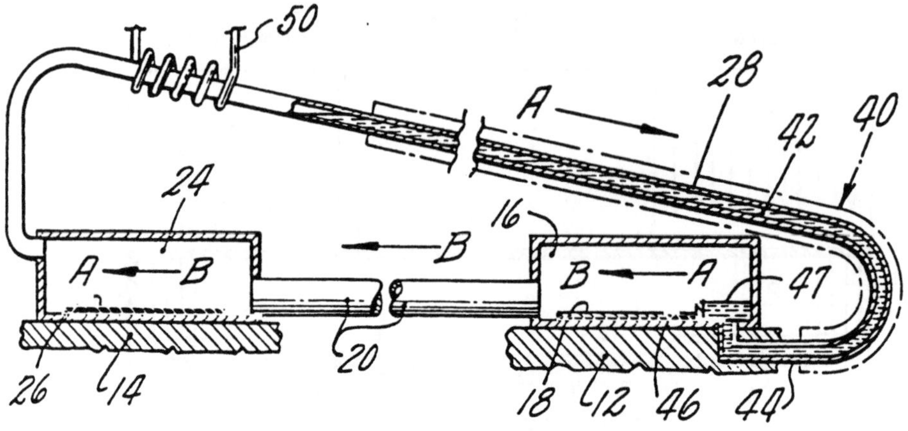

Fig. 13

flow about the circuit. Provision may be made for selectively reversibly storing reaction product B or reactant A to accommodate differences in time between heat production at the source 12 and heat demand at the sink 14.

Avoiding impurities collection

During the operation of a heat pipe impurities may deposit at an end of the heat pipe, which may disturb its operation. To avoid these difficulties Koch, H. and Breeb, H. of Dornier System GmbH., (47.2), proposed to have a heat pipe ready to be filled with the heat transferring fluid provided with a vacuum valve which has to be disposed near the condensating area. A predetermined amount of the heat carrying fluid is directly introduced in the heat pipe, which amount being larger than needed for the operation of the heat pipe. During the operation of the heat pipe rest gases and impurities together with the surplus of heat carrying vapour will be collected at the condensating area and discharged by the vacuum valve until the weight of the heat pipe together with the heat carrier fluid has reached a predetermined value.

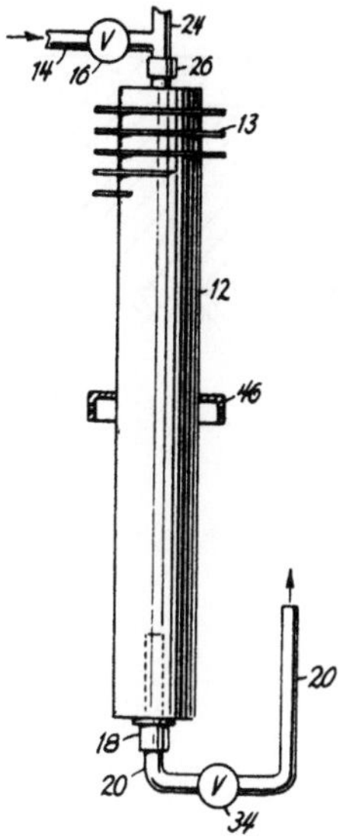

Fig. 14

In this connection, the method of Stockman, R.F. of Combustion Engineering, Inc., (33), is deemed to be of interest. In Fig. 14 a heat pipe 12 is shown into which a quantity of vaporizable fluid 10 is admitted through an inlet assembly to displace residual air in the heat pipe for exhaust through vent 24. When the heat pipe is full of fluid, The inlet assembly including valve 16 and fitting 24 is removed and coupling 26 plugged. Pump 26, having a section connection to the heat pipe, is actuated and fluid is withdrawn from the heat pipe through check valve 18 leaving an evacuated space to form above a predetermined quantity of vaporizable fluid. A cylindrical standpipe 38 above check valve 18 precludes the evacuation of vaporizable fluid from the heat pipe lower than the upper rim of the standpipe.

Auxiliary evaporator

The heat pipe designed by Faccini, E.C. (57), and adapted to be used in roadways, bridges, etc., includes an auxiliary evaporator formed

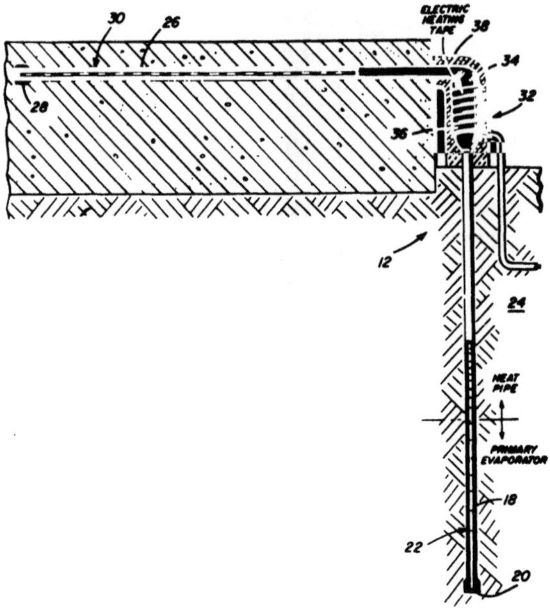

Fig. 15

concentrically with the upper end of a vertically disposed primary evaporator portion. The auxiliary evaporator portion comprises an annular sleeve disposed about the upper end of the primary evaporator portion and arranged between the primary evaporator portion and the condenser part of the heat pipe so that all of the condensed working fluid returning to the primary evaporator portion must enter and overflow the auxiliary evaporator portion prior to return to the primary evaporator portion. The auxiliary evaporator is provided with a heater whereby the auxiliary evaporator may function even in the absence of heat pipe function by the primary evaporator.

Fig. 15 shows the Faccini heat pipe.

1.2 Heat pipe claddings or coatings

Some inventors have directed their attention to claddings or coatings of heat pipes to protect them from corrosive influences of some fluids into which they have sometimes to be placed.

Usually, heat pipes are being made of steel, aluminium, copper or their alloys, refractive metals and so on. The choice of construction materials is mainly determined by the operation temperature and the corrosive action of the fluids surrounding the heat pipes. Possible reactions between a construction metal and water, if this is used as the heat conducting fluid, are for example:

$$Me + 2H_2O \longrightarrow Me(OH)_2 + H_2$$

Hydrogen will collect in the form of a gas cushion which detrimentally affects the metals from which the heat pipe has been made. Some researchers have tried to find means to avoid the above disadvantages in the last ten years.

Willmann, G. of Dornier System GmbH., (47.6), proposes to coat the inner wall of a silicon infiltrated silicon carbide heat pipe by annealing and degassing the heat pipe at a temperature range from 700°C to 1400°C, preferably at about 1000°C, in an atmosphere having an oxygen partial pressure of above 10^{-8} bar. The annealing time is preferably between 0.5 to 10 hours.

Eastman, G.Y. of Thermacore Inc., (181.1) suggested to form an aluminium oxide surface layer on an aluminium bearing steel alloy by heating the steel at 700°C for one hour.

Eastman, (181.10), also proposes to cover the steel casing of a heat pipe used in a fluidized bed of limestone and coal, with a thin aluminium oxide layer to prevent hydrogen permeation into the casing and have a plasma sprayed coating of ceramic on the portion within the fluidized bed to prevent erosion of the heat pipe casing by the solid particles within the bed.

1.3 Heat transferring liquids

In the period covered by this prior art study some inventors paid their attention to the kind of liquids to be used as heat transferring fluid in heat pipes.
Although usually water is preferred in that its latent evaporation heat is high with respect to that of many other liquids, quite often it is necessary to make use of those liquids dependent upon the working temperatures of the heat pipes.
If it is desired to transfer heat energy from one liquid to a second liquid having other properties, then a heat exchanger has to be used where by means of mechanical separating walls the heat energy is transferred from the first or primary liquid to the second one.

According to Henning, E., (88), the primary medium is evaporized in the heat source and condensed in the heat exchanger, so that the primary medium is continuously kept at its boiling point or somewhat therebelow. In the heat exchanger the heat energy is more quickly transferred to the secondary medium in that the primary medium in gaseous form is fed into the heat exchanger and condensed, and is returned to the heat source at elevated temperature. As a primary medium, preferably freon, ammonia or the like is used.

Vanadates

Hashimoto, M. of the Showa Aluminium Corporation, (163.3), uses an aqueous solution of a vanadate as a working liquid. Examples of suitable vanadates are ammonium, sodium, potassium and lithium metavanadate. The aqueous vanadate solution preferably has a concentration of from 0.1% by weight to saturation concentration.

Hara, Tsohitsuga et al, of Hitachi Ltd., (89.2), provided a heat exchanging system containing a liquid having a low evaporating or boiling point and a non-condensable gas, the liquid and the gas being stored within a container which is divided into a heating area and a cooling area by an adiabatic member to have the liquid be evaporated or boiled at temperatures higher than the desired temperature.

Magnetically susceptible working fluid

A heat pipe containing a magnetically susceptible liquid as the working fluid has been provided by Mayer, A.H. of the US Secretary of the Air Force (191.4).

The heat pipe is surrounded by an electro-magnet or parmanent magnet, which produces a magnetic field gradient which interacts with the magnetically susceptible liquid to produce an artificial body force field analogous to, but which may be substantially greater than, the gravitational force field, for separation of the liquid from gaseous phases of the magnetically susceptible working fluid within the heat pipe.

Fig. 16 shows the principles of the Mayer heat pipe.

As magnetically susceptible working fluid certain organic liquids such as benzene or hydrocarbons of the methane, ehtylene and the acetylene series of homologous organic compounds may be used.

The heat pipe combination of Arcella, F.G. of Westinghouse Electric Corp., (210.2), consisting of a common condenser section with evaporator sections at either end, contains two working fluids of different vapour pressures to form two heat pipe sections within the same cavity, to support an amplifier mode of operation. As working fluids methanol and water may be used.

Fig. 17 shows the construction of the Arcella heat pipe.

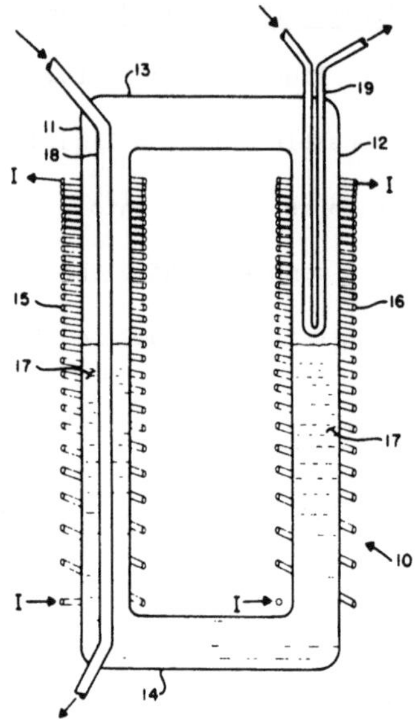

Fig. 16

A typical hydrogen getter proposed by Lamp, T.R. of Hughes Aircraft Company, (90.5), is tantalum and like metals, added to a water/ferrous metal heat pipe in the form of a foil, gauze or wire. The foil may have holes for vapour communication.

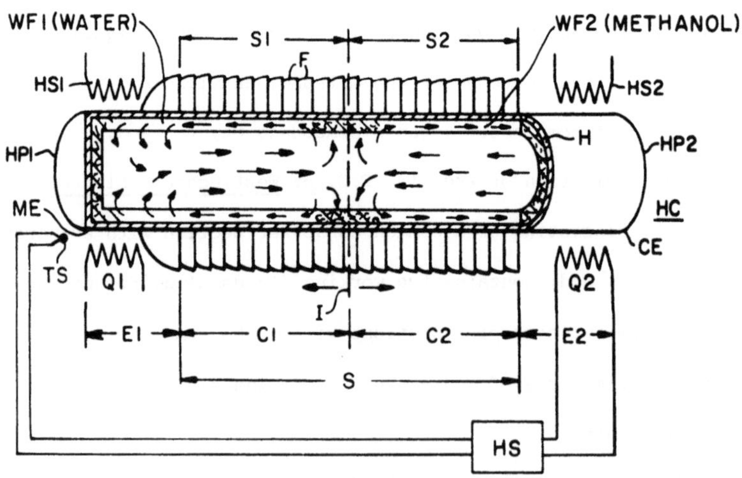

Fig. 17

1.4 Sealings and joints

Applegate, G. of Curwen & Newbery Limited, (38.4), designed a pinned heat pipe to be used in a heat exchanger. The heat pipe is provided with external transverse fins, and a sealing construction having an aperture of substantially the same shape and dimensions as the fins, through which aperture the pipe extends, and a sealing substance or body connecting st least two adjacent fins in the region of the wall and disposed between them, so as to form a seal or part of a seal between the heat pipe and the wall. Preferably, the sealing substance is plastic or putty.

The heat pipe housing wall through passage of Bährle, F. of Daimler Benz AG, (40.11), has particularly been designed for automotive engines. The passage is formed by an aluminium alloy flange moulded around the heat pipe after having machined off the outer ribbing in the flange area. The heat pipe consists of a string-pressable aluminium knead alloy and preferably, by intermediate ribs, contains several fluidly interconnected chambers. By mechanical treating a tight wall-integrated outer ribbing is formed.

The heat conductive heat pipe fitting made by Speidel, K. of Dornier System GmbH, (47.8), has, according to Fig. 18 a first construction part 1.3 which is flexibly sealed in a bushing 6 projecting into working fluid 4. Slit 14 between bushing 6 and construction part 1.3 is filled up by solder 13.

To improve the heat transfer of a heat pipe to a heat sink, Franke, J. of Erno Raumfahrtechnik GmbH, (53.1), designed a socket adapted to receive a heat pipe by a plug-in connection.

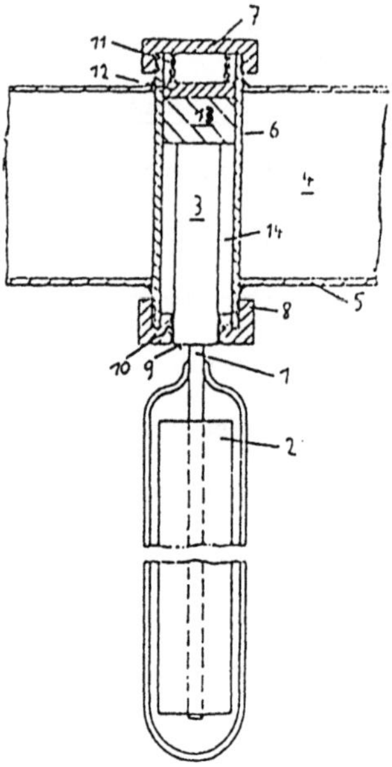

Fig. 18

This socket contains a sealing and the intermediate spaces thereof are filled up by a heat conducting material such as tin, solder etc.

Many heat pipes are manufactured by closing one end thereof by a sealing cap which is mounted after the pipe has been filled with a working liquid and has been evacuated.

To avoid this time-consuming and costly step of the manufacturing process, Jermyn, T.,(91.1) proposes to close at least one end of the heat pipe by an integral deformation of the open end thereof.

In a vacuum environment such as space, conductive ball and socket members are fixed to respective heat pipes for permitting orthogonal movement of one heat pipe relative to the other and the gap between the ball and socket members is maintained under a light gas pressure with the low pressure gas forming a low thermal impedance path across the gap.

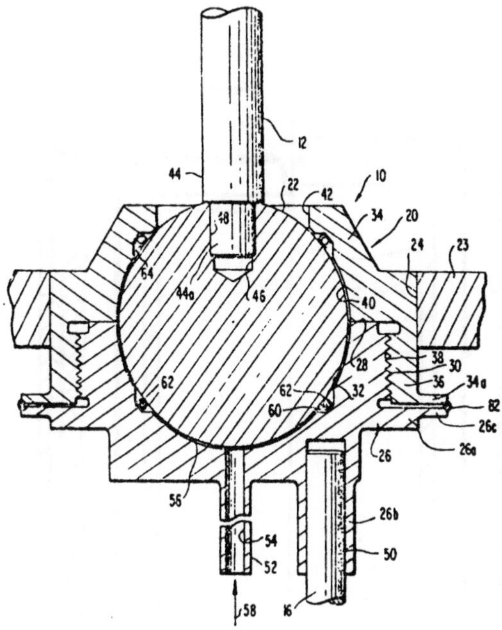

Fig. 19

Such a gas-filled swivel joint which particularly has been designed for cryogenic heat pipes, has been invented by Noyoryta, R.B. and Dawson, F.W. of Martin Marietta Corporation, (116.1).
Fig. 19 shows a cross section of the respective joint.
A multiple section heat pipe with swivel junctions between the sections has been developed by Ernst, D.M. and Copenhaver, R.L. of Thermacore Inc., (181.8).
The individual sections are independent individual heat pipes configured to interlock with each other at rotatable joints filled with high heat conductivity liquid. The clearance between the sections is maintained small to establish capillary forces that maintain the conductive liquid in place and also to minimize the temperature differential across the rotating joint. When used in conjunction with several right angle heat pipes, the combination furnishes multiple possibilities for end to end relationship of a heat pipe system.

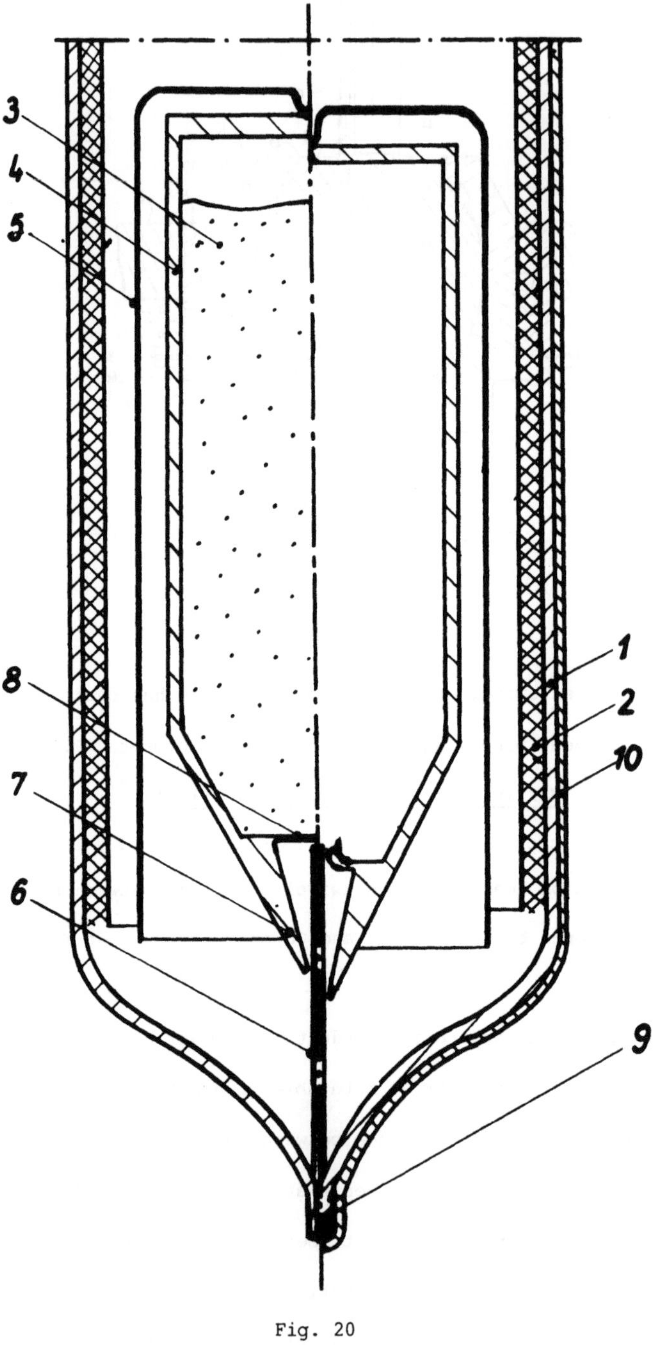

Fig. 20

For filling and sealing a heat pipe. Stiller, N. of Buderus AG, (25.1), found a method according to which filling of the heat pipe with a working fluid is effected so that the filled heat pipe can be heat-treated without disturbing of the heat transferring medium. To this end Stiller does not directly feed this medium into the heat pipe where it would be disturbed by high temperatures, but instead thereof it is fed through the container wall (Figure 20), before its direct contact with the tube wall.

In Figure 20, 1 represents the heat pipe, 2 a capillary structure, 3 a vapourisable medium, 4 a container, 5 a protecting sheath of aluminium foil, 6 a needle onto which container 4 is bearing; 7 a guider, 8 a very thin wall which easily can be penetrated.

CHAPTER 2

INVENTIONS MORE SPECIFICALLY DIRECTED TO CAPILLARITY

2.1 General aspects

As is known in the art, in heat pipe structures it is usual and quite often necessary to make use of the capillarity of various materials, or to have the inner surfaces of heat pipe be provided with ribs or grooves so as to roughen these surfaces and thereby providing a capillary system.
This section particularly deals with some inventions which are directed to combinations of capillary means and other structural measures to improve the heat transport efficiency of heat pipes.

Curven & Newbery Limited, (38.1), found that vibrating the heat pipe and its wick or wicks considerably improves the efficiency of the heat pipe.

According to Kreeb, H. and Koch, H. of Dornier System GmbH, (47.1), the inner wall surface of a heat pipe is covered with a capillary structure such as a wire screen, a sintered layer or the like which, in the direction of heat transport, is connected to and merges into a longitudinally grooved capillary structure of the heat pipe.

In order to prevent the vapour flow within a heat pipe from resisting the return of condensed, heat transfer liquid towards the evaporator section, the heat pipe of G. Rattcliff and P.I. Patrickson of G.B.P. Holdings Limited, (73), is arranged so that it has separate

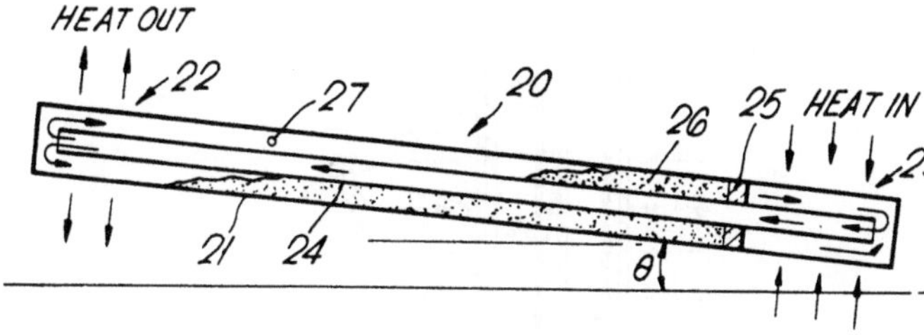

Fig. 21

paths for the vapour flow and liquid return, this being achieved by means of an open-ended tube 24 (Fig.21) located coaxially within the heat pipe. The tube 24 is supported by means of a semi-porous plug 25 which causes the liquid to collect as at 26 and slowly drain into the evaporator section. Numerous modifications of the heat pipe, including one where the condenser section is at a greater angle to the horizontal than the evaporator section, are also disclosed.

Where a wick for condensate return is provided, this may be sandwiched between the wall of the heat pipe and a coaxial inner tube to increase the area of plain wettable surface contact with the wick, to improve the wicking performance.

In Fig. 22 a heat pipe is shown, which has been designed by Basiulis, A. of Hughes Aircraft Company, (90.6). The heat pipe has an evaporator section 14, a condenser section 18, and within both sections are independent capillary wicks or grooves 20 and 22.
It is necessary that these wicks are separated; therefore, wick 20 ends at 24 and wick 22 at 26, thereby forming an adiabatic section 18.

Jermyn, T., Rattcliff, G. and Kemp, R.S., (91.2) proposed to have a heat pipe be provided with two concentric wicks, one of which (the inner one) extending the evaporation zone of the pipe, while the outer wick extends the whole length of the heat pipe.

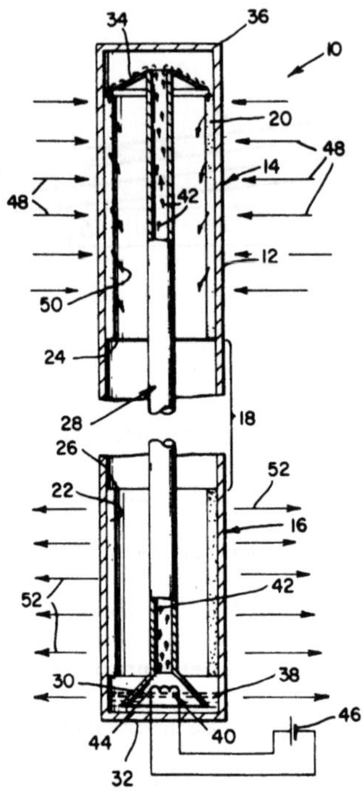

Fig. 22

Fig. 23 shows such an arrangement.

A heat pipe having porous metal wick parts formed by metal deposition, thus forming an enclosure and providing both a metallurgical bond to the wick and a hermetic seal, is subject of an invention by Token, K.H. and Garner, E.C. of McDonnell Douglas Corp., (121.10).

The heat pipe enclosure is formed by bridging the pores in the porous

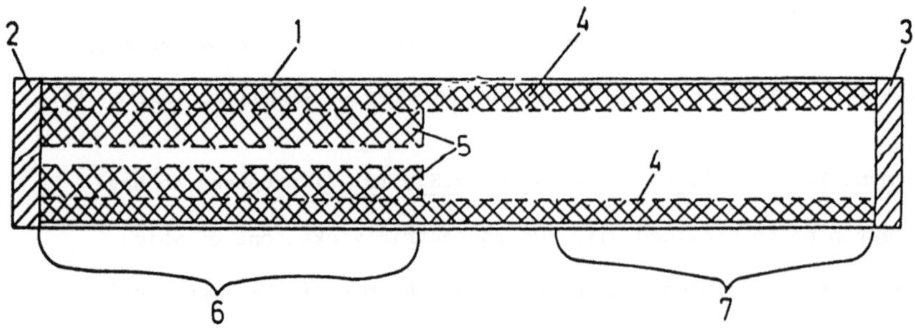

Fig. 23

materials and the joints between the metal pieces. A continuous metal hermetic seal is formed over the entire exterior of the heat pipe and increases its structural integrity. This hermetic container forms a one-piece case around all exterior surfaces of the wick and other parts and provides a metallurgical bond between the porous wick and the case so formed.

A heat pipe for fin coolers for cooling molten cones of glass drawn from a fibre glass manufacturing bushing, and having a wick containing a liquid therein, has been designed by Gunter, C.E. of PPG Industries, Inc., (140).

Figure 24 shows an embodiment of Gunter's heat pipe system, wherein 51 represents a header having means for the circulation of a cooling

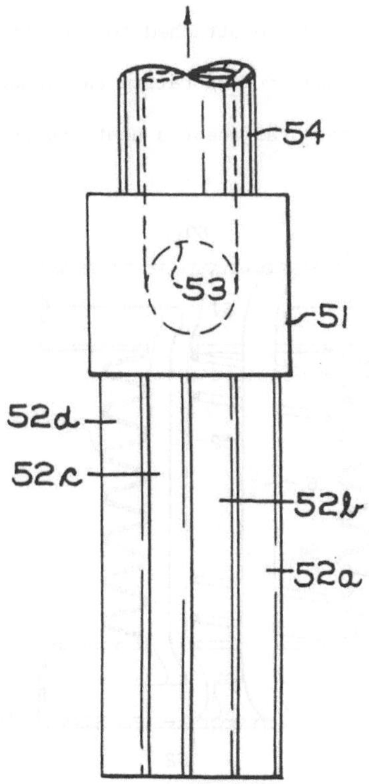

Fig. 24

fluid. Projecting from the header are sealed, platelike fins 52, in parallel rows, constructed and arranged to be positioned between rows of fibre glass bushing orifices or tips to thereby provide a flat surface on each side of a fibre glass bushing.

A thermal switch for controlling the dissipation of heat and having a bellow which defines an expansible vapour chamber, is invented by Wolf, D.A. of the National Aeronautics and Space Administration of the United States, (195.4).

Referring to Figure 25, inside the bellows (12) is located a coiled retaining spring (54) and four axial metal mesh wicks (46, 48, 50, 52), two of which (46, 48) have their central portions located inside of the spring (54) while the other two (50, 52) have their central portions located between the spring (54) and the side wall of the bellows (12). The wicks are terminated and are attached to the inner surfaces of the outer end walls (60, 62) of evaporation and condensation chambers (14, 16), respectively, located adjacent a heat source and heat sink.

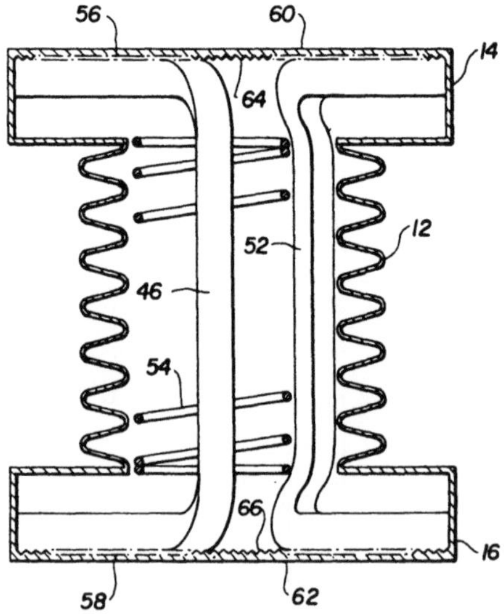

Fig. 25

The inner surfaces of the end walls furthermore include grooves (64, 66) to provide flow channels of the working fluid to and from the wick ends. The evaporation and condensation chambers are connected by turnbuckles and tension springs to provide a set point adjustment for setting the gap between an interface plate on the condensation chamber (14) and the heat sink (22).

According to Wulf, H. of Daimler Benz AG, (40.13), a heat pipe is provided which has a flexible wall 7, (Fig. 26), which at its ends is connected to a heat receiving section 1 and a heat dissipating section 13. Within the wall or flexible pipe 7 a capillary wick 8 is disposed which is spirally wound, so as to follow deformation of the wall during the operation of the heat pipe.

2.2 Details, wicks, wall surface ribs or grooves

In the following sections of the present literature study a number of inventions will be discussed which more specifically are directed to the construction, properties and materials of wicks, wall surface ribs or grooves.

2.2.1 Inventions directed to the construction and materials of wicks

Shinji Sawata et al. of Agency of Industrial Science & Technology, (2.1 and 2.3), designed a wick extending throughout the heat pipe wall and being disposed so that the resistance offered thereby to the flow of liquid gradually decreases from the heat input side to the heat output side.

The invention of Applegate, G., (7), resides in the fact that the heat pipe wick is formed from a number of layers in contact with, or close to the internal wall surfaces of the heat pipe. Each layer

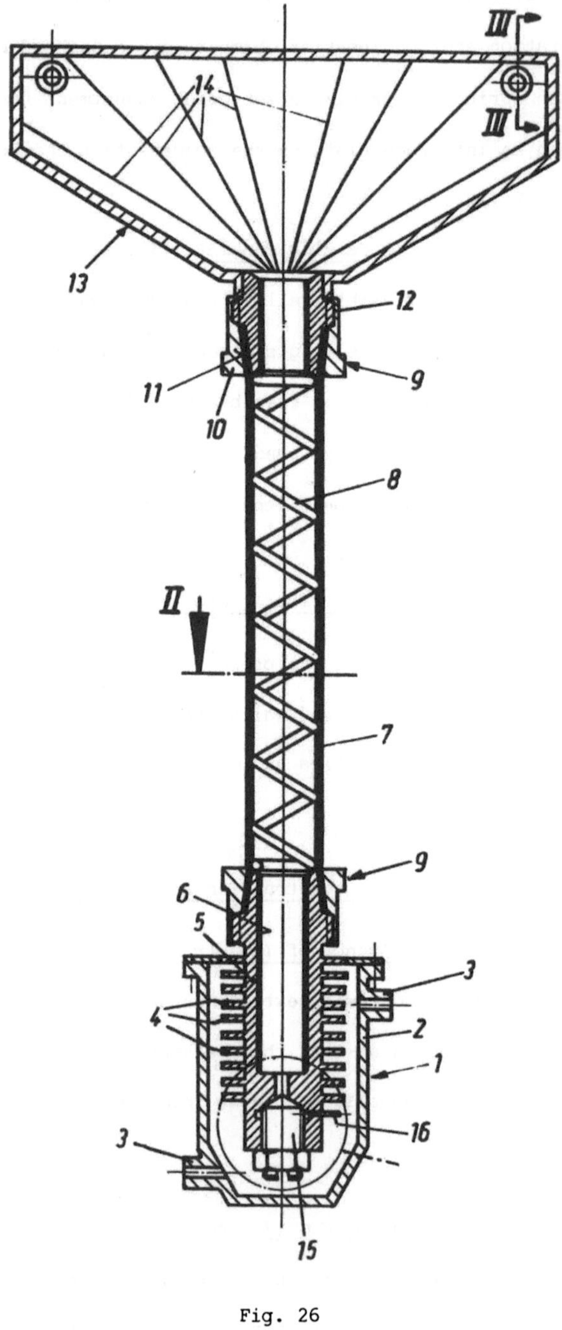

Fig. 26

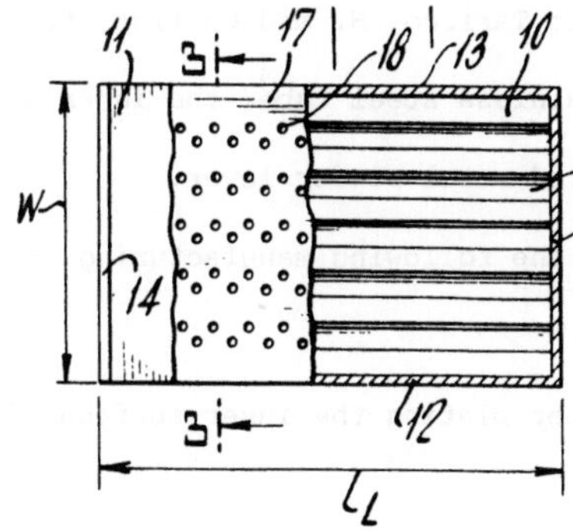

Fig. 27

may be of gauze, fabric, or non woven fibrous material and may be formed of metal fibres, glass fibres and other non-corroding material.

To avoid the formation of vapour bubbles during the operation of a heat pipe which at its inner surface is provided with grooves, Kreeb, H. and Koch, H. of Dornier System GmbH.,(47.11), use a capillary system covering the grooves so that a very large thermic flow occurs between the capillary covering and the contacting points thereof with the grooves.

An integral screen-wick structure providing a relatively large contact area between the internal working fluid or condensable medium and the heat input is an invention of Wayne, P. of E-Systems Inc., (54). The basic principles of Wayne's structure are shown in Figures 27 and 28 , the latter being a cross section taken at the line 3-3 of Figure 27.

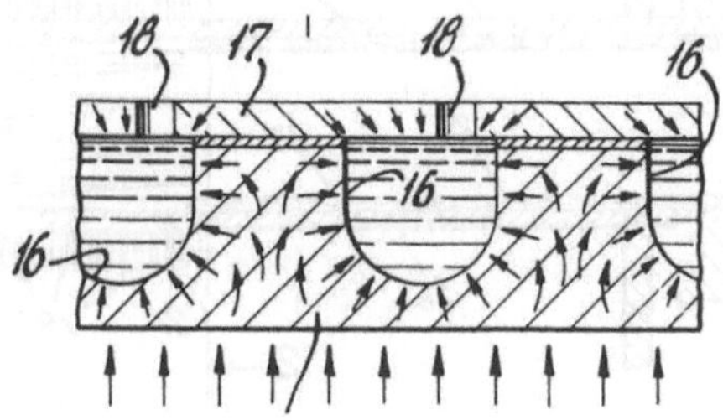

Fig. 28

The heat pipe according to Taricco, S. and Cottino, F. of Fiat SpA, (61.1), consists of a stainless steel tube, the inner surface of which being covered by a sintered metal powder layer.

To make such a heat pipe the following manufacturing steps have to be carried out:

a) electrolytically copper plating the inner surface of a stainless steel tube;

b) staining off the tube;

c) introducing and positioning a prestained core of non-corrosive steel into the internal space of the tube;

d) introducing and compacting the metal powder;

e) sintering.

In the heat pipe of Corman, J.C. of General Electric Company, (76.1 and 76.4), the portion of the wicking material which receives the liquid condensed at the condenser surface at the heat output end of the heat pipe is provided with a plurality of large openings extending from the surface adjacent the condenser surface to the opposing surface thereof to avoid condensate accumulation adjacent the condensation surface. Such accumulation blocks access of the condensation surface

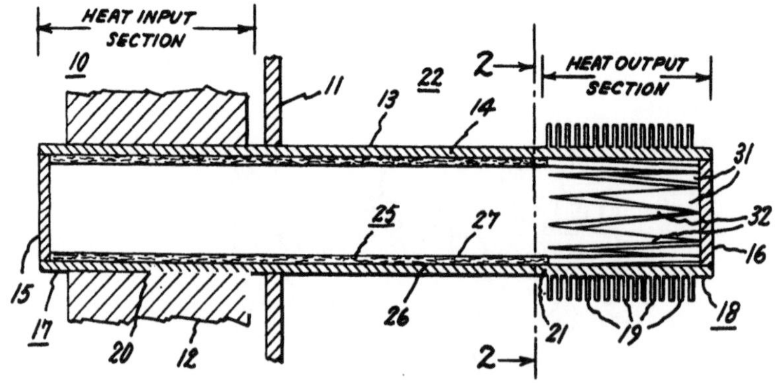

Fig. 29

to vapour and hence increases the thermal impedance of the heat output end of the heat pipe.

Fig. 29 shows the Corman heat pipe.

The arterial heat pipe of Kosson, R. of Grumman Aerospace Corp., (82.1), is self filling. It comprises a porous structure that is disposed around a hollow core, such as is shown in Fig. 30.

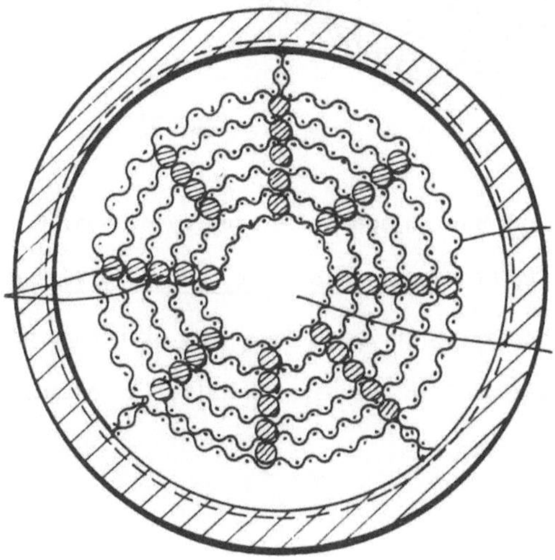

Fig. 30

Kosson together with Roukis, J.G. and Westell, J.R., (82.2), developed still another wick artery comprising a plurality of screen layers which are separated by spacers.

A quite similar construction made by Leinoff, S. et al. of Grumman Aerospace Corp., (82.4), has an artery arranged so that the warp and wave of the wire mesh are at about a 45° angle with respect to the axis of the heat pipe.

Fig. 31 shows a perspective view of Leinhoff's heat pipe.

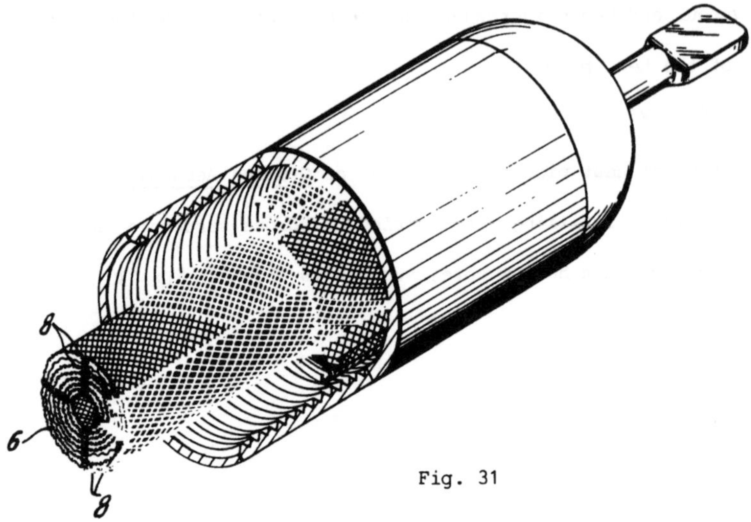

Fig. 31

Still another wick system again provided by Kosson, R.L. together with Quadrini, A. of Grumman Aerospace Corp., (82.6), comprises a mesh screen wick having at least one artery structure used in conjunction with a closed or open system while in contact with the wall capillary.

See Fig. 32.

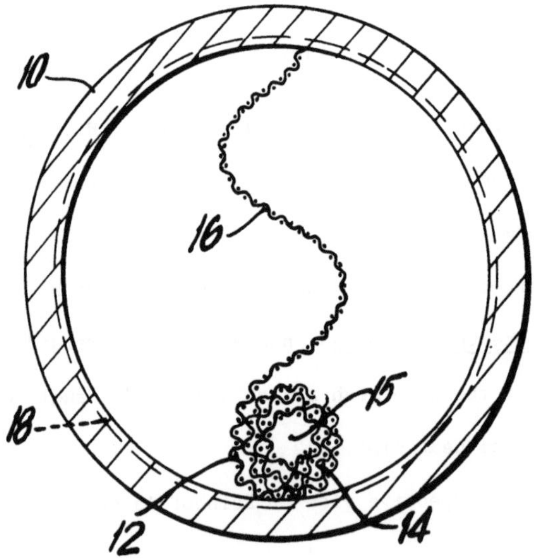

Fig. 32

Hydrogen getter

The wick proposed by Lamp, T.R. of Hughes Aircraft Company, (90.5), is in the form of a foil, gauze or wire made of a material selected from the group consisting of tantalum, titanium or niobium which have an affinity for hydrogen gas. The foil may have holes therein for vapour communication.

A method of preparing a capillary heat pipe wicking structure comprises the steps of (a) disposing an inner surface of a tubing in contact with a solution containing a salt of a thermally conductive metal and (b) depositing the metal from the solution upon the surface so as to form a porous metallic layer thereon constituting the capillary heat pipe wicking structure.
This method has been developed by Kiyoshi Inoue of Inoue Japax Research Corporation,(92).

In another gettering structure of N.V. Philips Gloeilampenfabrieken, (136.9),use is made of a metallic cloth 6,(Fig. 33), for example of chrome-nickel steel. This cloth locally has portions 9 of a gettering substance. These portions may be constituted by lanthanium, yttrium or scandium.

Glass beads

A heat pipe with annular grooves engaging a composite wick having a porous sheath containing a capillary core of glass beads and arranged to provide at least one axial vapour passage is of Calhoun, L.D. of McDonnell Douglas Corporation, (121.). The working fluid projects through the screen sheath to expose the meniscus to the heated surface at the evaporator end of the heat pipe to provide thin film boiling.

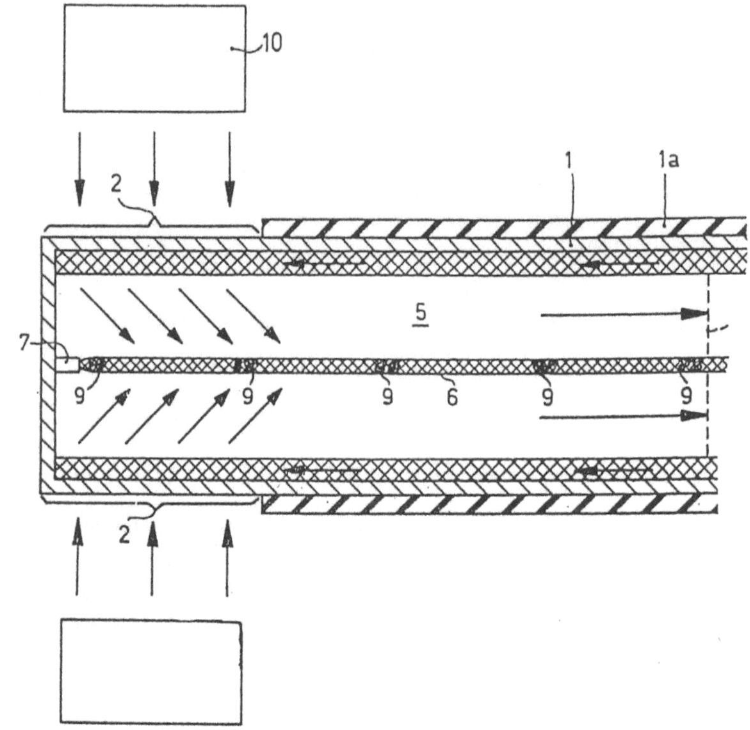
10
1
1a
2
5
7
9
6
2

Fig. 33

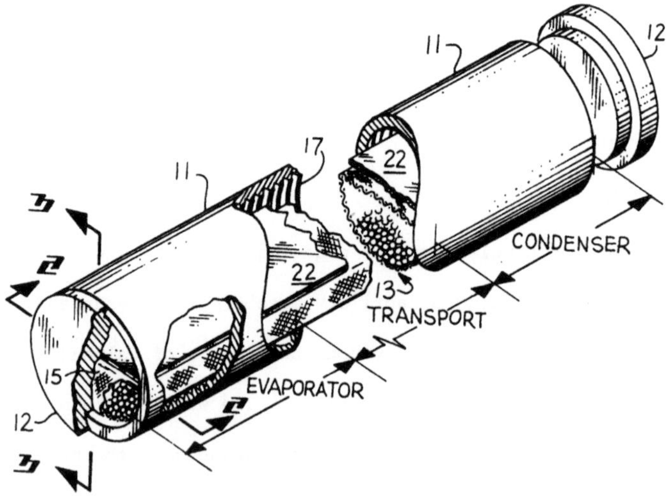
12
11
17
22
CONDENSER
13
TRANSPORT
EVAPORATOR
15
12

Fig. 34

The annular grooves conduct the vapour to the vapour channel.

Fig. 34 shows the Calhoun construction.

The heat pipe of B.E. Nelson of The Perkin-Elmer Corporation, (135), has the usual central wick for transporting a liquid, such as a cryogenic liquid, from end to end. A secondary wick transports liquid radially between the central wick and the wall of the heat pipe. This secondary wick is formed with lobes for contacting the heat pipe walls. In this manner, much of the wall surface is free of any contact with the wick, thereby providing improved heat transfer characteristics.

Metal whiskers

According to Schladitz, H., (152), a capillary system is provided having poly-crystalline metal whiskers at, at least, both ends of the heat pipe.

Whiskers of the type used by Schladitz are described in "Zeitschrift für Metallkunde (1968), 1, pp. 19 to 22.

The wick structure of Dean, D.J. of The Secretary of State for Defence, (158.1), comprises a prefabricated cushion having a continuous outer cover, of which at least part is formed of mesh or gauze wick material, filled with a mass of resilient fibrous material of open structure, the cushion being compressed within the enclosure so that the mass of resilient fibrous material urges the wick material towards the wall of the enclosure.

By fabricating the wick structure prior to assembly of the heat pipe, the integrity of the continuous outer cover can be tested, and any faults rectified, before the heat pipe enclosure is sealed.

It has been found in the past that in a heat pipe fluid circulation can substantially be improved by using wicks consisting of two layers, one of which having a structure consisting of a material having large pores, while the other layer is formed of a material having small pores.

In this connection Fries, P. of Siemens AG, (164.1 and 164.2), developed a wick in which the layer having large pores, at least at its face adjacent to the evaporating zone, is provided with a layer having the small pores. The pore diameter of the small-pored layer preferably is smaller than one half the diameter of the large-pored layer.
Figure 35 shows a cross section of an embodiment in which the large-pored layer 8 of sintered material is covered by a small pored-layer 9.

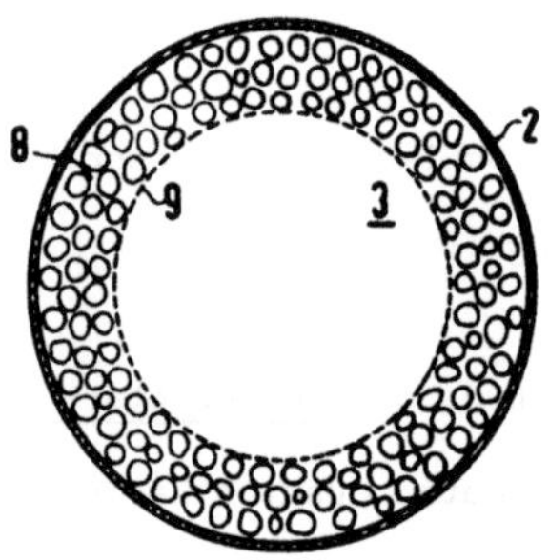

Fig. 35

A heat pipe wick which can survive freezing of the heat transfer fluid within the heat pipe and return to full operation automatically was invented by G.Y. Eastman of Thermacore, Inc., (181.2).
A flexible high lift wick is used with a limited liquid inventory to prevent damage from freezing. The limited amount of liquid is completely retained in the wick at all times by capillary forces preventing puddling at the lowest point of the heat pipe, thus avoiding damage to the casing by expansion.

Another wick structure of Eastman of Thermacore, Inc., (181.3), consists of sintered metal which is directly formed into the heat pipe casing and contains simultaneously formed tunnels which permit the flow of large quantities of liquid with low pressure loss.

Still another capillary wick of Eastman, (181.7), is constructed from a metal cylinder formed in close contact with the inner diameter of the heat pipe casing, and containing longitudinal grooves on the wick's inner surface adjacent to the vapour space. The grooves provide a longitudinal capillary pumping action while the high capillary pressure of the sintered wick provides liquid to fill the grooves and assure effective circumferential distribution of liquid in the heat pipe.
Figure 36 shows part of 9 cross sections through Eastman's wick structure.
Buturlia, J.A. of Thermo Electron Corp., (Patent 182.6), provided a method of manufacturing a heat pipe wick by bonding particles to the interior wall of the heat pipe. These particles are also bonded to each other and this bond is obtained by making a suspension of particles in a suitable binding agent. The suspension is applied to the heat pipe wall, for example by means of a brush or by pulverising. The wall thus covered is first heated at a relatively low temperature during a short period of time so as to eliminate the binding agent. Then the wall is heated at a relatively high temperature so as to adhere the particles to each other and to the wall surface.

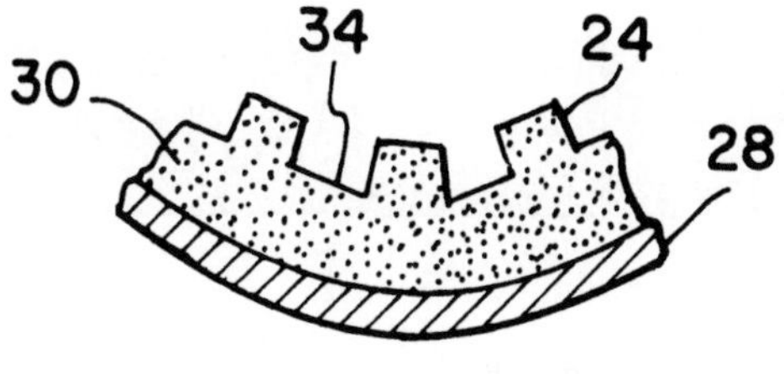

Fig. 36

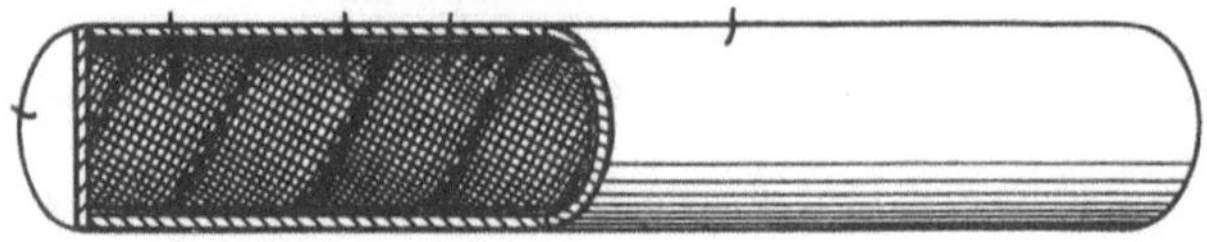

Fig. 37

The wick of Masayoshi Kata of Tokico Ltd., (185), is made up of a ribbon-shaped material of relatively narrow width wound in the form of a helix and is fabricated by winding the wick material preferably in alternately opposite helical directions on a mandrel, which is thereafter extracted to leave a hollow tubular wick. See Fig. 37.

Heat pipes containing graded-pore non-arterial wicks have substantially improved reliability when compared with arteries which are used.

Heat pipes having wicks which are optimally graded in pore size in an axial direction, with the pore size decreasing from the condenser to the evaporator end are subject of an invention of Marcus, B.D. and Edwards, K. of TRW Inc., (188). These graded pore size wicks yield more than twice the capacity of axially uniform pore size wicks having similar geometries.

Figure 38 shows the respective construction, while Figure 39 graphically represents the increase in the reciprocal of the wick pore size per length of the heat pipe.

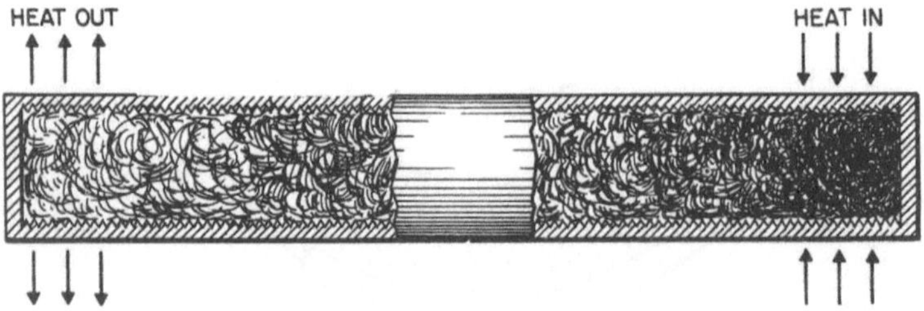

Fig. 38

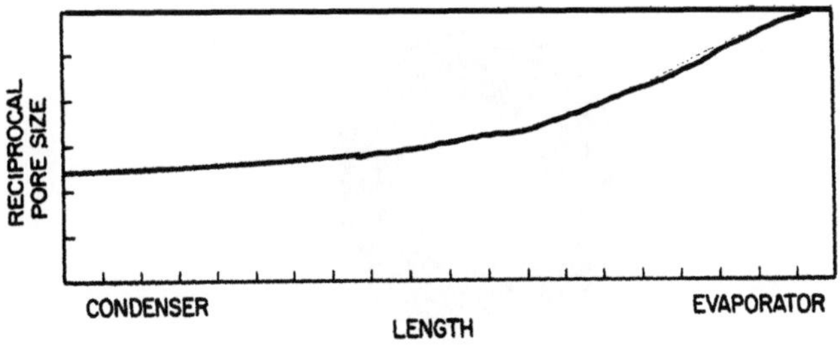

Fig. 39

A method of forming a tubular wick for a heat pipe has been proposed by Arcella, F.G., et al. of United States National Aeronautics and Space Administration, (195.1). The method is characterised by the steps of forming a wick blank of a predetermined thickness, comprising a plurality of superimposed layers of stainless steel mesh screen, wet rolling the blank for reducing the thickness thereof, wrapping the blank about an inner mandrel, compressing the blank into a rigid tubular structure, removing the tubular structure from the mandrel and sintering the tubular structure. See Fig. 40.

Fig. 40

Fig. 41

Somewhat comparable therewith is the wick construction of Masayosho Usui of Usui International Industry Ltd., (199).

Here, the inner tubular member is formed by winding in a spiral shape from a web of sheet material with a narrow continuous space between the adjacent surfaces of a thus-shaped inner member so that a capillary action may occur in the narrow space. A multiplicity of fone recesses extending longitudinally of the inner and/or outer members may be formed in the inner and/or outer circumference of the inner member and/or in the inner circumference of the outer member. See Fig. 41.

Ichiro Honda, et al.of Suzuki Metal Industrial Co. Ltd., (178.1), developed specific materials for a heat pipe wick and for securing it to the inner wall of the pipe. The inner tube includes a solid drawn metal tube and the wick consists of tows or yarns of metal fibres of 0.004 mm to 0.03 mm in diameter. The material for securing the wick to the inner tube includes yarns of metal fibres or metal wire. The outer tube includes a metal tube and a metal hoop capable of being laterally curled or wound while welding the edges of the wound hoop by the use of a tube-forming machine.

According to a further aspect of Ichiro Honda's invention, (178.2), the metal fibres are heated to 350°C - 900°C in an oxidising atmosphere before

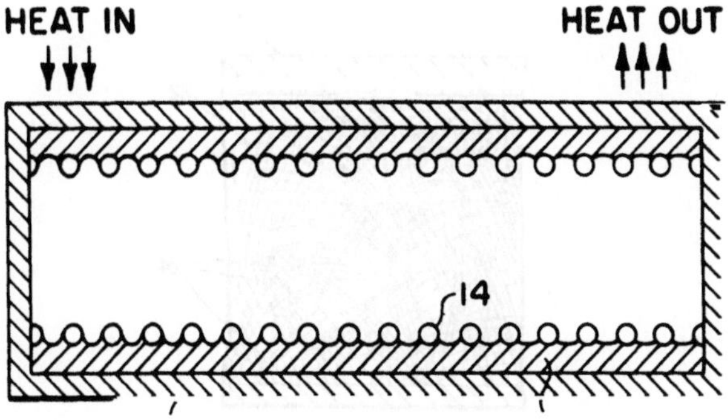

Fig. 42

sealing the outer tube at both ends, to form a film of their oxide thereon, thereby improving the fibres wettability properties with the working liquid.

Figure 42 shows a heat pipe designed by Morris, J.F. of NASA, (195.3), which is developed for its use in terrestrial and space energy processing. The device comprises a sealed container or envelope which contains a capillary wick 14. The temperature of one end of the heat pipe is raised By the input of heat from an external heat source which is extremely hot and corrosive. A working fluid of a corrosive alkali metal, such as lithium, sodium or potassium transfers this heat to a heat receiver remote from the heat source.

In accordance with the invention the container and wick are fabricated from a super alloy containing a small percentage of a corrosion inhibiting or gettering element. Lanthanum, scandium, yttrium, thorium and hafnium are utilized as the alloying metal.

2.2.2 Inventions directed to wall surface ribs, grooves and other capillary structures.

To improve the capillary action of heat pipes it has been proposed to provide the inner surface walls of the pipes with ribs, grooves or

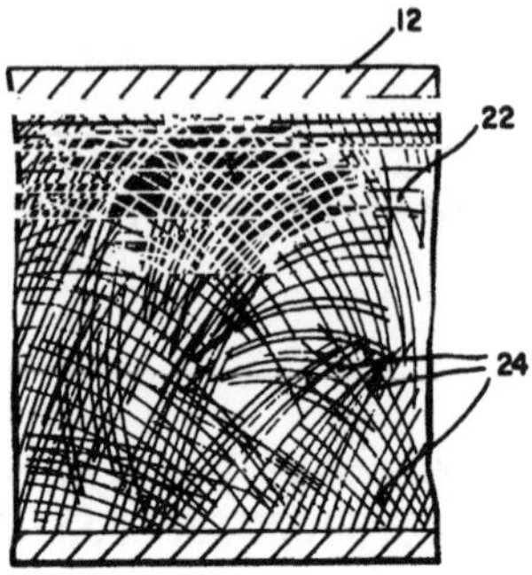

Fig. 43

other irregularities, or to use spirally wound obstructions inserted into the heat pipe so as to avoid bubbling of the heat transferring liquid.

According to Hughes Aircraft Company, (90.10), the wall surface is provided with thin grooves which, with respect to the heat pipe axis, are radially directed, such as shown in Fig. 43.
Furthermore, the respective heat pipe has an additional channel or tube resting on the bottom of the heat pipe.
The heat pipe designed by Franklin et al. of The Boeing Company, (14.3), has an outwardly extending conduit (125), (see Fig. 44) disposed in both the evaporator section (135) and condenser section (140) and provides fluid communication between the vapour tube (120) and the return tube (130). Circumferential v-shaped grooves (145) terminate at a slot-like opening (165) formed in each of the conduits (125).
A cap member (170) traverses each opening (165) and coacts with the grooves (145) to form a plurality of fluid passageways (175).
Another heat pipe of Franklin, (14.4), capable of being primed under gravity conditions, comprises a sealed envelope (95)(see Fig. 45) defining an evaporator (110) and a condenser (115). An arterial tube (125) of large liquid carrying capability is provided with an axial

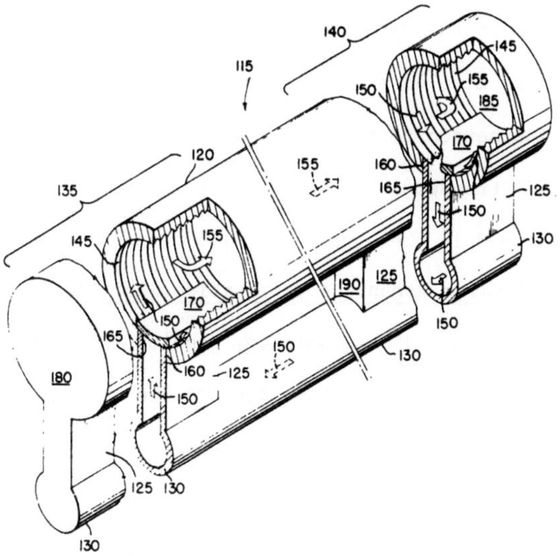
140
115
120
135
145
155
150
185
170
160
165
125
130
190
180

Fig. 44

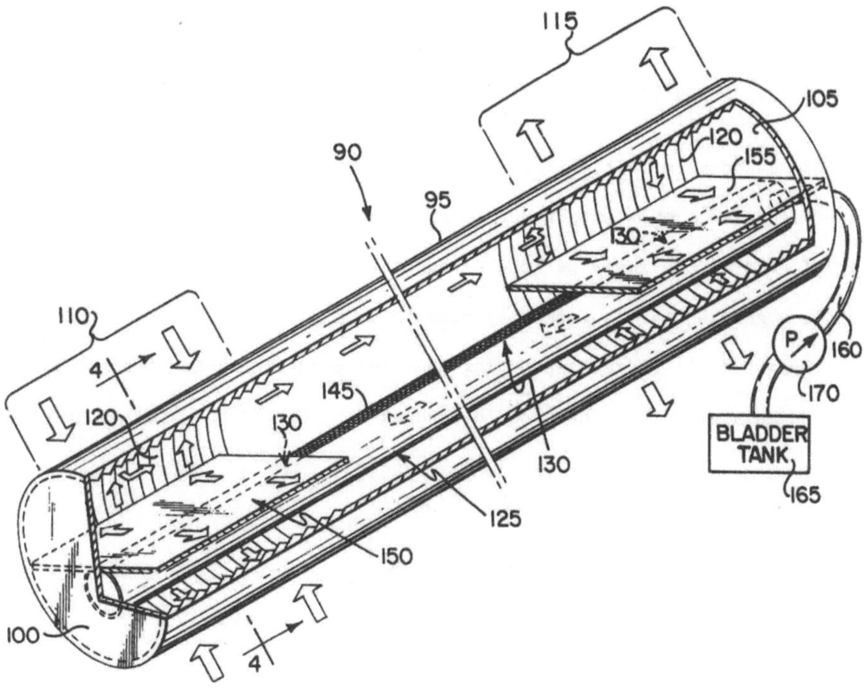
115
105
120
155
90
95
130
110
4
145
P
160
170
BLADDER TANK
165
125
150
100

Fig. 45

slot (130) formed on the upper surface of a segment of the tube (125). The slot (130) has beveled surfaces (135), (140) which increase the capillary pumping action to transverse wicking bodies (150, 155). The bodies (150, 155) transport working fluid to or from circumferential grooves (120) formed in the evaporator (110) and condenser (115).
Leszak, E.H. of Grumman Aerospace Corporation, (82.5), has proposed a method for making longitudinal capillary grooves in the inner wall of a heat pipe by passing a rotating tool through the internal passage of the tube such that a swaging ball carried by the tool flattens the lands between the grooves to thereby produce a capillary configuration in which the width of the opening of the grooves is narrower than a cross-section of the groove and the lands between the grooves have a mushroom-shaped cross-section. Deformation of the lands by the swaging ball is essentially of a limited-load point-contact type to avoid a destructive collapse of the lands during the swaging process.

Michio Yanadori of Hitachi Ltd., (89.4), designed a heat pipe (Fig. 46) provided with a multiplicity of longitudinal deep grooves (2) and ridges (3) formed in the inner peripheral surface thereof. A plurality of shallow grooves (4) is formed by plastic working in the top surfaces of the longitudinal ridges separating the deep grooves. Parts of burrs (5) formed as a result of the plastic working for forming the shallow grooves

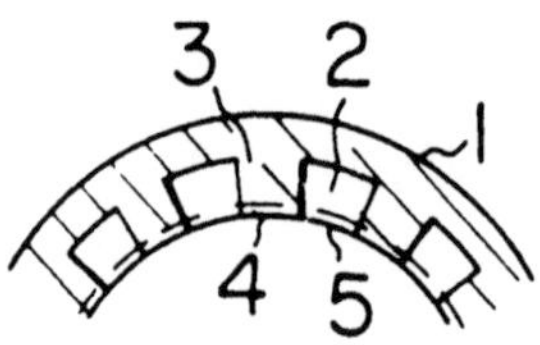

Fig. 46

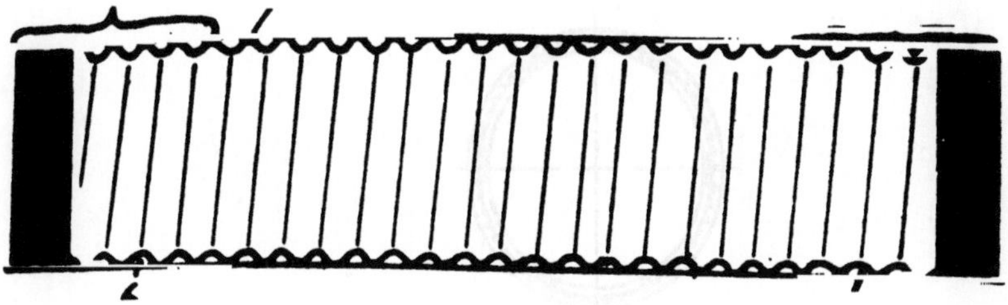

Fig. 47

are extended over the deep groove so as to form bridges connecting adjacent ridges over the deep grooves. The bridges (5) enhance the capillary action of the deep grooves (2).

To reduce manufacturing cost and to enhance mounting facilities of heat pipes, Schieferdecker, F.D. of Kabel- und Metallwerke Gutehoffnungshütte, (100.1), proposed to have a heat pipe constituted of at least two coaxial metal tubes, one of which having a screw lined corrugation as is shown in Figure 47.

Another double-walled heat pipe has been designed by Janssen, M. et al, of Mannesmann AG, (115).

To manufacture such a heat pipe, Janssen uses a metal sheet which he deforms to an inner wall and an outer wall of the heat pipe. As is shown in Figure 48a, one end of the inner wall is provided with rectangular holes, while the outer wall has longitudinal grooves, as shown in Figure 48b.

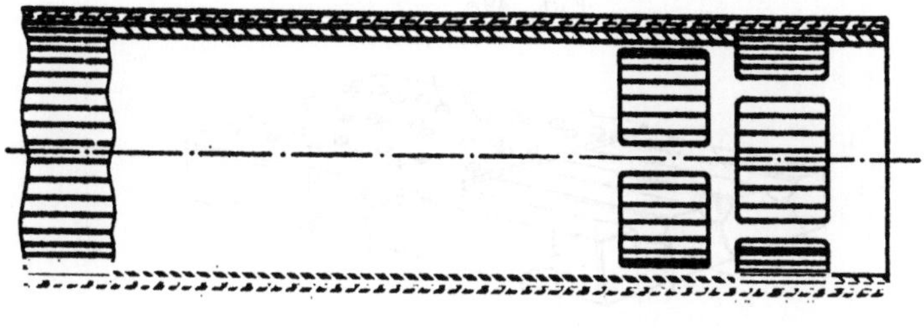

Fig. 48a

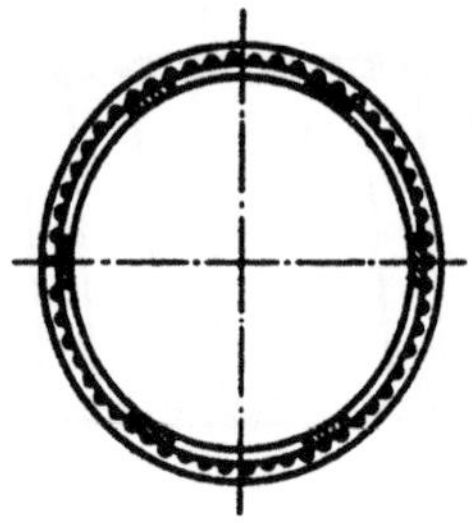

Fig. 48b

The heat pipe of Wayne, P.P. of NCR Corporation, (128), has an integral screen-wick structure which provides a relatively large contact area between the internal working fluid and the heat input. Some embodiments of Wayne's heat pipe are shown in Figures 49a and 49b.

Grover, G.M. of Q-Dot Corporation, (141.1 and 141.2), developed a method and apparatus for fabricating a circumferentially capillary grooved heat pipe. The capillary grooves in the side walls of the heat pipes transport the liquid phase vertically above the liquid level to increase the area of the liquid-vapour interface. Additionally, the solid metal strips which form the grooves, provide a low impedance

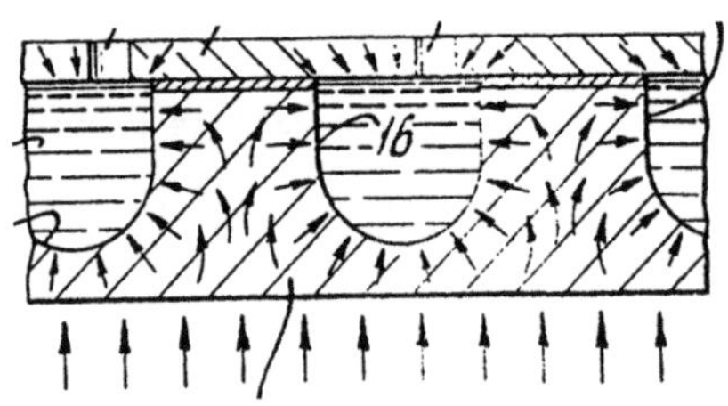

Fig. 49a

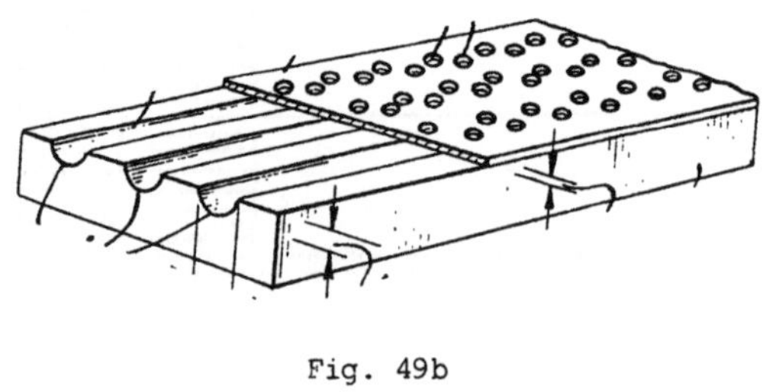

Fig. 49b

thermal path from the walls of the heat pipe to the liquid-vapour interface where evaporation and condensation occur. These two factors significantly increase the efficiency of the system. A divider plate having an X-shaped cross-section separates the liquid phase from the high velocity vapour phase to prevent slugging under high energy transfer conditions. The divider plate is operative when the unit is disposed in either of two horizontal positions.

The method and apparatus provide a means for fabricating a spiral capillary groove by cutting the metal from the wall of the tube and raising and folding the cut metal over to provide a groove having a narrow opening for a maximum capillary action. The cutting tool has a curved cutting edge formed by the intersection of a planar surface and a cylindrical surface.

Figure 50 shows an important part of Grover's heat pipe.

Another method of producing a heat transfer surface on the metal wall of a heat pipe is subject of research by Müllejans, H. of the

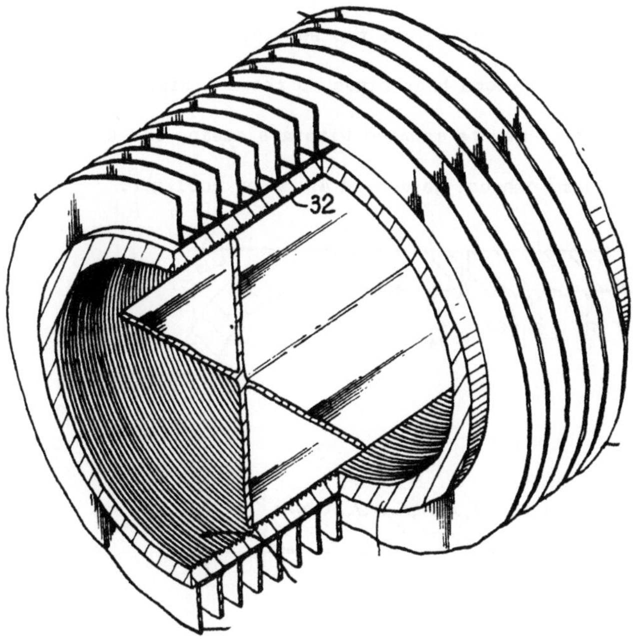

Fig. 50

Süddeutsche Kühlerfabrik Julius Fr. Behr GmbH & Co. KG.,(177) .
The respective method comprises first mechanically smoothing the wall and then making it rough by chemical etching to produce minute depressions or elevations.

Figure 51 shows a cross-section of a heat pipe invented by Klaschka, J.T. and Davies, M.J. of the UK Atomic Energy Authority, (190.1).
The internal surface consists of a number of equi-spaced longitudinally extending ribs 26 which define grooves 25 between them, and the ribs 26 may be of rectangular, semicircular, or triangular form. In use the bulk of the condensate in the thermal siphon 11 is pulled by surface tension effects into the corners of the grooves 25, and leaves a thin film of the condensate between the rivulets.
An assembly of heat transfer fins is attached externally, one at each side of the heat pipe for air flow therethrough transversely to the length of the heat pipe, and a number of such assemblies is clustered together with a silicon rubber sealing strip between them at about the centre of each assembly to divide a cluster into two sealingly separated portions. A number of modules of such heat pipes or thermal siphons together with the heat transfer fins may be mounted within a casing for use in heat transfer between warm exhaust gases and incoming cool air.

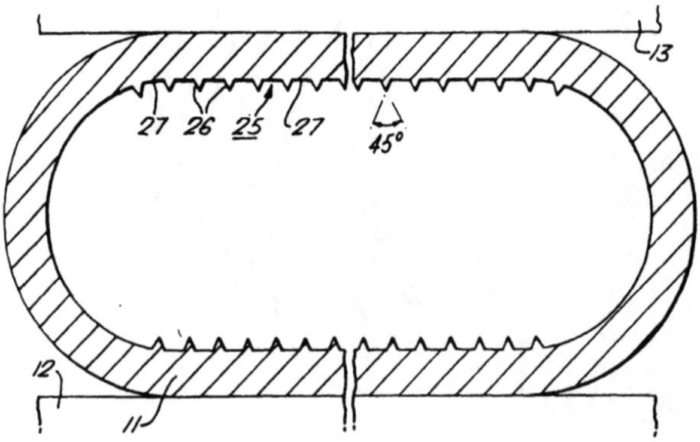

Fig. 51

CHAPTER 3

HEAT PIPES HAVING DEVIATING CONFIGURATIONS; FLEXIBLE HEAT PIPES.

This chapter and its sections deal with inventions which specifically are directed to the configurations of heat pipes. These configurations may differ from the usual tube like pipes having a circular cross-section either due to the mode of operation or to the use whereto the heat pipe has to be designed.

3.1 Spirally wound and serpentine-like heat pipes

According to Kabel- und Metallwerke Gutehoffnungshütte A.G., (100.4), a heat pipe is provided which has a configuration as shown in Fig. 52 and which can be used as a radiator which crosses hot and cold zones along its length.

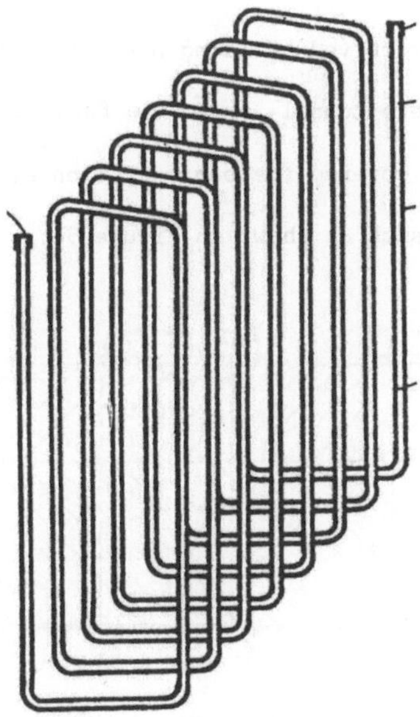

Fig. 52

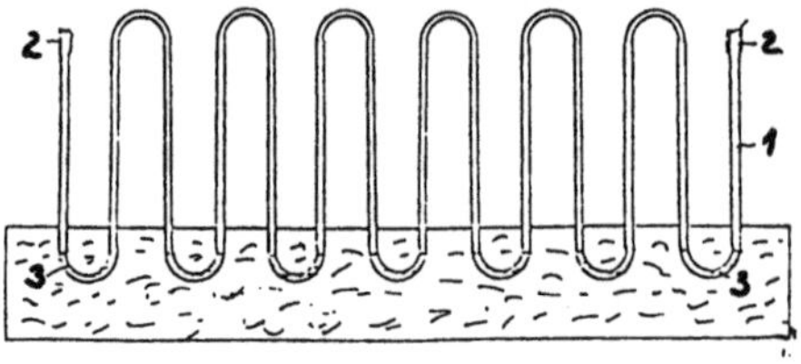

Fig. 53

According to a method for the series production of heat pipes Rohner, P., (100.5), of the foregoing company proposes to evacuate a large length of tube, and to fill it with a working liquid, after which the tube is closed at both ends (2, Figure 53), and formed to a serpentine like configuration, such as shown in the figure.

Each individual winding is heated at its lowermost point and cooled at its highest point. The individual windings are cut under vacuum to a multiplicity of finished heat pipes which can be interconnected to form a radiator.

Wiersma, T.G. of Smit Ovens Nijmegen BV, (166), has been assigned to design a heat exchanging element which can be manufactured as a module like sheet-shaped or curved building element. With such elements furnace walls or also complete tunnel- or muffle furnaces can be built.
To this end Wiersma covered the basis of each element by one or several curved heat tubes, such as shown in Figure 54.

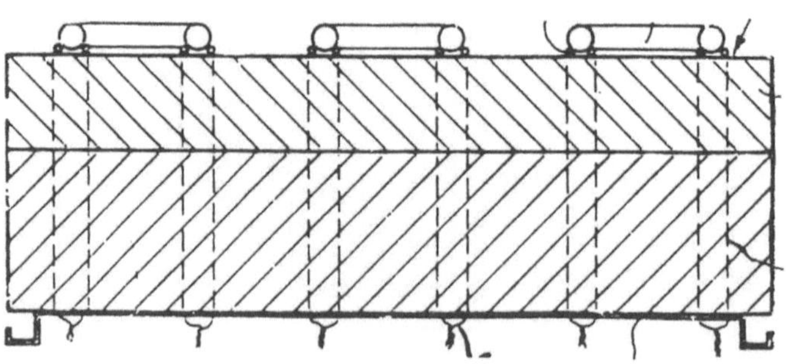

Fig. 54

3.2 Heat pipes having other configurations

T-form

Koch, H. and Bey, R. of Dornier System GmbH, (47.10), designed a heat pipe adapted to have a large heat receiving surface at its evaporating section and a large heat emitting surface at its condensation section. To this end they formed a large surface metal sheet integrally with a pipe, as shown in Fig. 55.

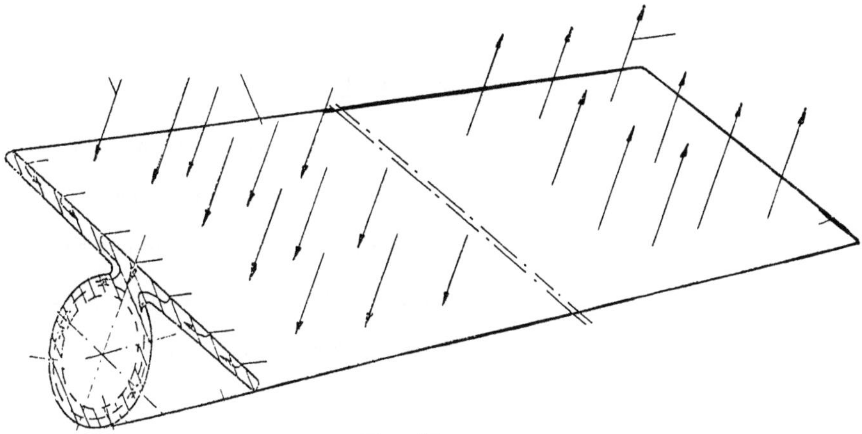

Fig. 55

Rectangular cross section

The heat pipe of Schlitt, S.R. of Erno Raumtechnik GmbH, (53.2), which particularly has been designed for spacecraft has a rectangular outer cross-section (5, Figure 56), the shorter side of which corresponding to the outer diameter of a heat pipe 6 which preferably has a square cross section. The longer side of tube 5 has a length equal to the sun of the shorter side and the second side of a hollow profile (7) having a square cross section. The outer side of the hollow profile, which side is directed from said profile , has to be disposed in the second surface of the structure, for example a honeycomb sandwich structure.

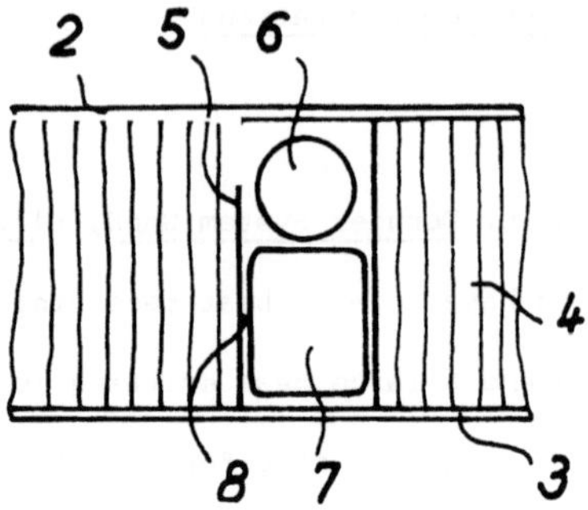

Fig. 56

Vapour conduit having different cross sections

The vapour conduit of a heat pipe provided by Busse, C.A. of Euratom, (55), has different cross-sections so as to influence the speed of vapour flow through the conduit. These different cross-sections may be obtained by disposing an obstructing member in the conduit, having a countour for example, such as shown in Figure 57.

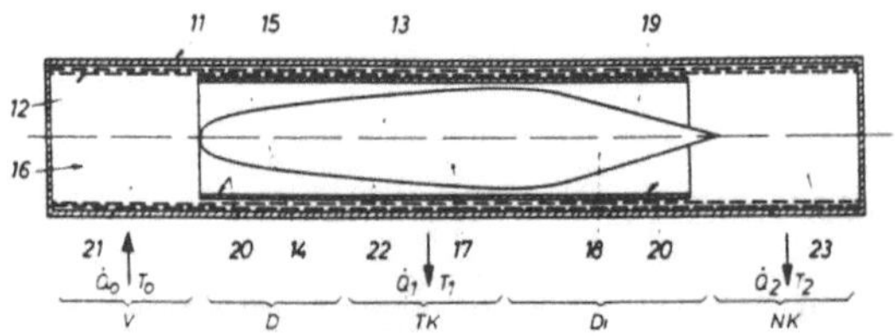

Fig. 57

Internal baffle

Another transverse variably conducting heat pipe has been designed by Edelstein, F. of Grumman Aerospace Corp., (82.3). It is based on the combination of a wicked closed casing having an internal baffle for changing the direction of vapour flow.

8-shaped cross section

Figure 58 shows a heat pipe having an 8-shaped cross section. This conception proposed by Alario, J.P. of Grumman Aerospace Corporation, (82.12), is arranged, so that the lower channel 20 (Figure 58), is the liquid phase channel, while the upper channel 18 acts as the evaporation section of the heat pipe. The heat pipe can be provided with an integral fin for rejection of heat by radiation to the ambient or for attachment to a source of heat in the evaporator region.

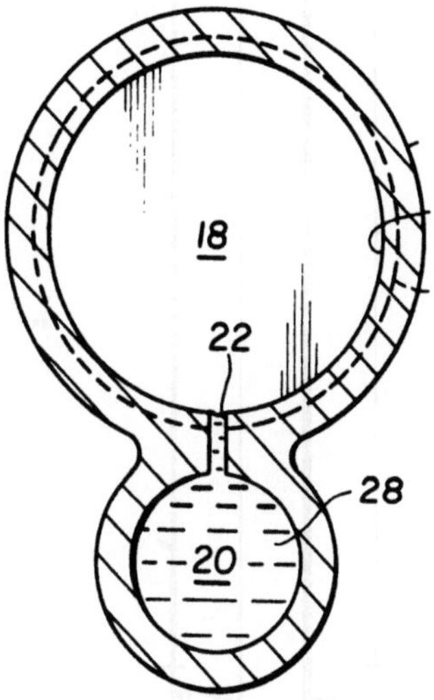

Fig. 58

H- or Y-shaped heat pipe

The heat pipe assembly of Tsu Hung Sun and Basiulus, A. of Hughes Aircraft Company, (90.3), is formed into an H-shape or a Y-shape. The H-shaped configuration comprises two heat pipes, each having condenser and evaporator sections with wicking therein coupled by a tube with wick at their evaporator sections. The Y-shaped configuration utilizes a common evaporator section in place of the two evaporator sections of the H-shaped configuration. In both configurations, the connection between the vapour spaces of the two heat pipes equalizes vapour pressure within the heat pipes. Although both heat pipes have wicks, they have sufficient fluid only to saturate a single pipe. If heat is applied to the condenser section of one of the pipes, this heat pipe becomes inoperative since all the fluid is transferred to the second pipe which can operate with a lower thermal load. Figure 59 shows the respective heat pipe assembly in its H-shape.

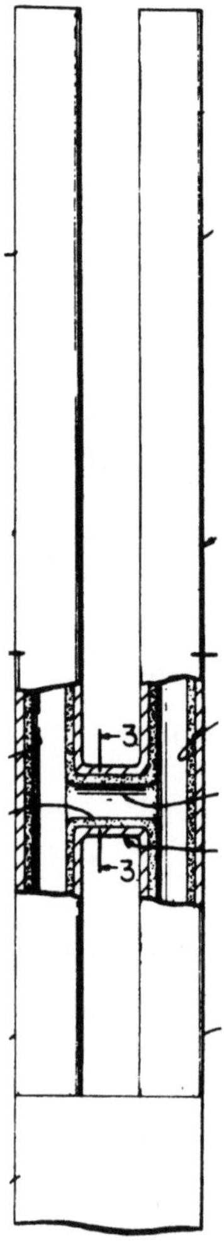

Fig. 59

Hexagonal and triangular configurations

Figure 60 shows a heat pipe assembly consisting of a hexagonal outer tube having at its inner wall heat pipes connected thereto, which pipes may have the usual circular cross section or may be of triangular cross-section. This assembly provided by Rohner, P. of Kabel- und Metallwerke Gutehoffnungshütte AG, (100.6), is adapted to its use as a cooling tube in, for example, power plants.

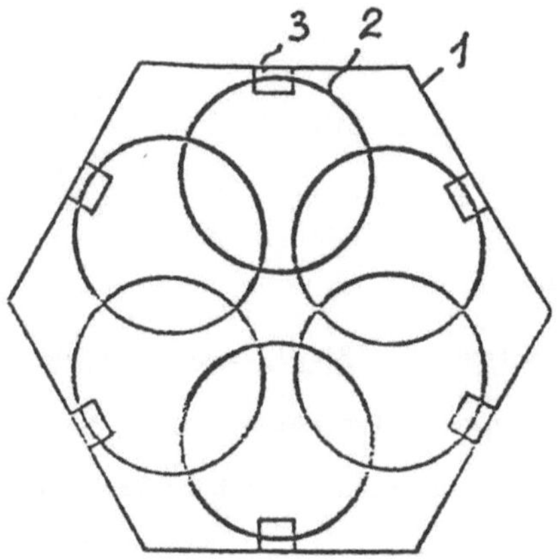

Fig. 60

1 represents the outer pipe;

2 represents a circular heat pipe;

3 represents a connector soldered to outer wall 1.

Conical configuration

In Figure 61a a heat pipe is shown, having a conical heating chamber 6 provided with an electric winding 8 and reversed conical obstruction walls 10 acting as condensate return flow walls. The respective heat pipe proposed by Kodaira, N. and Kusunose, M., (104), has several embodiments such as shown in Figures 61b and 61c.

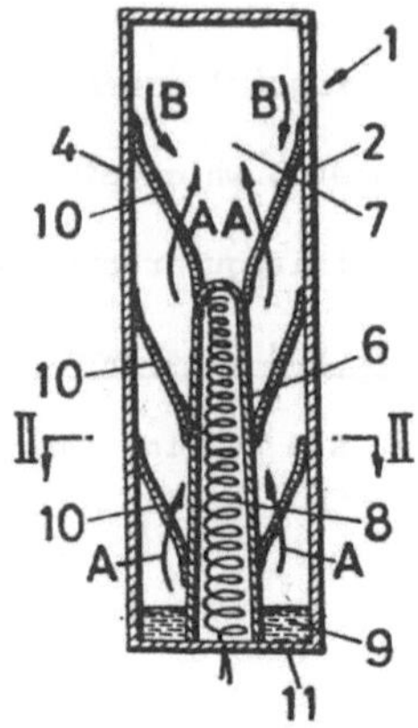

Fig. 61a

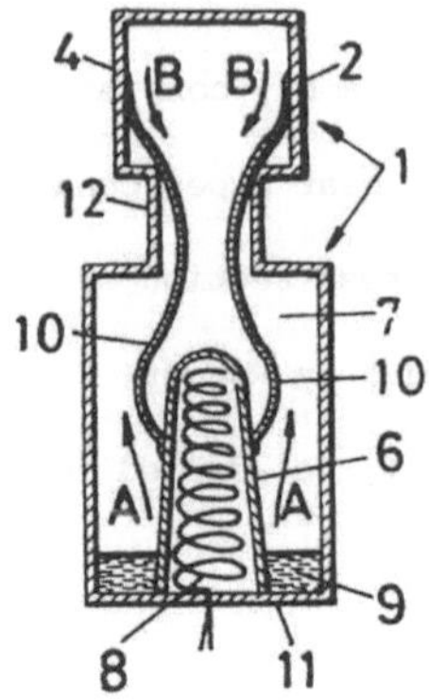

Fig. 61b

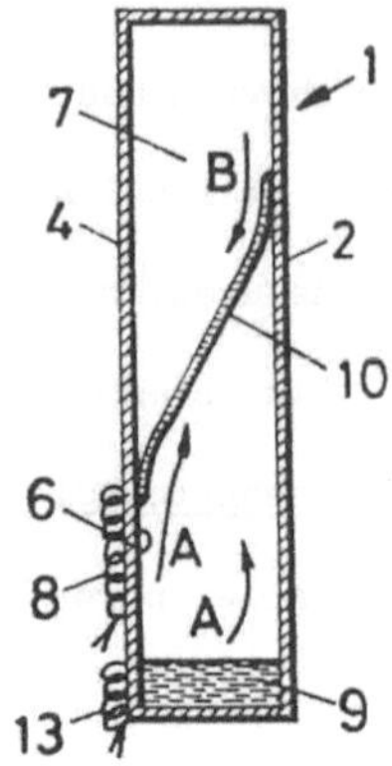

Fig. 61c

Fluted ribbed heat pipes

Applegate, G. of Curwen & Newbery, (38.3), uses in a heat pipe assembly fluted ribbed pipes. In Figure 62 two passageways 4 and 5 are arranged the one above the other and are connected to each other by heat pipes 1 having the respective fluted ribbed configuration.

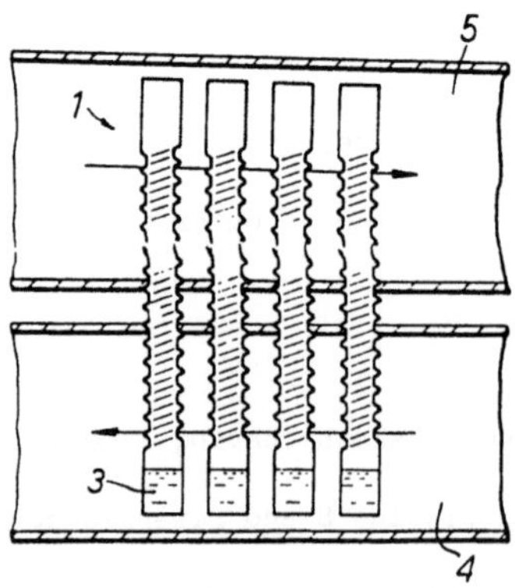

Fig. 62

Nodded heat pipes

A three-way heat pipe configured in a generally Y-shape having a condenser-evaporator or heat-sink end, an evaporator arm and a condenser arm, is subject of an invention of Cady, E.C. of McDonnell Douglas Corporation, (121.9).

In addition to conventional internal pumping the heat pipe is mechanically pumped externally from the heat-sink end to the evaporator arm. Heat transfer can take place between any two of the three elements by thermostatically controlling the external pump. Heat or cold may be stored in a suitable medium surrounding the condenser-evaporator end of the pipe for later transfer through the condenser or evaporator ends used as heat exchangers without a secondary circulation system. Heat transfer may also take place directly between the evaporator and condenser arms without storage. In the preferred embodiment of the heat storage configuration, a valve is placed in the condenser arm to prevent unwanted heat trans-

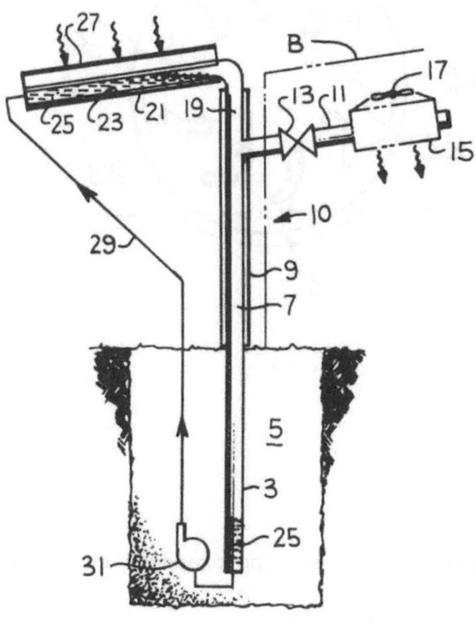

Fig. 63

fer. A system of three-way heat pipes can provide long duration thermal storage of heat or cold energy which can be later used for space heating and/or cooling of buildings.

Figure 63 shows Cady's heat pipe.

Blade shaped heat pipes

It has been proposed in the art to transport fluids and simultaneously exchange heat from the fluid to another medium.

In this respect Hayashi Kazuo of Nissan Motor Company, (129), designed a pump or ventilator like device which is characterised by a blade wheel rotatably mounted on a shaft and having at least one supporting element and a group of a large number of closed and evacuated tubes of heat conducting material in the form of blades, such as shown in Figure 64.

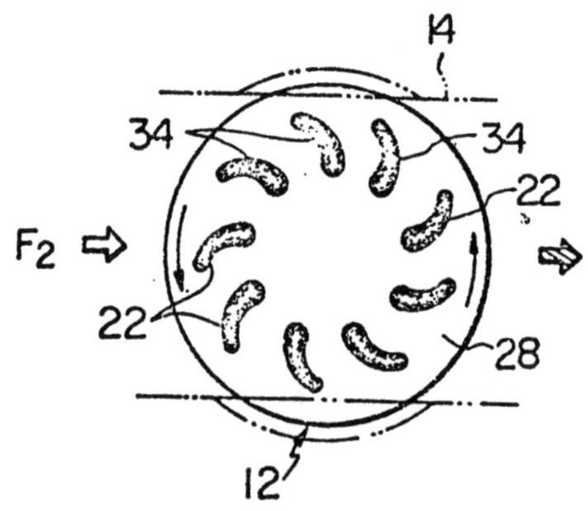

Fig. 64

Flat-plate heat pipes

Figure 65 shows a flat plate heat pipe proposed by Marcus, B.D. and Fleischman, G.L. of NASA, (195.2).

Such flat-plate heat pipes are made by enclosing metal wicking 5 between

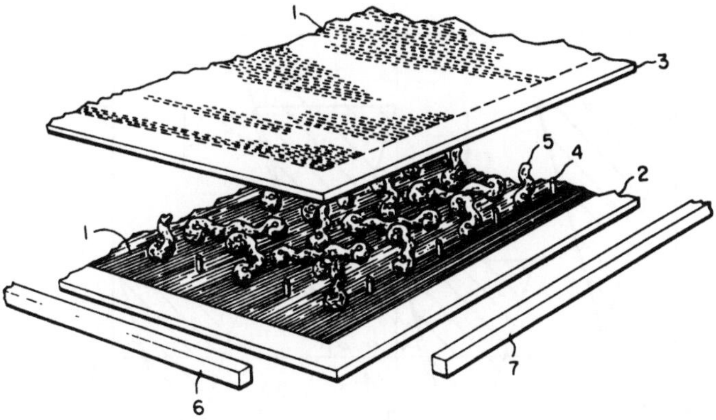

Fig. 65

two capillary grooved flat panels 1 and 2. These heat pipes have good capacity and conductance capabilities in zero gravity. When these flat-plate vapour chamber heat pipes are heated or cooled, the surfaces are essentially isothermal, varying only 3° to $5^{o}C$ over the panel surface.

Fin arrangements

A rotating heat pipe operating at low rotating speeds and providing a two-phase thermo-siphon has been constructed by Chisholm, D. and Finlay, I.C. of the Secretary of State for Industry (Great Britain), (159.1).

It comprises a fin arrangement located within the housing and extending from the condenser to the evaporator. The fin arrangement comprises a plurality of isosceles triangular cross sectional fins 15 (Figures 66a and 66b) symmetrical about the axis of rotation and located in the annular space between a sleeve 9 and the inner circumferential wall of the scaled housing 1, and having their respective vertices between the equal sides pointing in the same axial direction towards the evaporator 4.

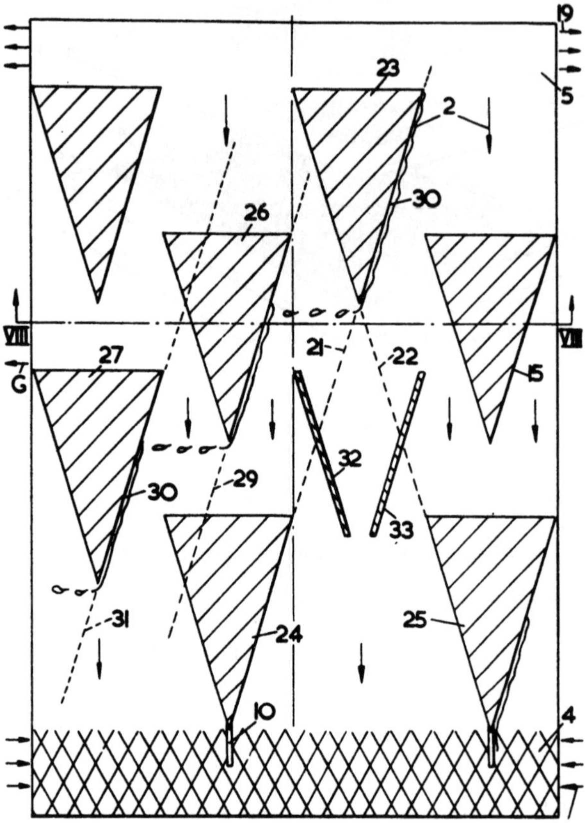

Fig. 66a

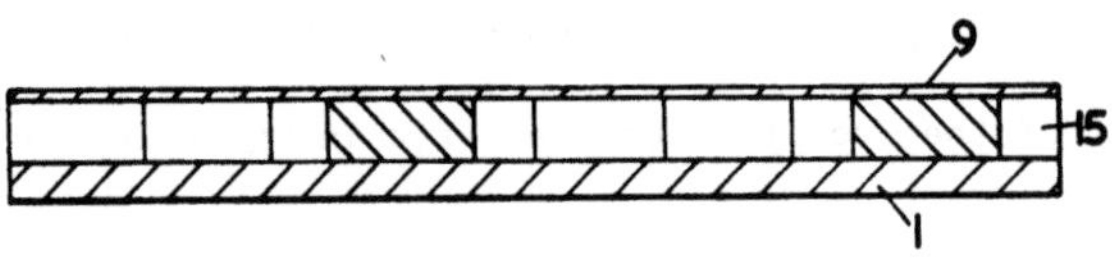

Fig. 66b

Flattened heat pipes

A heat pipe configuration for use in a magnetic field environment of a fusion reactor is subject of an invention of Werner, R.W., (207) of the US Department of Energy.

Heat pipes for operation in a magnetic field when liquid metal working fluids are used are optimized by flattening of the heat pipes having

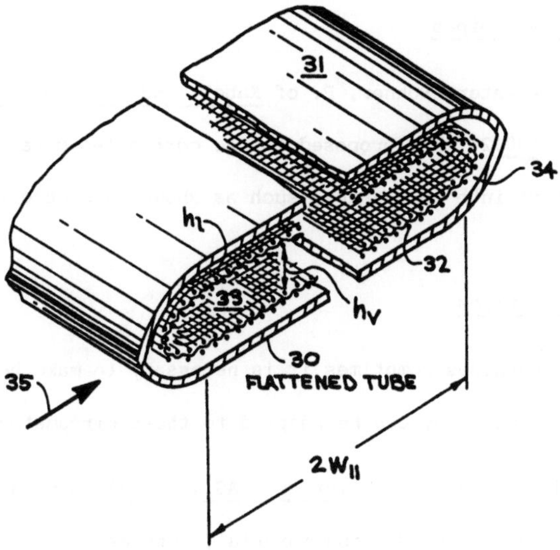

Fig. 67

an unobstructed annulus which significantly reduces the adverse side region effect of the prior known cylindrically configured heat pipes. The flattened heat pipes operating in a magnetic field can remove 2-3 times the heat as a cylindrical heat pipe of the same cross sectional area.

Figure 67 shows a perspective view of the Werner heat pipe.

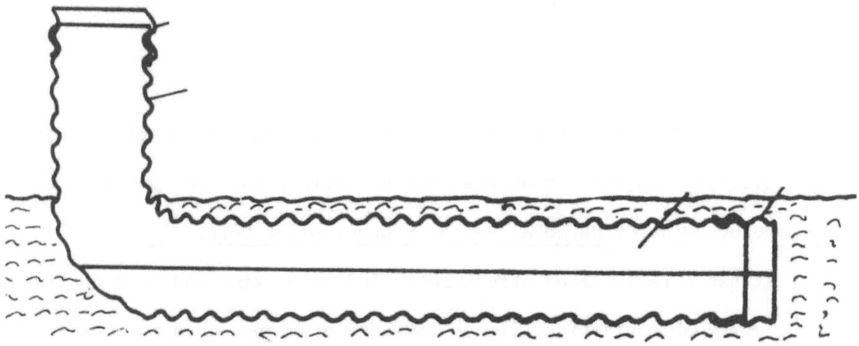

Fig. 68

Corrugated heat pipes

Tp cool waste water, Rohner, P. of Kabel- und Metallwerke Gutehoffnungshütte AG., (100.2), has proposed to use corrugated metal heat pipes which are bent into an L-shape, such as shown in Figure 68.

3.3 Flexible heat pipes

For various purposes sometimes it is necessary to make heat pipes flexible, so that they can be adapted to those circumstances.

To this end Wulf, H. of Daimler-Benz AG., (40.3), made a long length heat pipe of a heat conducting material, such as copper, aluminium or corrosion resistant steel. The respective heat pipe has a flat lens-formed cross-section, formed when the faces 1 and 2 (Fig. 69) are separated from each other by one or more spacing tubes or threads 7.

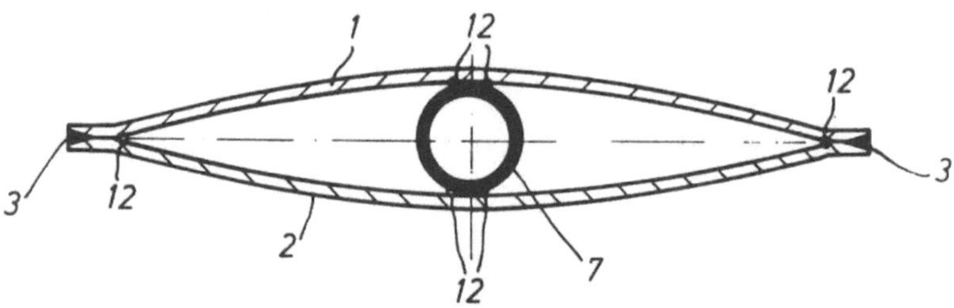

Fig. 69

A flat, flexible or conformable heat pipe assembly usable in conditioning clothing and other heat transfer situations has been provided by Gravveney, M.J. of National Research Development Corp., (127).
It has a reticulated structure including wicking and void continua, an impermeable plastics film or laminate envelope surrounding the structure and a valve in the assembly by which the assembly may be outgassed and evacuated and liquid introduced thereinto. See Fig. 70.

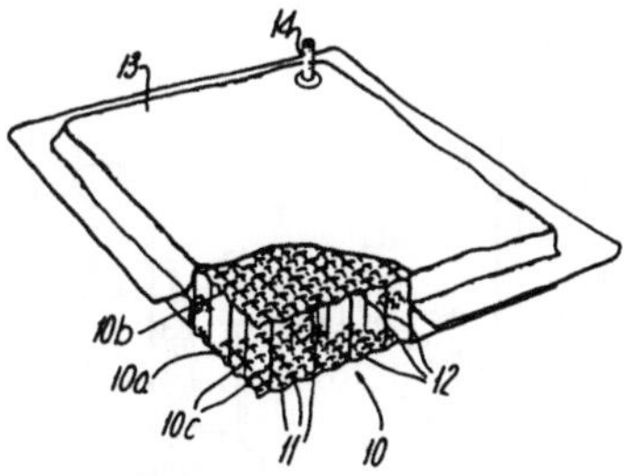

Fig. 70

The Secretary of State for Defence (GB), (158.6), also provides an apparatus for cooling a scalp during administration of drugs liable to cause cause Alopecia. The apparatus comprises a heat pipe in conformable sheet form and is shown in Fig. 71.

According to Bewley, A.D.; Gent, R.W. and Graveney, M.J. of The Secretary of State for Defence (GB), 158.3; 158.4 and 158.5) a heat pipe is provided which is principally a conformable garment. The assembly may include dismantleable connectors and a pump for evacuating the heat pipe interior and/or pumping the working fluid. According to the (158.3) embodiment the assembly may include a heat sink comprising a thermoelectric module array connected to a power source.

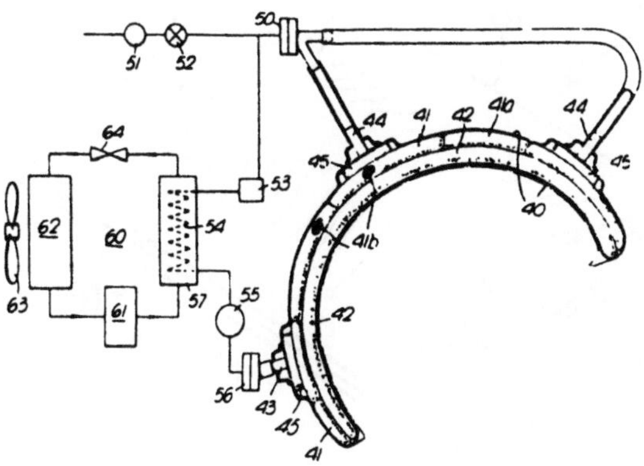

Fig. 71

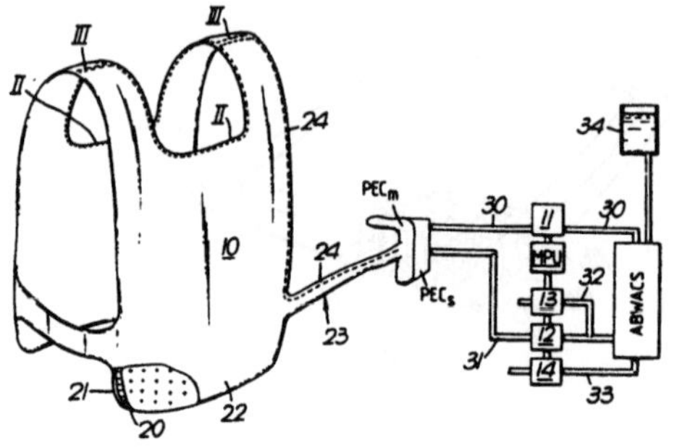

Fig. 72

Figure 72 shows a garment of the Bewley et al invention.

A heat pipe which can be rolled up for storage and automatically deploys when heat is applied, is subject of an invention by Eastman, G.Y. of Thermocore Inc., (181.4) (See fig. 73).

Two highly flexible parallel sheets are bonded together at their edges, thı permitting compact rolled storage. The inside portions of the joined edges form creases which act as capillary channels to move the heat exchange liquid from the condenser to the evaporator. A further embodiment involves multiple longitudinal cells which yield many more capillary channels and increases the structural strength of the deployed heat pipe, while maintaining the large surface area for heat transfer.

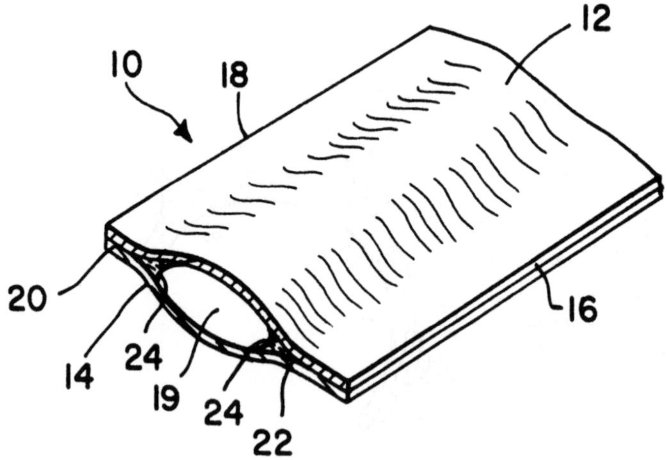

Fig. 73

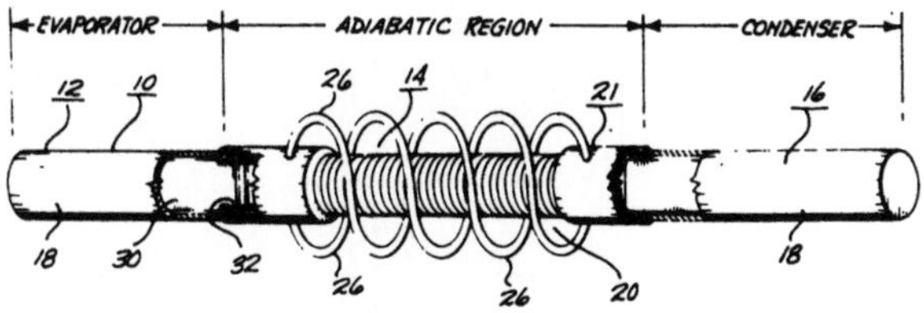

Fig. 74

A flexible heat pipe employing external tube arteries in the adiabatic region to transfer the heat pipe working fluid from the wick contained in the condenser portion to the wick contained in the evaporator section, has been developed by Franklin, J.L. and Pogson, T. of The Boeing Corp., (14.1).

Figure 74 shows a perspective view of the Franklin and Pogson heat pipe. Still another flexible structure developed by Pogson, T. together with Streck, (14.2), has equal characteristics, but its structure is as shown in Fig. 75.

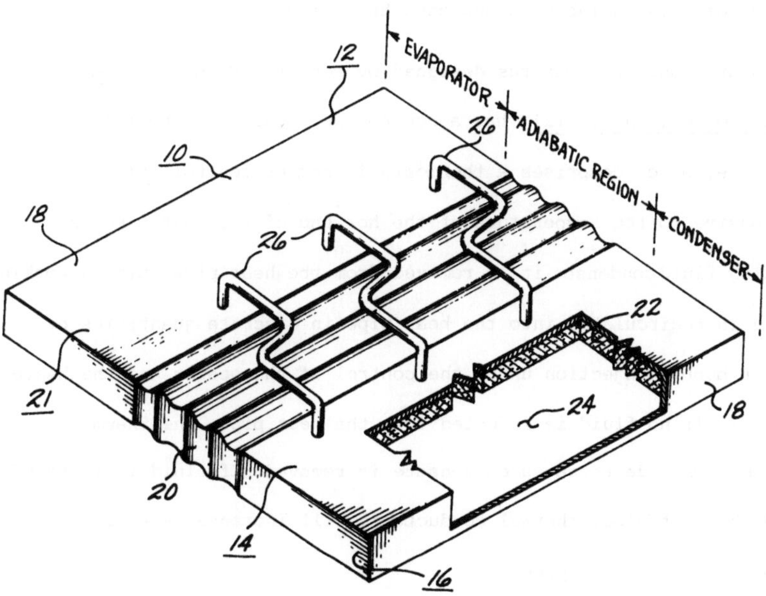

Fig. 75

CHAPTER 4

HEAT PIPE CONSTRUCTIONS MORE SPECIFICALLY DESIGNED FOR CONTROLLING PURPOSES

In many heat pipe applications it is advantageous to be able to control the thermal conductance of the heat pipe. For example, electronic equipment may generate more heat during certain phases of operation and, therefore, require the removal of more heat to maintain a fixed temperature. Alternately, it may be desirable to operate particular equipment at a variety of temperatures. It is possible to accommodate both of these situations by removing heat from the equipment through a controllably variable conductance heat pipe.

The heat transfer apparatus designed by Roberts, C. of Bell Telephone Laboratories, Inc., (12), has a controllably variable thermal conductance, which comprises a thermostatic heater control which is responsive to the temperature of the hot end of the heat pipe. As working fluid condenses it is removed from the heat pipe. Liquid working fluid is recirculated into the heat pipe in discrete quantities by vapour bubble injection under the control of the heater and the heater control. If no fluid is injected into the heat pipe, the thermal conductance will decrease as condensate is removed. If fluid is injected into the heat pipe, thermal conductance will increase as fluid is supplied to the heat pipe.

Figure 76 shows the Roberts' heat pipe in longitudinal section.

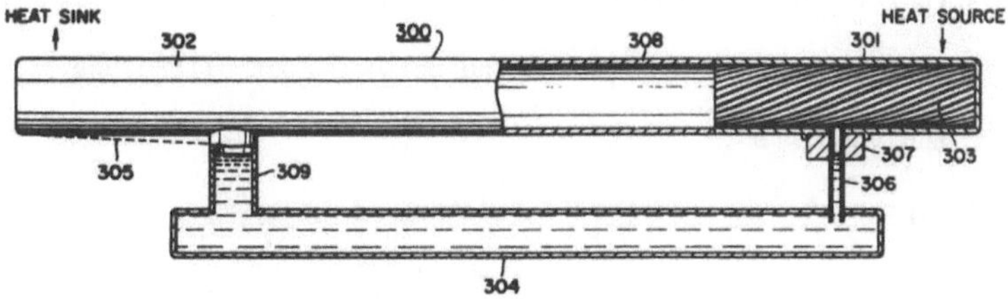

Fig. 76

It is known to improve the temperature constant in heat tubes by using a pressure controlling system, constituted e.g. by a gas reservoir having a constant pressure.

Under ideal circumstances there exists a correlation between the control gas pressure and the temperature of the heat pipe imposed by the vapour tension curve of the working liquid used in the pipe, but several causes affect these ideal circumstances. One of those affective influences results from the presence of controlling gas in the working liquid vapour, due to the solubility of the gas in the working liquid, particularly in regions where the partial pressure of the gas is high. To obviate this effect Busse, C.A. and Lebrande, J.P. of Euratom, (35.1 and 35.2), proposed to separate the closed evaporation-condensation circuit from the controlling gas by an auxiliary evaporation-condensation circuit. Preferably both closed circuits have a common evaporation zone.

Figure 77 shows the Busse and Lebrande proposal,

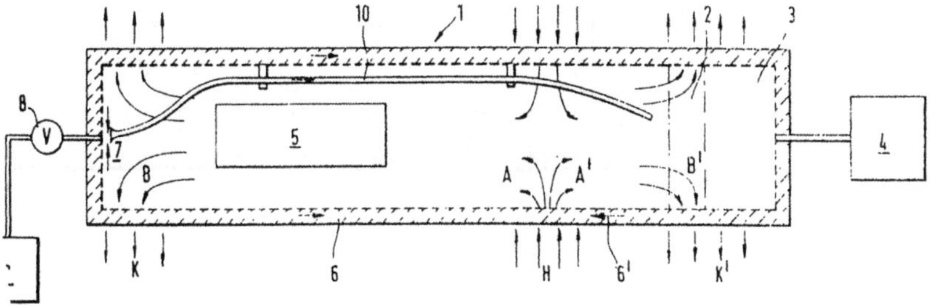

Fig. 77

erein 4 represents the pressure controlling system.

2 represents the transition zone between controlling gas and vapour B^1.

5 represents the isothermic treating chamber.

According to the embodiment of the Busse and Labrande, (35.1), heat pipe, the heating chamber formed by a closed cavity is constituted by at least two heat pipes mounted one behind the other. The temperature in each part of a hollow wall is adjusted by the controlling gas pressure in that cavity. The heat transmitting capacity along the hollow wall parts can be differentially controlled.

Figure 78 shows still another embodiment of the Busse and Labrande invention, (35.3).

In the figure reference number 7 indicates a cushion formed by a mixture of vapour and controlling gas, e.g. argon, into which a part of the vapour escapes. This cushion communicates a controlling system G which controls the pressure of the controlling gas. Another part of the vapour

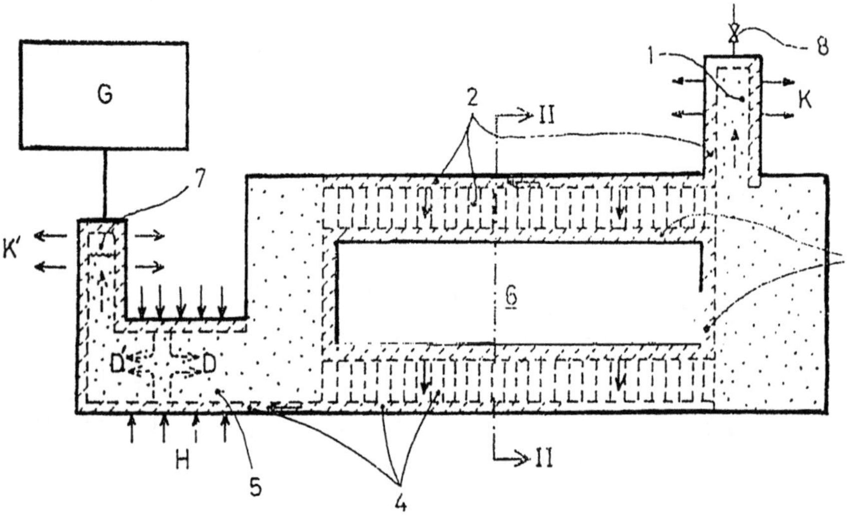

Fig. 78

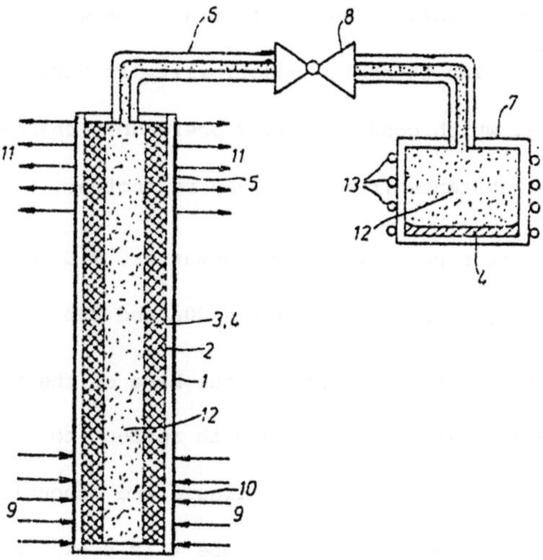

Fig. 79

follows arrows D and condenses in cooling zone K which by means of valve 8 can be brought in communication with a decompression chamber (not shown). The physical and technical phenomena behind the above structure is known to skilled persons. The invention resides more particularly in that the hot pipe has a cooling surface for the production of condensate and being connected to the outer wall of the isothermic chamber so as to clean the latter from impurities by a directed flow of liquid.

In figure 79 a controllable heat pipe is shown, which is developed by Kögler, K.; Koch, H. and Molt, W. of Dornier System GmbH., (47.9). The heat pipe is controllable in that the quantity of working fluid 4 is adjusted. The cool zone 5 of the heat pipe is connected to a container 7 to which the vapour of the working fluid can flow and in which container the liquid phase is wholly or partly stored. The container 7 can be heated or cooled, and in one embodiment its volume can be varied. According to Molt, W. of Dornier System GmbH., (47.12), a heat pipe can be controlled by a variable thermic resistance disposed between a heat source and the heat pipe, which thermical resistance is provided with a solid liquid or gaseous substance placed between the heat pipe and the heat source. The thermal conductibility is variable as a function of the temperature, which conductibility preferably changes during the transition of the substance from one phase to another.

The controllable heat pipe proposed by Macmaster, M.D. of Emhart Industries Inc.,(50), is shown in Figures 80a and 80b.

The heat transmission structure forming the base of the Macmaster heat pipe is conduct 2 which is sealed with respect to the ambiant atmosphere and is partly filled with a liquid and vapourised refrigerant referred to by 4. The configuration of tube 4 is shown in figure 80a. The tube may be sub-divided into two general parts being condenser 6 disposed along radiation board 28 and evaporator 8. The main function of condenser 6 is detecting the mean temperature in alimentation com-

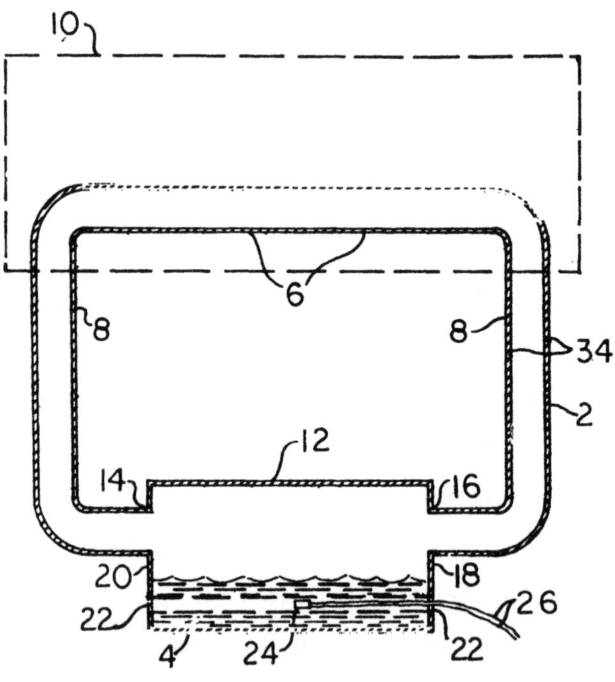

Fig. 80a

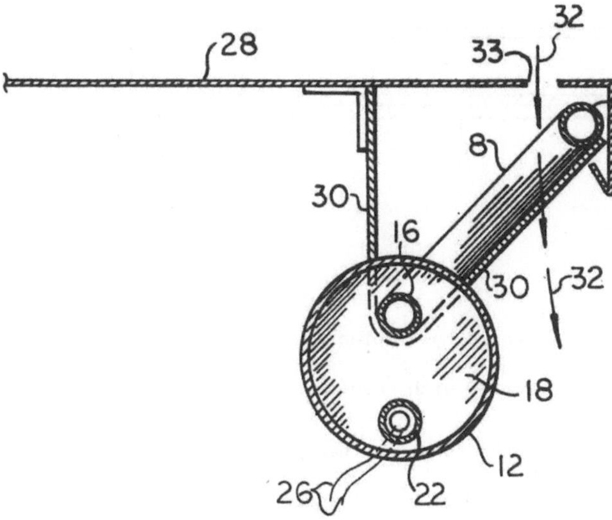

Fig. 80b

partment 10 and transmitting the detected information to the detection apparatus disposed in the evaporator 8 beneath screen 28 (figure 80b).

Okamoto, Eiji of Gadelius K.K., (71.2), found a method of controlling a heat pipe comprising a heat transmission zone and a controlling zone. Between the transmission zone and the controlling zone a conduct is formed through which only vapour of the working medium is permitted to flow, thus not its condensate. To control the amount of working medium present in the heat transmission zone, the temperature of the controlling zone is controlled by means of an external heat source, so that thereby at least one of the characteristics of the heat transmission capacity of the heat transmission zone, the temperature of said zone and the internal pressure of the heat pipe can be controlled.

The controllable heat pipe of Figure 81 is subject of an invention of Nakao and Takahashi of Babcock-Hitachi K.K., (89.7).

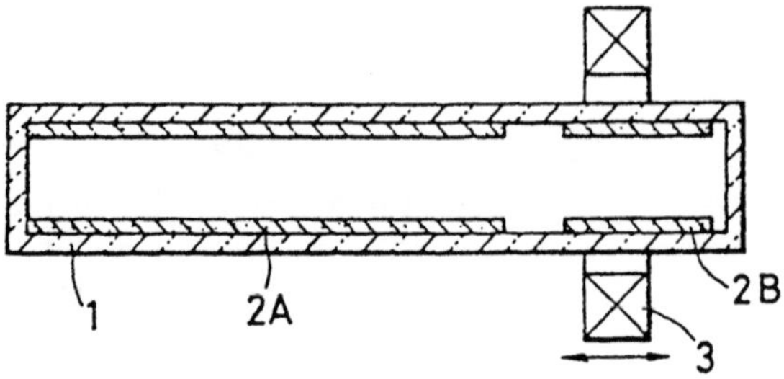

Fig. 81

Controlling can be effected by displacing electro-magnet 3, so as to displace wick portion 2B with respect to wick portion 2A.

A heat valve device in the form of a heat pipe, especially suitable for controlling heat flow from a heat source to aheat sink is subject of an invention of Basilius of Hughes Aircraft Company, (90.2). This device includes an elongated tubular housing of low thermal conductivity being closed at both ends. A capillary wick lining the inner lateral surface of the housing is adapted to convey a volatile working fluid such as water from one end to the other. A sealed working fluid reservoir adapted to being heated is in fluid communication with the housing.

Basilius' heat valve is shown in Figure 82.

Brost, O.; Schubert, P.; Groll, M. and Zimmermann, P. of Institut für Kerntechnik und Energiewandlung, (93), has improved the heat pipe operation by stabilising the operating temperature thereof with high precision and independent upon the direction of the heat flow. By their method an important additional effect of gas buffering, i.e. by filling the total cooling zone of a heat pipe with controlling gas during switching on and off of the heat pipe independant upon the operating temperature, can be obtained.

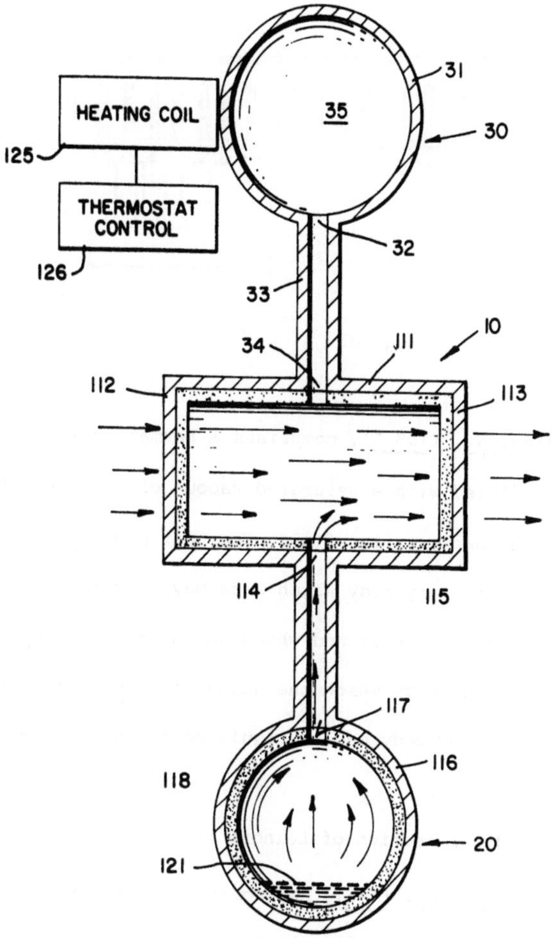

Fig. 82

The thermodynamic phenomena of temperature stabilising with heat pipes at changing heat flows is described in document EUR 4634d of the European Communities.

Brost et al. found that the above requirements can be obtained by using a tube extending from the controlling gas reservoir into the heat pipe housing, which tube has openings in the cooling and heating areas of the heat pipe.

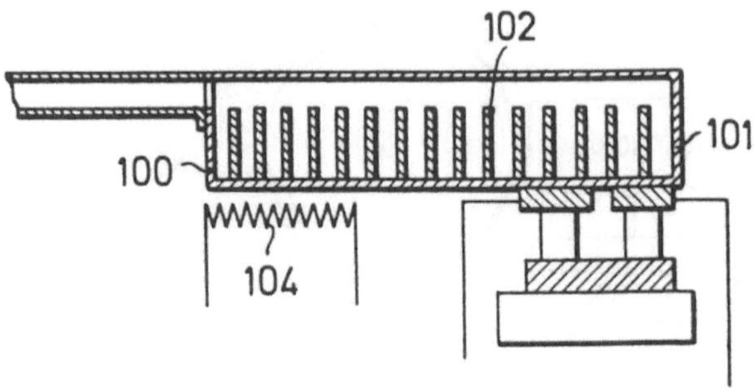

Fig. 83

The heat pipe of Laing, N., (108.1), comprises a hermetically sealed hollow body which is filled with a saturated vapour of and a small amount of the condensate of a heat carrying fluid. The heat pipe communicates with the inside of an auxiliary body which also may contain a small amount of the condensate and vapour and which has associated therewith a heat sink and a heat source by which the amount of vapour in the auxiliary body and in the hollow body may be regulated to control heat flow in the heat pipe.

Figure 83 is a longitudinal section of Laing's heat pipe.

The heat pipe of Roberts, C.C., (147), having controllably variable thermal conductance, comprises a reflux heat pipe having a capillary wick therein, an evaporator and a condenser zone, a circulating supply of working fluid which is controlled by a bendable section in the heat pipe body and an external device to control the bend in the heat pipe body, thereby controlling the amount of liquid returning to the evaporator. As returning liquid condensate is trapped in the center of the deformed tube, less liquid is available for evaporation in the evaporator. This results in a low liquid level that saturates only a portion of the evaporator wick. This results in higher thermal impedance. If the heat pipe body is deformed so that all liquid is trapped in the deformed portion,

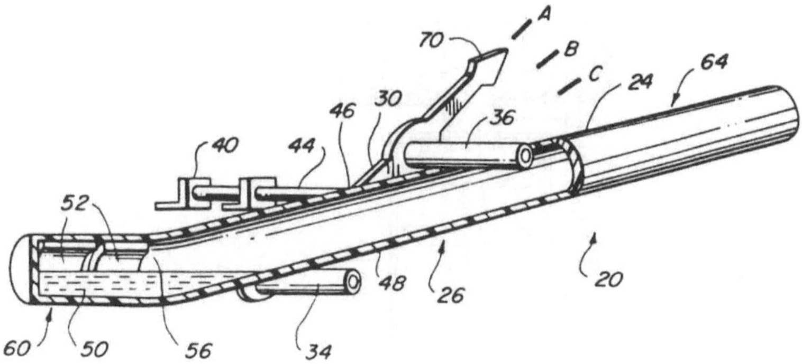

Fig. 84

no liquid is available in the evaporator and there is a high thermal impedance. See Figure 84.

The UK Secretary of State for Industry, (159.5), provided an evaporation/condensation system comprising a hermetically sealed vessel containing a liquid and its vapour together with a quantity of inert gas, the vessel having an evaporator, a main condenser, a supplementary condenser capable of holding a quantity of the liquid, means for restricting the return flow of liquid from the supplementary condenser to the evaporator, and a reservoir for inert gas, the main condenser being located in a flow path between the evaporator and the supplementary condenser and the supplementary condenser being located in a flow path between the main condenser and the reservoir for inert gas, whereby, in use, when heat is transferred by supply to the evaporator and withdrawal from the main condenser, a body of the inert gas tends to accumulate in the reservoir, the extent, if any, to which the said body extends into the supplementary condenser being dependent on the rate of heat transfer. See Figure 85.

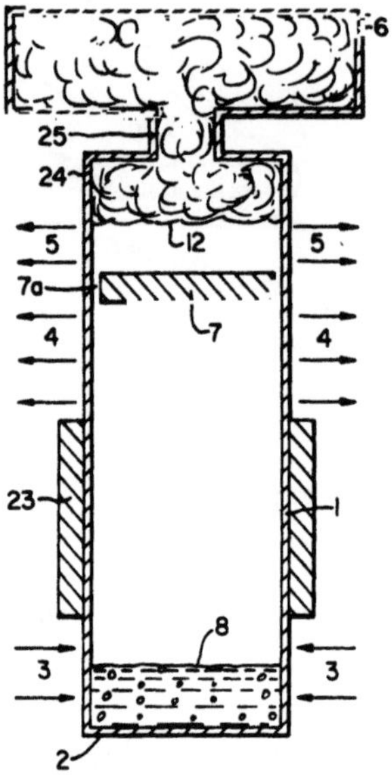

Fig. 85

The heat pipe of Martorana, R.T. of US Secretary of the Navy, (196.2), has a check valve 31 (Fig. 86), which is operated by very low pressure, and is placed in the vapour channel of a heat pipe to allow vapour to flow in a forward direction from the heat source to the heat sink. In the event that the heat sink becomes hotter than the heat source, vapour flow will reserve in direction, but will be blocked by the check valve.

The invention of Ward, C.E. of Varian Associates, Inc., (201), is a method for controlling the temperature of a primary heat-generating source in an environment whose temperature may vary widely. A heat-pipe is used to carry the heat from the source to a sink at ambient temperature. A separate control source applies heat to the heat pipe. The condensed liquid returning in the heat pipe is partly evaporated by the control heat, whereby the flow of liquid available for evaporation by the primary heat source is reduced and the heat

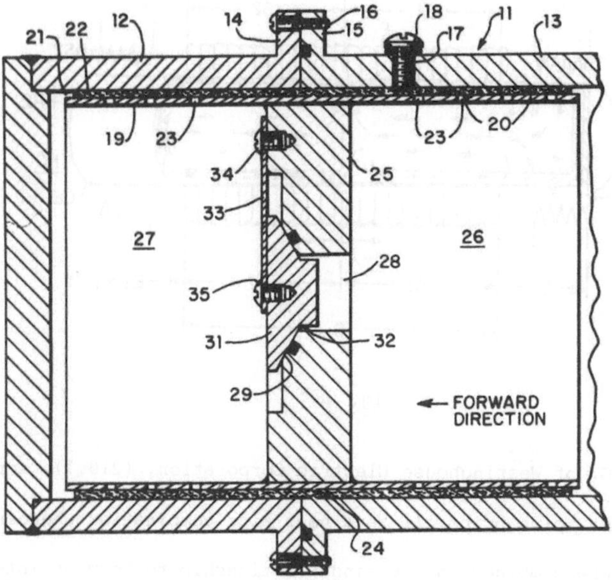

Fig. 86

conducted away from the source by the pipe is consequently reduced. The control source heat is regulated by a servo circuit to maintain the temperature of the primary source. The efficiency is higher than prior-art controllers using a constant thermal impedance between source and sink.

Figure 87 shows the principles of Ward's method.

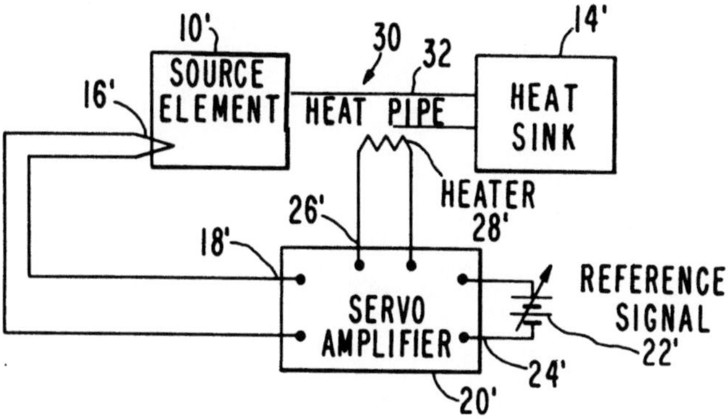

Fig. 87

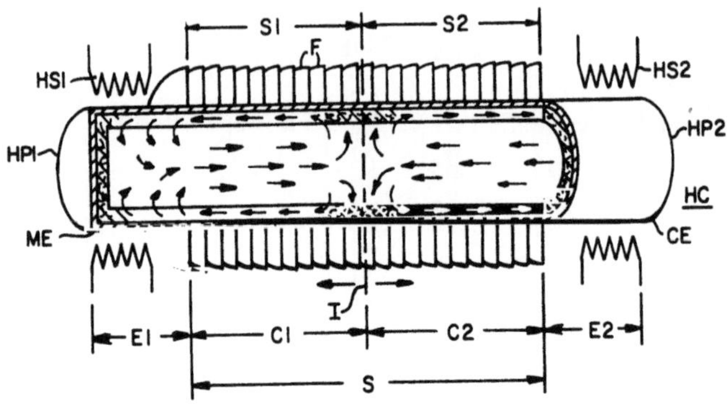

Fig. 88

Arcella, F.G. of Westinghouse Electric Corporation, (210.1), employs the basic heat pipe principle in a heat pipe combination wherein two heat pipes are combined in opposing relationship to form an integral unit, such that the temperature, heat flow, thermal characteristics, and temperature-related parameters of a monitored environment or object exposed to one end of the heat pipe combination can be measured and controlled by controlling the heat flow of the opposite end of the heat pipe combination.

Figure 88 is a basic sectioned schematic illustration of the Arcella heat pipe.

CHAPTER 5

APPLICATION OF HEAT PIPES, HEAT PIPE EQUIPPED HEAT EXCHANGERS

This chapter deals with heat exchanging apparatus, in which the heat pipe construction is of secondary order. In most of the heat exchangers described the heat pipes and the operation thereof can be adapted to the heat exchanger construction.

5.1 Multiple heat pipe exchangers comprising parallelly arranged heat pipes, e.g. flat radiators.

The heat pipe heat exchanger of D.E. Maxson, (Barber-Colman Co.), (10.1 and 10.2), has a plurality of heat pipes which are interconnected so as to permit fluid communication between the pipes at least during charging so that the heat pipes may be simultaneously filled with the heat transfer fluid.
Filling of the heat exchanger panels is of extreme importance in connection with the heat transferring capacity of the exchanger.

H. Klein, (Buderus AG), (25.2), recognised this problem and took it into consideration in designing a heat pipe panel as shown in Fig. 89a.
Fig. 89b of patent 25.2 shows a bottom view of the heat exchanger of Fig. 89a.
According to this patent, the heat pipes are simultaneously filled, evacuated and sealed.

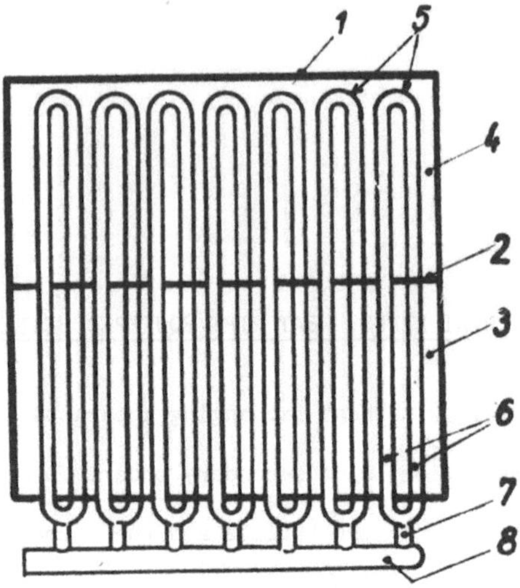

Fig. 89a

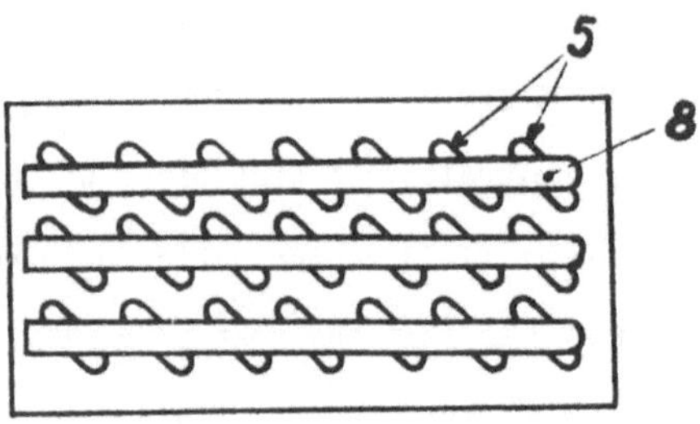

Fig. 89b

For connecting heat pipes to a heat pipe assembly, Ikin, D.D. et al, (Alcoa of Australia Ltd.), proposed in Patent (5) a connector, adapted to embrace a part of the heat pipe. At least one extremity of the connector can be connected to the pipe in a fluid-tight manner.

Fig. 90 shows an embodiment of the Ikin-connector.

The heat exchanger of R. Andres, et al., (Daimler-Benz AG) disclosed in Patent (40.9), consists of a bundle of parallelly arranged heat pipes, which may be subjected to an air flow.

A connecting box extends transversely to the pipe bundle.

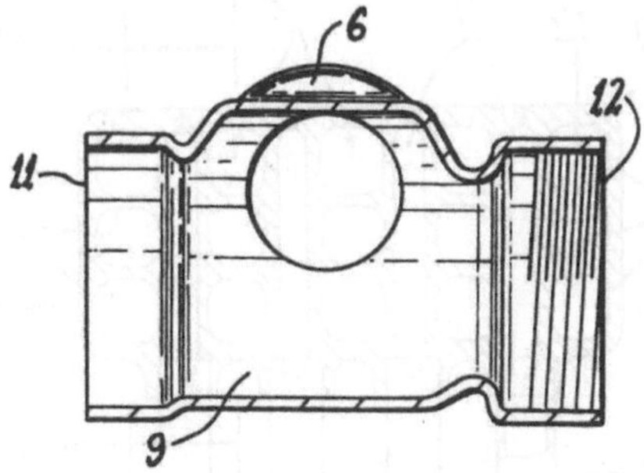

Fig. 90

Through this box a heat carrier fluid is flowing. The box is preferably made of a pressure-casted profile, the length of the profile being parallel to the rows of pipes and being provided at their frontal sides with conducting covers over the division and return of the heat carrier medium flowing through the conduits present in the said profiles. By dividing the conducting covers into two separate independent sections and by using two connecting boxes, (one being used for cooling and the other one for heating), the air flowing through the pipe bundle can be dried. The Andres-heat exchanger is shown in cross-section in Fig. 91.

In another heat exchanger, revealed in Patent (40.15), Andres et al, use a combination of the following characteristics:

1) the tubes and cavities of the heat pipes of one tube array (see Fig. 92) form a first unitary heat exchanging panel and extend into the air current, up to the area of the exchanger head, the tubes being constructed of conduits formed in the panel;

2) the heat exchanging panels are in parellel arrangement with the air current;

3) each first heat exchanging panel is connected to a second panel,

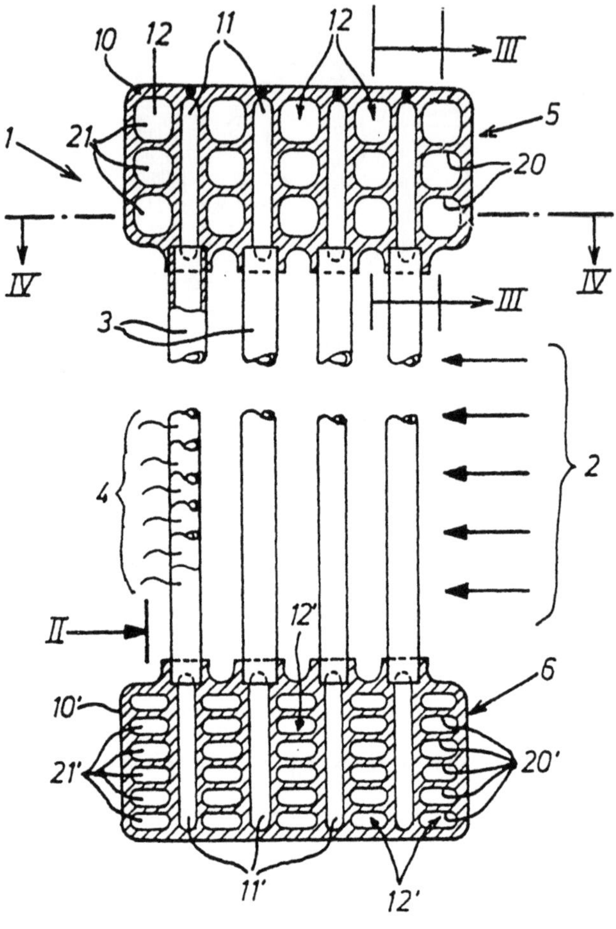

Fig. 91

containing heat carrying conduits and ends at the heat exchanger head, while the adjacent panels are soldered so as to form heat conductors. See Fig. 92.

According to still another patent of Andres et al, (40.8), triple layered bodies are provided with two separate, sealed heat pipes, forming spaces or channels (8, 9 in Fig. 93) the said bodies can be manufactured at low cost and at high production rate. In this construction several separate channels can be formed, between which a heat conducting separating wall is located.

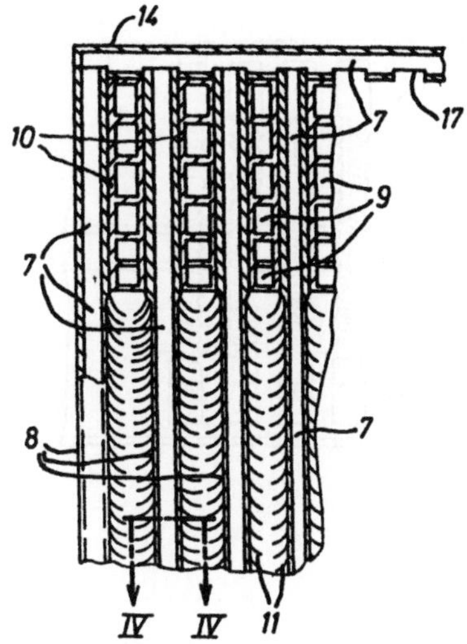

Fig. 92

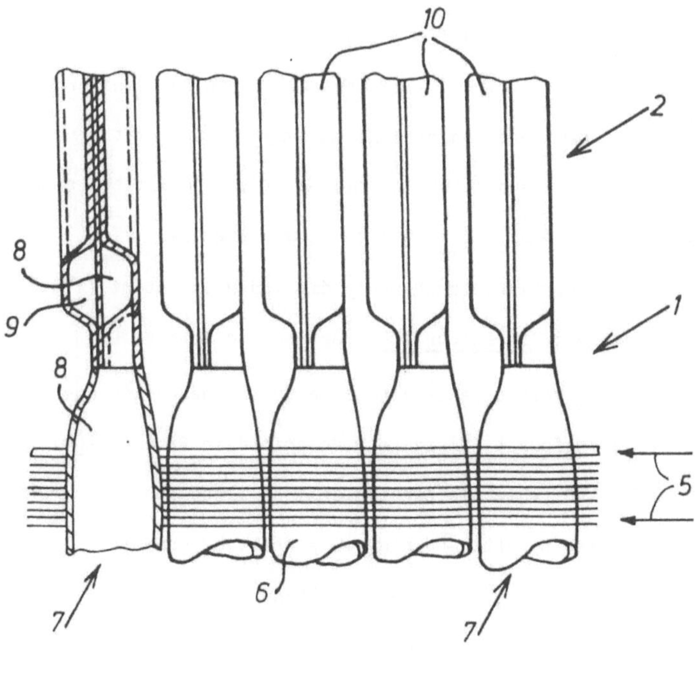

Fig. 93

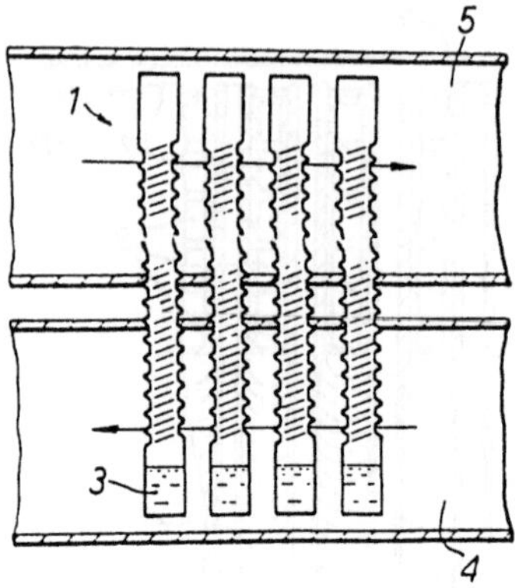

Fig. 94

The heat transfer element proposed by Q. Applegate (Curwen & Newbury Ltd.), in Patent (38.3) comprises a pipe 1 provided with spiral fluting ribs, enclosing a quantity of a vaporisable heat transfer liquid 3. Heating of one end of the pipe 1 by a hot fluid stream in one passageway 4 will cause the liquid 3 to evaporate and migrate to the other end of the pipe 1. The heat will then be transferred to a cold fluid stream in a separate passageway 5, causing the vaporised liquid to condense and return to the original end via the spiral passageways formed by the fluted ribs. See Fig. 94.

An essentially similar structure is disclosed in Patent (173.1) of Dews and Hazard of Stein Surface, represented in Fig. 95.

In this respect reference is also made to the heat exchanger of Knödel, P. of Werner & Pfleiderer, described in Patent (208) and shown in Fig. 96, which is particularly designed for the purification of exhaust gases by after-burning in a reaction chamber.

According to the invention of R. Basilius, Patent (90.8), (Hughes Aircraft Co.), a selected number of heat pipes, located in the front rows of a plurality of otherwise operable heat pipes, which are disposed between intake and exhaust ducts, have liquid-trap sections,

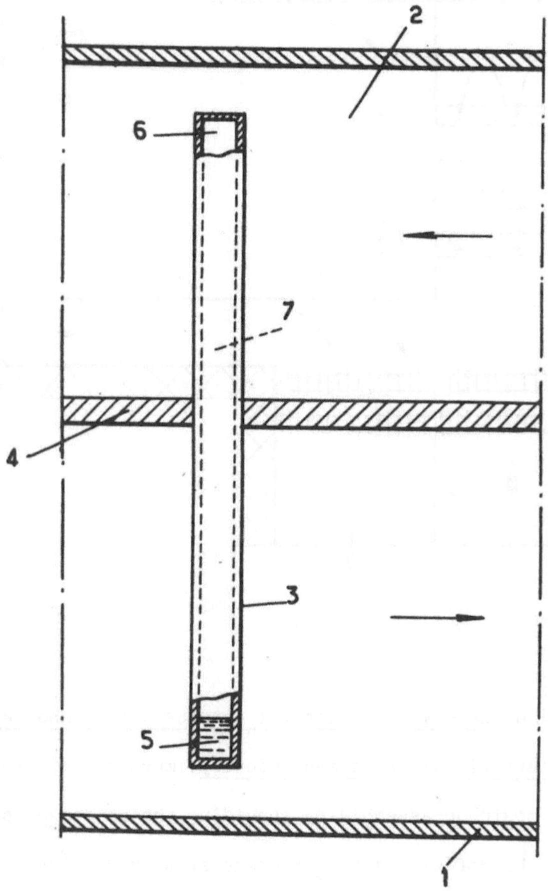

Fig. 95

extending into a switching section. During normal operation reservoirs in the switching section are dry and the plurality of heat pipes operate in a conventional manner. However, if some of the heat pipes in the exhaust duct becomes frosted or over-cooled, due to excessive cold in the intake duct, thermostatically or command-controlled valves or louvres cause the fluid in the exhaust duct to warm up or defrost the excessively cooled heat pipes therein. Prevention of excessive cooling is used to avoid frost build-up in air conditioning equipment, or solidification of solids and condensation of corrosive liquids.

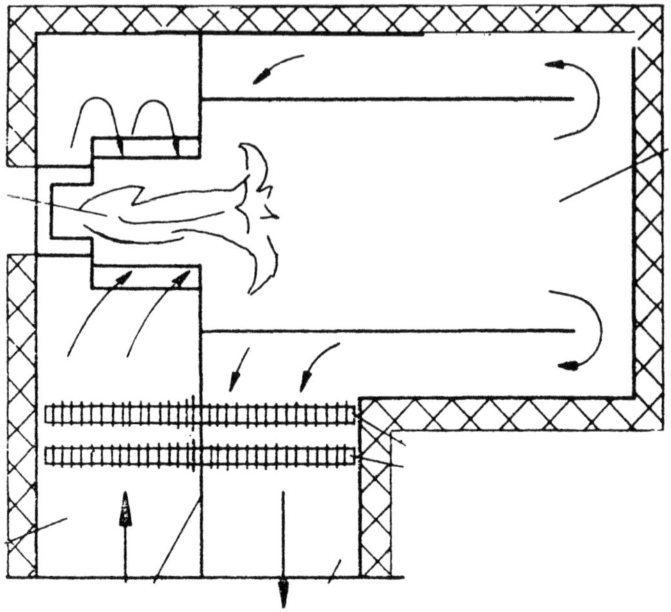

Fig. 96

Masao Shiravshi of Agency of Industrial Science and Technology; Ministry of International Trade and Industry improves in Patents (2.1 and 2.2) a thermosiphon assembly by providing the condenser section on the outer wall thereof with a heat storage section and further providing the condenser section in the lower portion thereof and adjacently to the heat storage with a fluid reservoir for receiving the working fluid which has been condensed in the condenser section. In this thermosiphon, the working fluid which has been vaporized by the absorbed heat is condensed on the inner wall of the condenser section with liberation of part of its heat to the exterior of the thermosiphon. The remainder of the heat is accumulated in the heat storage section.

See Fig. 97.

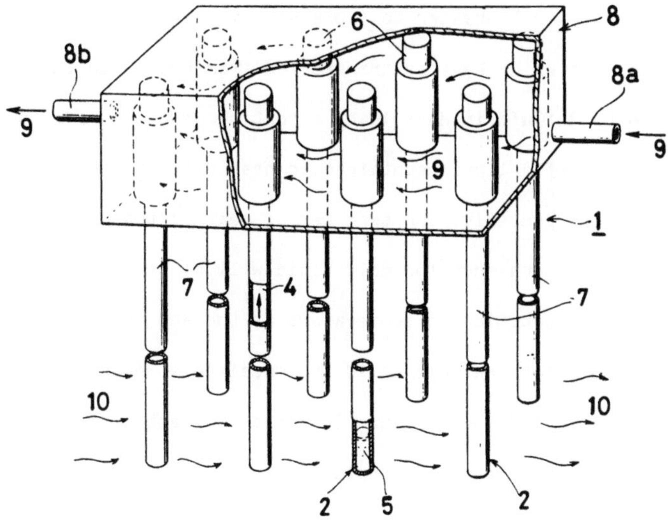

Fig.97

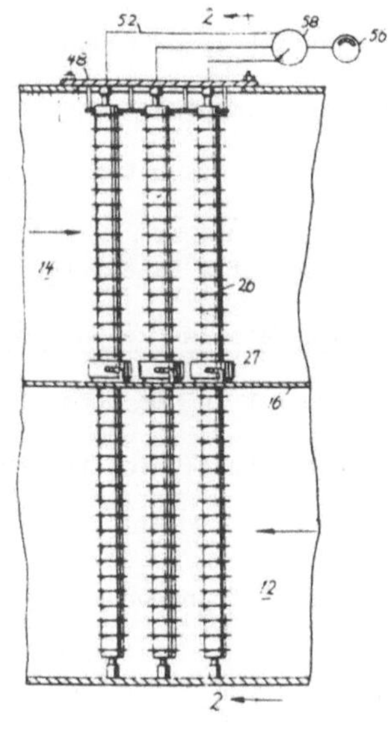

Fig. 98

R.F. Storkman of Air Preheater Co., presents in Patent (3) a heat pipe heat exchanger in which heat is transferred from hot fluid in duct 12 (Fig. 98) to cold fluid in duct 14 by a continuous cycle of evaporation and condensation of a liquid contained in heat pipe 26. Should the heat pipe not function properly, the transfer of heat to that portion of the heat pipe lying in the duct for cold fluid will be reduced partially or completely. This reduction in temperature at the end of the heat pipe lying in the duct for the cold fluid is immediately sensed by a thermocouple 5 and indicated upon a gauge 56 which is adapted to identify a particular heat pipe. Repair or replacement of a faulty heat pipe will return each heat pipe to its maximum effectiveness. See Fig. 98.

The heat exchanger of P. Paikert of GEA Luftkühler Ges., described in Patent (74.1), consists of a number of externally ribbed heat pipes, filled with methylene chloride and transversing a separating wall, which separates two fluids, having different temperatures. According to Paikert such heat exchanger can be improved by constructing the heat pipes of an evacuated heat pipe hose, which is closed at both ends and is supported by the separating wall. As an important advantage of the invention it is claimed that the hose has to be evacuated and sealed only once.

A similar idea is the subject of a Patent (100.3) of P. Rohner and K.D. Schipple of Kabel- und Metallwerke Gutehoffnungshütte AG., according to which the heat pipes are not in the form of a multiple bent hose, but are formed of a multiple bent tube which traverses the separating wall several times.

Fig. 99 shows a so-called heat transformer, which can be powered by any type of heat source.

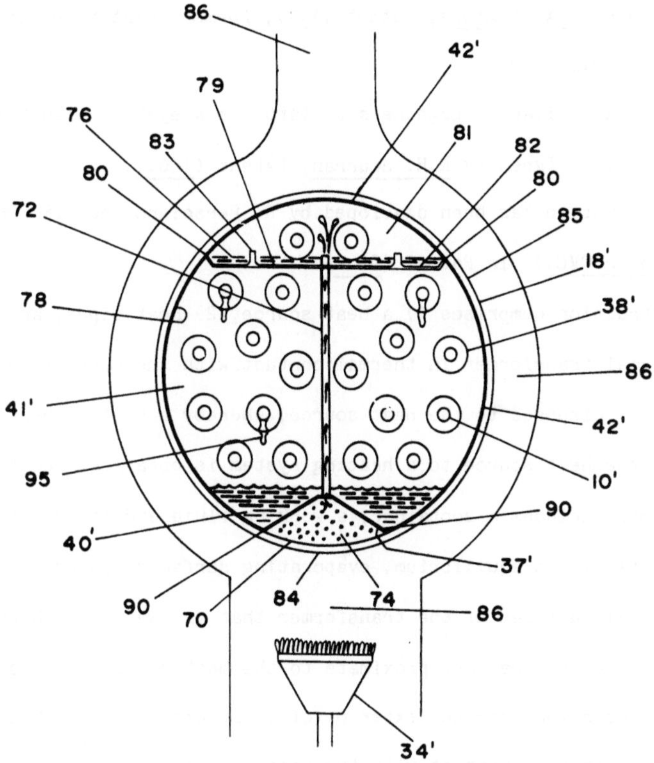

Fig. 99

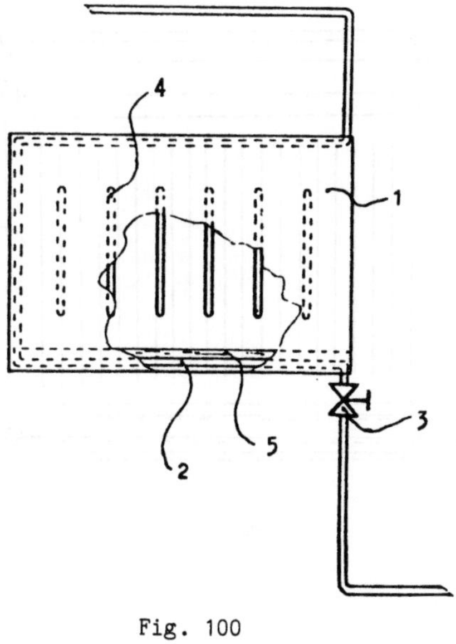

Fig. 100

A radiator of J.X. Lognone, Patent (111), forms a heat tube panel as shown in Fig. 100.

More or less similar to Lognone's radiator is a system, illustrated in Fig. 101, designed by N.H. Scurrah, Patent (156.2).

This construction has been developed by R. Persolano and R.B. Rhodes, (Isothermics OVC.), in Patent (99.1).

The construction comprises 1) a heat source, 2) heat pipes, and 3) a thermal transformer in thermal contact with the heat pipes and proximately disposed to the heat source, whereby a uniform heat transfer from the heat source to a heating system is obtained. The thermal transformer includes a container, a working fluid and its vapor which are essentially in equilibrium, evaporating surfaces depending from the internal surfaces of the transformer that are wetted with the working fluid and that are proximate to the heat source, and condensing surfaces depending from the exterior of those areas of said heat pipe that are in thermal contact with the said transformer.

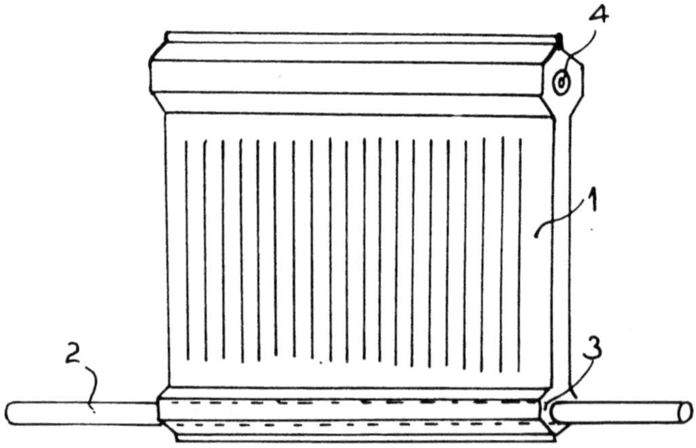

Fig. 101

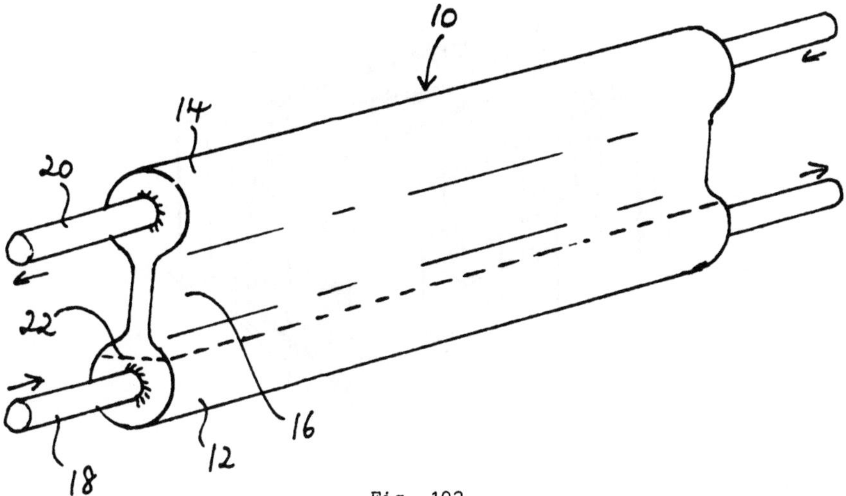

Fig. 102

Another heat exchanger of N.H. Scurrah, (156.3), shown in Fig. 102 comprises a waisted, elongate, sealed jacket 10 having elongate upper and lower bulbous portions 12, 14 above and below the waist 16 and through which upper and lower pipes, respectively 10, 18, extend with clearance. The sealed jacket 10 contains a refrigerant or similar vapourising fluid 22, the liquid phase of which is vapourised by contact with the lower pipe when a heated fluid is passed through, thence to condense on the upper pipe through which a cooler fluid is passed, thereby transferring heat to said cooler fluid, the condensed fluid dropping back to the bottom of the jacket for revapourisation. A gas conveying duct-type heat exchanger has been designed, having two gases passing therethrough, preferably in counterflow relationship. An easily assembled, sectioned gas separator is provided to divide the duct into separate gas conduits. The separator sections are designed to fit with each other and to support heat pipes extending between adjacent conduits for transferring heat from one conduit to the next. The structure is adapted for ready on-site assembly and modification,

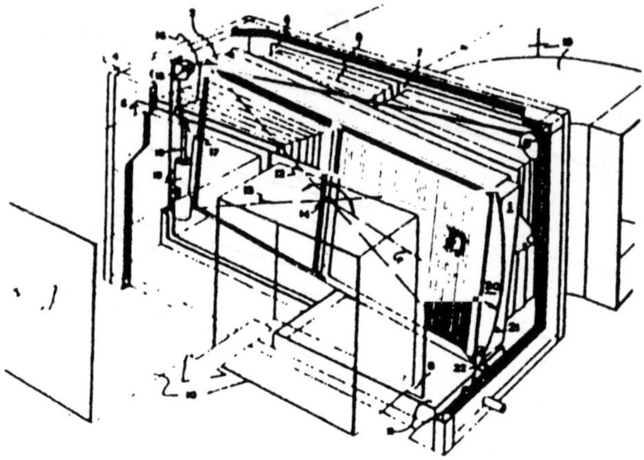

Fig. 103

thus providing facility in relating the constructed device to the heat load in a given installation and permitting shipment of a unit in disassembled condition. This heat exchanger has been developed by E.D. Waters, in Patents (121.5 and 121.7), assigned to the Mc.Donnell Douglas Corp.

Figures 103 and 104 illustrate two embodiments of a counter-current type heat exchanger, invented by W.E. Schönmann, (Patents 153.1 and 153.2)

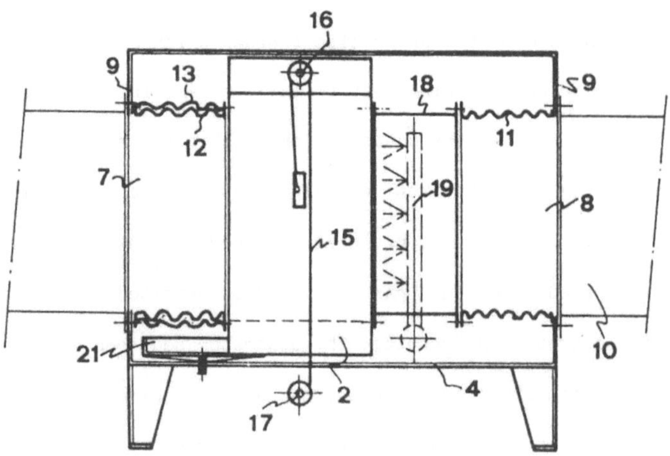

Fig. 104

containing heat exchanging registers, which are tiltably mounted. Register 3 consists of an evaporator 1 and a condenser 2, which are completely embraced by a housing 4.

A cable pull system effects the tilting of the register to a predetermined position.

For realising high heat transfer rate and uniform heat transfer such as for laminators, D.S. Sarcia, (Seal Inc.), developed in Patent (157) a heat transfer roller, comprising a heat pipe 10, in Fig. 105, which is mounted on a shaft 16, rotatably supported by bearings 32 and 33.

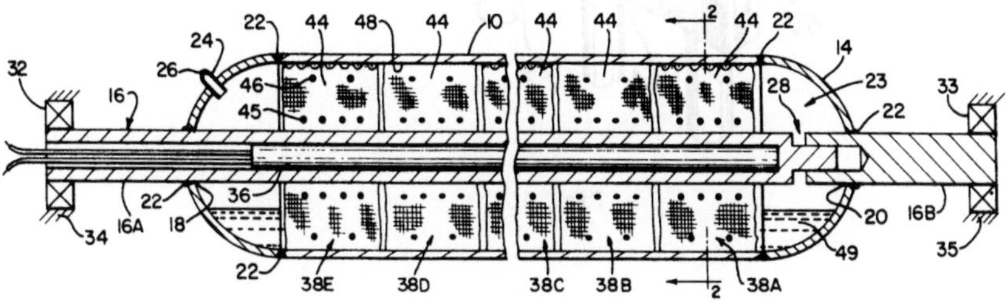

Fig. 105

M.J. Gravency, (assignee: The Secretary of State for Defence) presents in Patent (158.2) a heat exchanger suitable for use at a range of temperatures and pressures and with a pumped heat pipe cooler assembly comprising a pressurisable vessel 10 having a working fluid inlet 20, a working fluid outlet 21, and a rigid heat exchanger; a wick 15 in sheet form outside the vessel and contiguous with the heat exchange wall, and a liquid supply means 16 for supplying liquid to the wick. Air from a blower 25 evaporates the liquid on the wick to provide a cooling effect which condenses vapour entering the vessel at inlet 20. See Fig. 106.

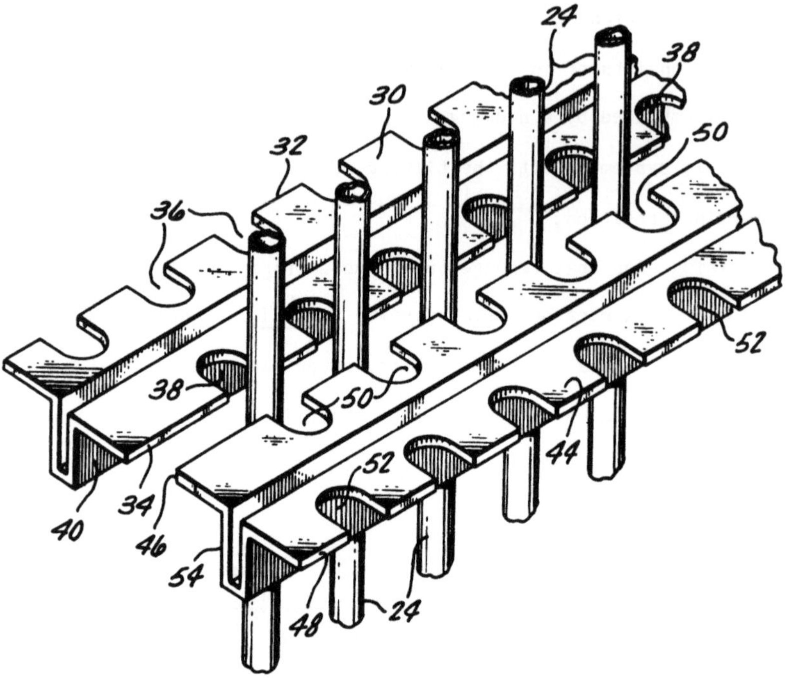

Fig. 106

In the closely coupled two-phase heat exchanger of G. Eastman, (Thermacore Inc.), Patent (181.5), the heat is transferred between gas streams by multiple intermediate fluid paths (Fig. 107).

Each of the several intermediate fluid paths passes alternately between the hot and the cold gas stream many times during the travel from the pump. The heat exchanger is designed so that the intermediate fluid vaporizes in the hot and condenses in the cold section each time, thereby minimizing the quantity of liquid necessary to transfer the heat. The pumping action for the intermediate fluid exchanger is accomplished by either mechanical or vapor pumps. An added feature of the heat exchanger is automatic control of the minimum temperature to which the hot gas is cooled. This is accomplished by shutting off one of several intermediate fluid paths at the cool end of the gas

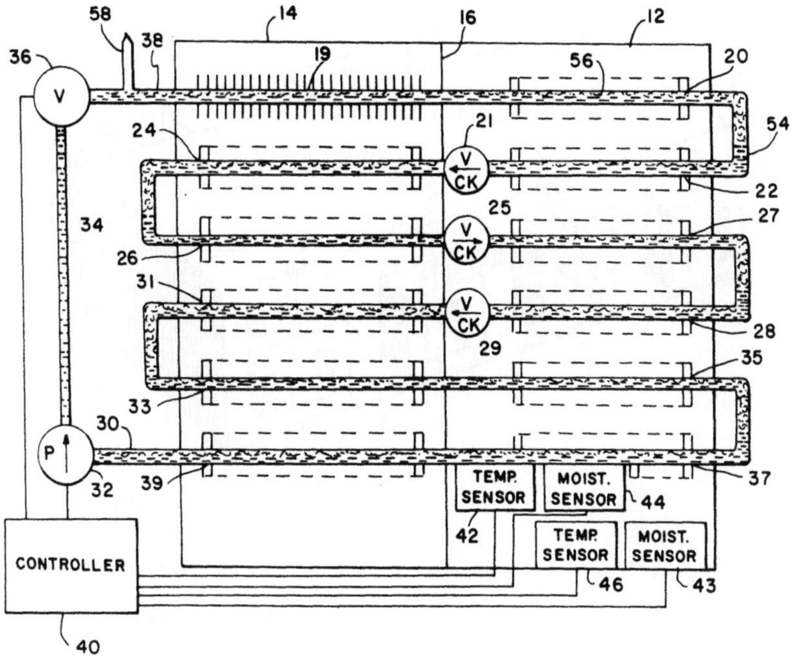

Fig. 107

being cooled by valves 21. The shutoff is accomplished either on the basis of temperature of the cooled gas or on the formation of condensates at the cool end of the gas flow path. See Fig. 107.

The heating device designed by Asselman and van der Leegte, (Patent 197.6), assignors to U.S. Philips Corp., comprises a plurality of heating chambers, each chamber being bounded by the heat-transmission wall of a reservoir in which a heat transport medium is present which completes an evaporation/condensation cycle. The reservoirs are connected, via a common reservoir which also contains heat transport medium, to the same common heat source. Fig. 108 is a perspective view of Patent (197.6).

The heat exchanger of G. Köster, (Weisstechnik GmbH.), revealed in (205), comprising parallelly arranged heat pipes and a separating wall, separating the hot and cold sections of the heat exchanger,

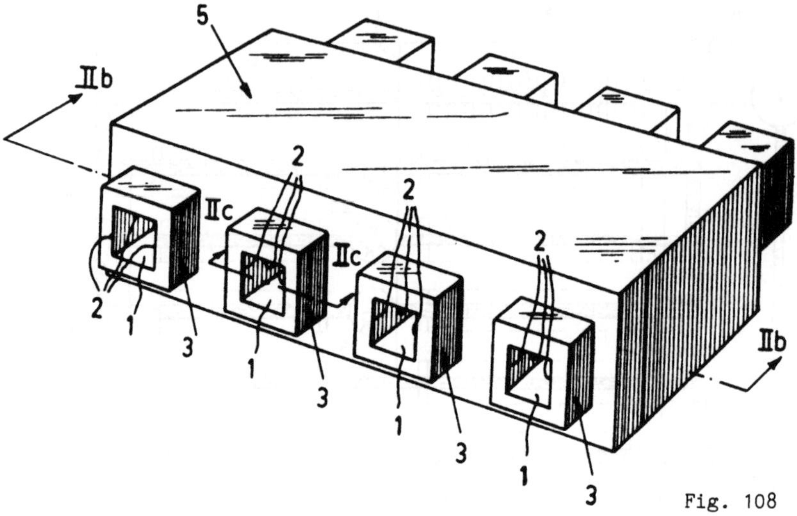

Fig. 108

has been so designed that the heat exchanger can be easily cleaned. To this end Köster provided a plurality of heat pipe registers enclosed in a heat exchanger frame. The registers can shift sideward and inward out and into the frame. Köster also proposed the use of an additional washing register, which is connected to a source of washing liquid and is provided with spraying nozzles.

See Fig. 109.

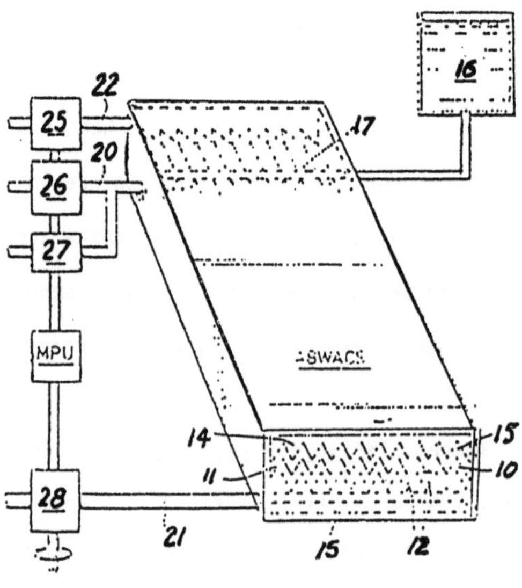

Fig. 109

5.2 Heat exchangers consisting of heat pipe bundles, e.g. rotary heat exchangers.

In order to realise a compact heat exchanger structure of moderate space requirement and optimum dynamic conditions for the heat transferring fluid currents, H. Lerch, et al. of Balcke-Dürr AG., proposes in Patent (9) a heat exchanger, wherein two fluid currents of different temperature are guided through separated channels (in Fig. 110: 1 and 2), traversed by a multirow matrix 8 of heat pipes 7. The matrix 8 is arranged on a polygon, a circle or a star, around which radial fluid currents are flowing. Fig. 110 represents a longitudinal section of the proposed structure and Fig. 111 the section II - II of Fig. 110.
A rotary heat exchanger, revealed by G. Applegate of Curwen & Newberry Ltd., in Patent (38.2), with two fluid systems, comprises: a rotary unit disposed in a casing, each fluid system having an aligned inlet and an outlet to the casing, and including an array of a plurality of heat pipes arranged around an axis, each pipe containing a thermal fluid, and means for rotating the unit about the axis, with the respective two end sections of each pipe of the unit spaced along

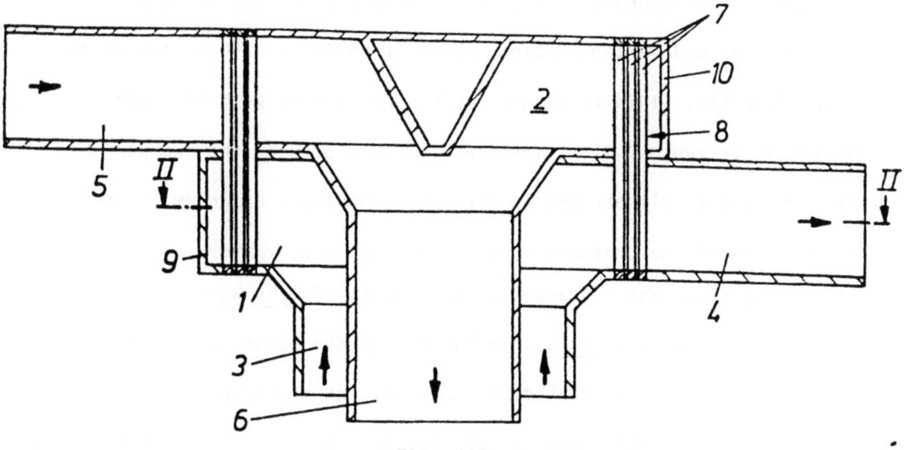

Fig. 110

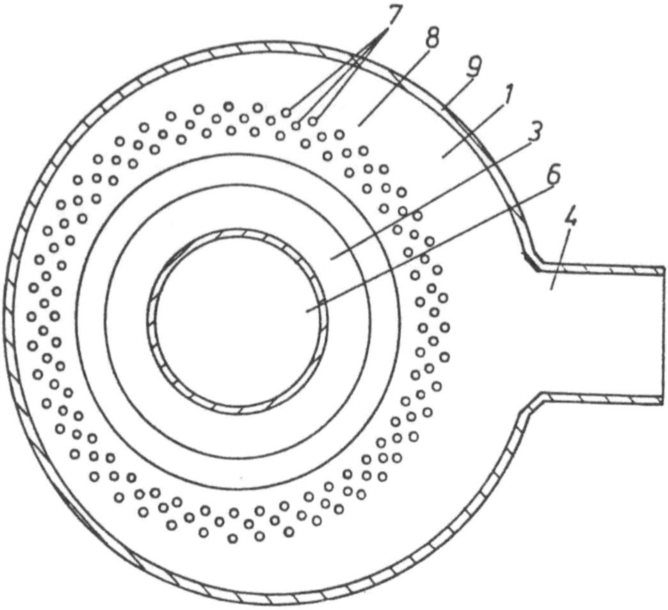

Fig. 111

the axis and exposed respectively one to each of the two fluid systems, the dimension to each inlet which extends orthogonal to the axis of rotation and transverse to the axis of the inlet being substantially greater than half the transverse dimension of the array. Preferably the apparatus includes a rotary drive means for turning the unit at a controlled, and variable speed, about its axis, and there may be at least one rotary fluid seal surrounding the unit between the ends thereof to prevent or limit movement of fluid between the two streams. The invention is illustrated in Fig. 112, showing a diagrammatic perspective view of the proposed rotary heat pipe regenerator.

A condenser-type heat exchanger of K.H. Junke, (Funke Wärme-Austauscher Apparatebau K.G.), in Patent(69) and shown in Fig. 113 comprises a closed casing subdivided by at leat one transverse bulkhead 16 into two cylindrical chambers 18, 20, each having a fluid inlet (e.g.

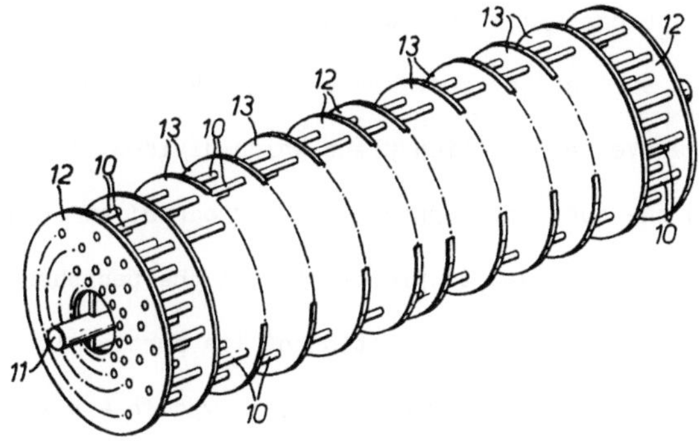

Fig. 112

22 or 24) and a fluid outlet (e.g. 26 or 28), and heat transfer tubes 40 (e.g. heat pipes) sealed into the bulkhead(s) and extending therefrom into each said chamber, the tubes acting to transfer heat between two fluids (e.g. oil and cooling water) flowing respectively through the chambers 18, 20. The bulkhead may comprise an elastic disc, sandwiched between two plates each comprising a chamber end-wall. Baffles 36 provide tortuous flow paths for the fluids.

According to Patent (70), assigned to the Furukawa Electric Co. Ltd., the space requirements of heat exchangers can be reduced and a uniform gas supply can be ensured by a construction, comprising: a cylindrical

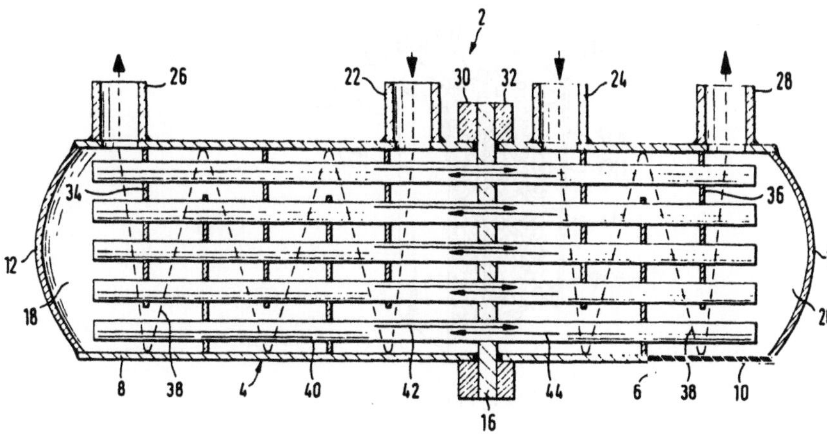

Fig. 113

or polygonal tubular casing; a duct which extends from the circumferential wall of the casing so that the duct and casing together have the form of a volute; a transverse partition plate which divides the interior of the casing and the duct into upper and lower parts; an array of heat pipes vertically arranged to pierce through the peripheral area of the partition plate, with the exception of its portion located in the duct, and in an annular configuration as viewed in a plan view thereof. The upper and lower parts of the partition plate inside the casing and the helical duct are arranged to form flow passages in a counter-flowing manner with respect to each other, the part of the heat pipe array above the partition plate being arranged to serve as a heat radiating portion which allows a low temperature gas to be heated to flow there and to come into contact therewith, and the part of the heat pipe array below the partition plate being arranged to serve as a heat receiving portion which allows a high temperature gas to flow there and thus to come into contact therewith. Fig. 114 is a perspective view of an arrangement of the heat pipes, while figures A to D show different configurations of the heat pipes of Fig. 114.

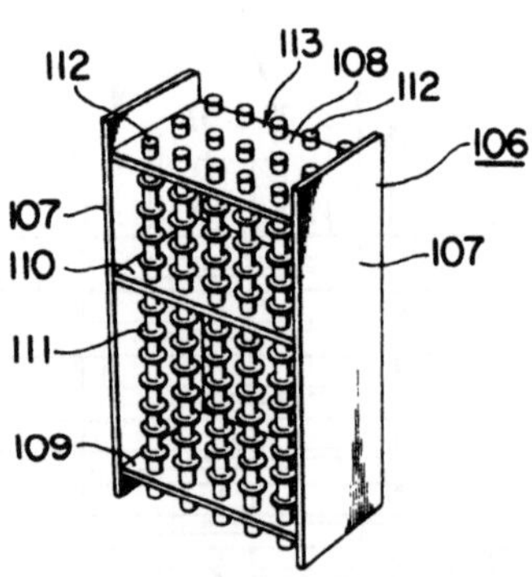

Fig. 114

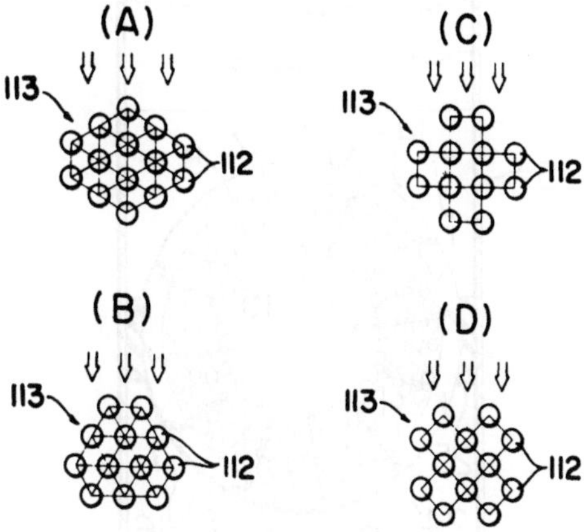

Fig. A to D

The rotary type heat pipe heat exchanger of Eiji Okamoto of Gadelius K.K., revealed in Patent (71.1), and shown in Fig. 115, comprises a first flow path (13, 14, 15, 16, 17), along which a high temperature fluid flows, a second flow path (18, 19, 20, 21, 22) along which a low temperature fluid flows, and a rotor 28 formed by a plurality of heat pipe 8 disposed across the two flow paths with such intervals that the high temperature fluid or the low temperature fluid can pass therebetween and transferring heat from the high temperature fluid to the low temperature fluid.

The rotor 28 has a separating member 9 fixed to the rotor and separating the first flow path (14, 15, 16) therein from the second flow path (19, 20, 21) and is rotatable. By this, corrosion, decreased wall thickness and clogging of the heat pipe is prevented from being localized in a particular portion thereof and the maintenance work of the heat exchanger can be facilitated.

In another patent, (71.3), Eiji Okamoto et al, (Gadelius K.K.), provide a fixed-type heat pipe heat exchanger, which can be used to recover heat from the combustion gases containing dust, sulfur oxide

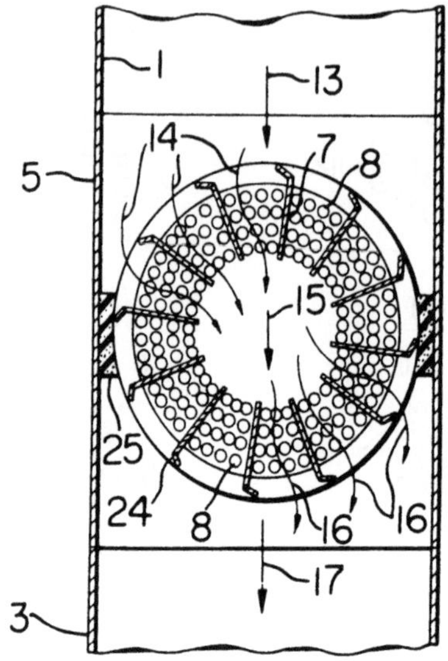

Fig. 115

(SO_x) and the like at high concentration, exhausted from a large-sized boiler of a steam-power plant and to utilize the heat effectively, and wherein the gas passages are not blocked up by dust and yet the heat pipes are not corroded by the sulfuric acid content in the exhaust gas. More particularly, in the proposed heat pipe heat exchanger groups of heat pipes are arranged within a box, the center of which is partitioned; a high temperature fluid flows on one side, a low temperature fluid flows on the other side, thereby transferring the heat of the high temperature fluid to the low temperature fluid via fluid-sealed heat pipes. In the heat exchanger, a plurality of heat pipes which are bare tubes at least on the high temperature fluid passage side, are arranged across both passages, and the heat pipes on the high temperature fluid side are inserted into pipes without fins. At least the outer surfaces of the pipes without fins are treated for corrosion resistance. Since the pipes without fins and the heat pipes are joind by a thermally conductive substance, the heat

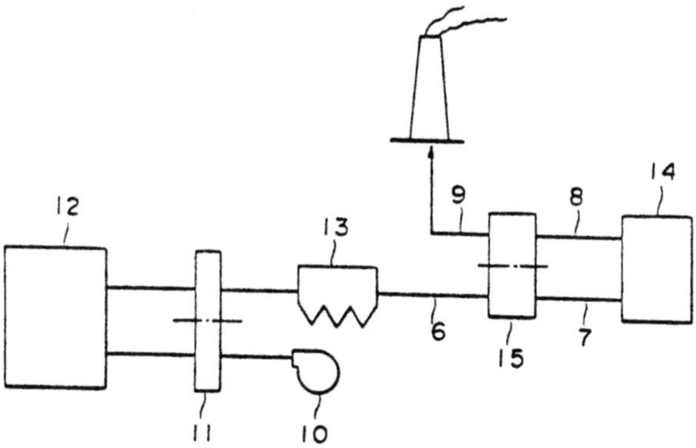

Fig. 116

pipes can be easily removed, and accordingly, the efficiency of the heat exchanger can be increased and decreased as required.

Fig. 116 is a flow sheet of an embodiment of the invention, wherein 15 indicates the proposed heat exchanger.

A rotary heat exchanger apparatus designed by W. Pozsicsányi, (Wilhelm Gebhardt GmbH.), disclosed in Patent (75.1) and illustrated in Figures 117 and 118, comprises: a rotor 6 mounted within the casing and carrying a plurality of blades 10, 11, each provided with at least one heat pipe 12, a partition wall 7, subdividing the

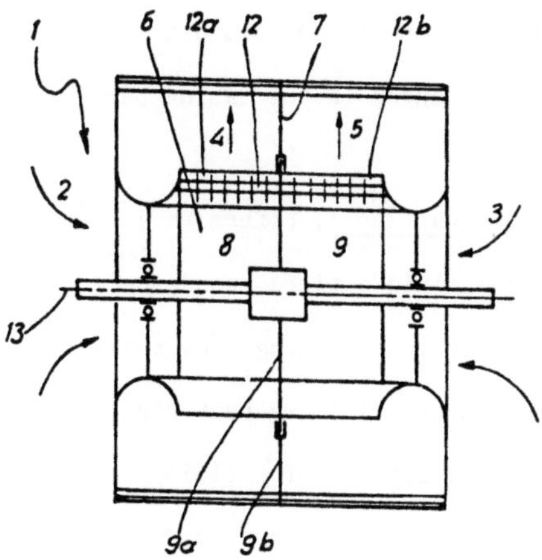

Fig. 117

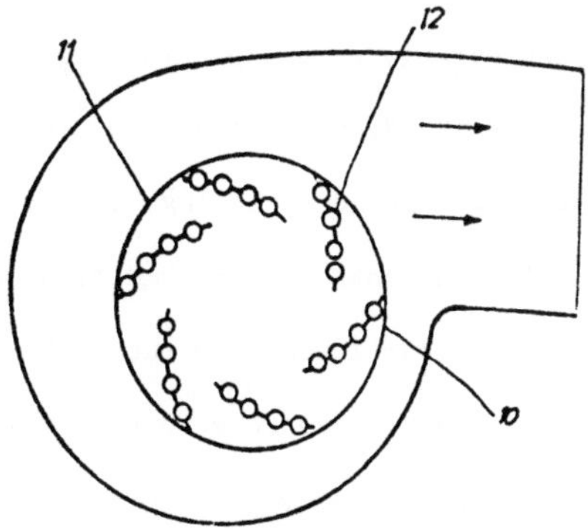

Fig. 118

casing into two separate fluid flow passages 8, 9, for first and second heat exchange fluids, the heat pipe 12 each having one end extending into said first fluid flow passage 8 and the other end extending into the second fluid flow passage 9 whereby heat is transferred between the first and second fluids via the heat pipes. The construction is relatively inexpensive, since no separate heat exchanger, pumping unit and guide channel systems are necessary, as all these are contained in a single piece of apparatus; other advantages of this invention are its compact construction, minimal space requirements, reduced assembly requirements and, above all, safety in operation requiring little or no maintenance.

G. Waldman of Gifa Planungs-Ges. für Industrie und Forschungs-Anlagen mbH., provides in Patent (78) a heat pipe, which can be inserted in a partition wall of a heat exchanger in a pressure-resistant and fluid-sealed manner, the partition wall and the pipe-system being arranged in a casing and the pipes being rigidly welded to the casing. The heat exchanger composed of such heat pipes is suitable for operating with heat exchange fluids of high pressure difference. Fig. 119 illustrates the invention, indicating with 1 the partition wall, two rooms 2 and 3, for the fluids, divided by the partition wall 1, and heat pipes 4, arranged in the casing perpendicularly to the partition wall.

Patent (90.4) of A. Basiulis, (Hughes Aircraft Co.), refers to a heat exchanger for dry cooling towers, air and fluid coolers, condensers, pollution control, air conditioning, chemical processing, and the like, in particular, to such a heat exchanger utilising heat pipes with attached gas reservoir in which solidification of a fluid to be cooled or processed is prevented. More particularly, the fluid to be cooled flows through a first duct, while the cooling fluid passes

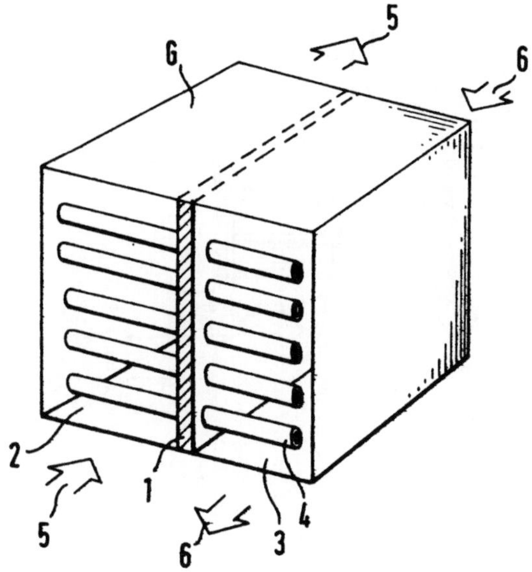

Fig. 119

through a second duct. Heat from the fluid to be cooled in the first duct passes into the heat pipes and is withdrawn in the cooling fluid second duct. A gas reservoir at the end of the heat pipes at the second duct controls the temperature of the heat pipes, maintaining a constant pressure and temperature and preventing solidification of the fluids in the first duct of the heat exchanger, while also maintaining the gas front of the working fluid in the heat pipes within that portion of the heat pipes, residing within the second duct. Fig. 120 illustrates an embodiment of the invention, utilising a gas reservoir common to the heat pipes with a diagram of changing temperature conditions, showing movement of the gas from of the working fluid. Another Patent (90.7) assigned to Hughes Aircraft Co. (inventor: W.R. Iriarte), relates to a heat pipe heat recovery unit for unidirectional, but reversible, and temperature controllable transfer of heat between at least two ducts. In the proposed heat exchanger heat pipes are in and thermally coupling two ducts, so that the fluid

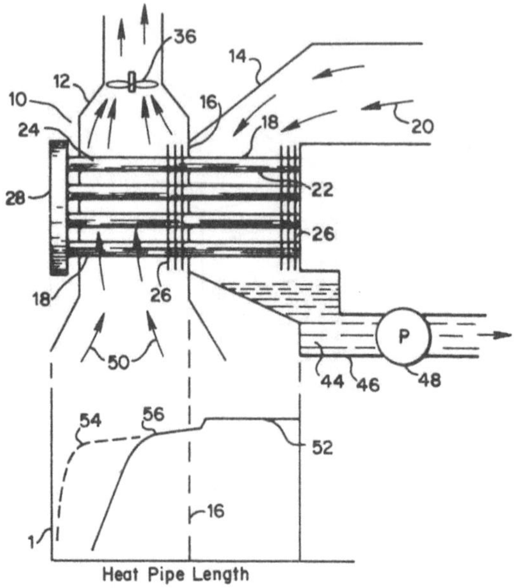

Fig. 120

in the ducts flows through the heat pipes. These pipes are angularly movable about an axis, extending through and normally to the ducts. It is important, that the heat pipes have a fixed angular offset from the said axis. Furthermore, in order to operate at higher capacity, the heat pipes are inclined at an angle to a line, normal to the gravitational force, for utilising the force of gravity in returning the working fluid to the evaporator end. Fig. 121 is a cross-sectional view of a pair of ducts of the proposed heat exchanger.

W. Schuster of Herbert Kannegiesser KG, presents in Patent (101) a rotary body for the production and/or treatment of bands, foils, webs, etc. (mainly of textile material), the rotary body being combined with a heating device, and displaying annular pockets, mounted at the ends of the rotary body and destined for receiving the evaporating fluids, the pockets having an outer diameter, surpassing the

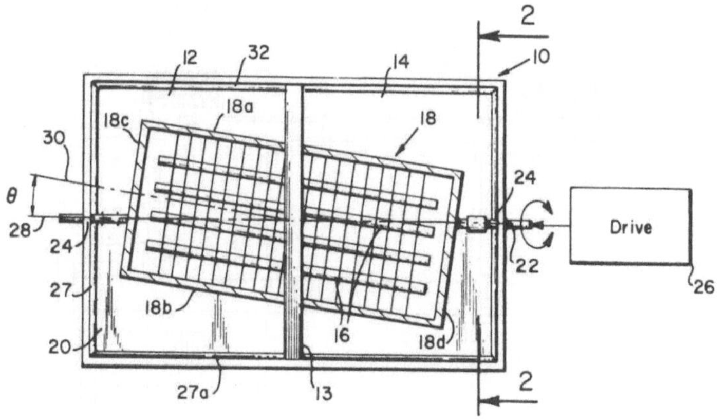

Fig. 121

outer diameter of that part of the rotary body, which contacts the bands, foils, webs produced. In the range of the annular pockets heating elements are arranged, while the pockets are filled with steel wool, displaying a spongy effect. Due to the steel wool filling, the evaporation surface of the fluid entering the pockets is considerably increased, while at the same time increasing the amount of fluids, which take part in the rotation of the rotary body, intensifying the evaporation process and improving the temperature distribution at the rotary body's surface. The heating elements are operated by electric current and are provided with heat and pressure sensors.

Fig. 122 illustrates the described rotary body, with annular pockets. The heat pipe heat exchanger provided by Hayashi Kazuo of Nissan Motor Co., in Patent (129), features an impeller wheel, comprising an axis around which it rotates, furthermore at least one support member and a set of several closed and evacuated pipes, made of a heat-conducting material, which are spaced off from each other and are carried by the support member on a ring in such a way, that they are in parellel with

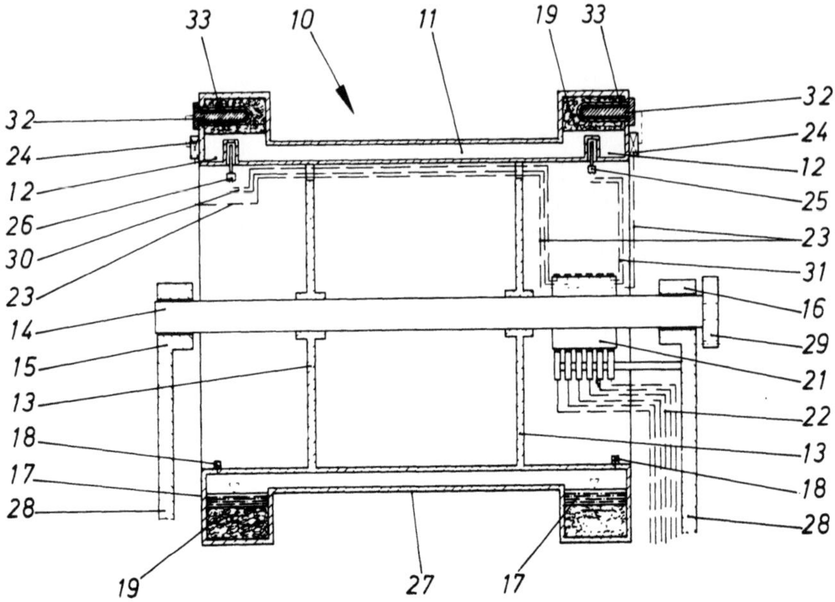

Fig. 122

the said axis and form the impellers (vanes) of the wheel. The inner face of each pipe is coated with a porous material, which is saturated with a fluid, which can evaporate in the pipe at a temperature, that is lower than the temperature of the fluid entering a sector of the pipes. Since a sector of the pipe is exposed to a first fluid, while another sector thereof - to a second, cooler fluid, heat exchange takes place by evaporation of the fluid in the first sector and condensation of vapour in the second sector.

Fig. 123 illustrates a lateral view of the proposed heat exchanger, (12 - impeller wheel;14 - casing; 16, 18 - separate fluid ducts; 20 - rotary axis; 22 - vanes).

G. Reisinger discloses in Patent (145) a system of axially or radially activated heat pipes, which can be cleaned by means of a scraper, the required motion of the pipes necessary for the cleaning being realised by an external drive or by the axial stroke, developing

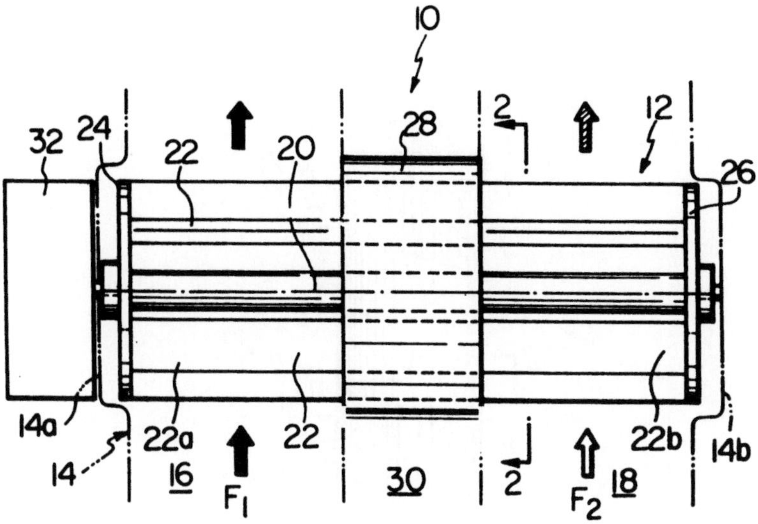

Fig. 123

during temperature fluctuations. The rotating motions of the pipes permit a uniform moistening (by the condensate) and a uniform tempering of the inner evaporation surface, thereby allowing the use of pipes of larger diameter. The inner condenser surface has a screw-like structure, which intensifies the condensate transfer and increases the inner heat exchange surface. The device can also be used as transport aggregate and by varying the rotation rate of the pipes, the heat transfer can be closely controlled.

5.3 Heat exchangers having their heat pipes in alternative arrangement

J.F. Maidanik, et al, (Otdel Fiziko-Tekhnicheskikh Problem Energetiki Uralskogo Nauchnogo Tsentra Akademii Nauk SSSR.), disclose in Patent (132.2) a heat exchanger, represented in Fig. 124. The heat exchanger comprises: an evaporation chamber 1, with a coaxial evaporator 3, made of a capillary material, that is saturated with a heat carrier medium, in contact with a heat source, furthermore a condensation chamber 4.

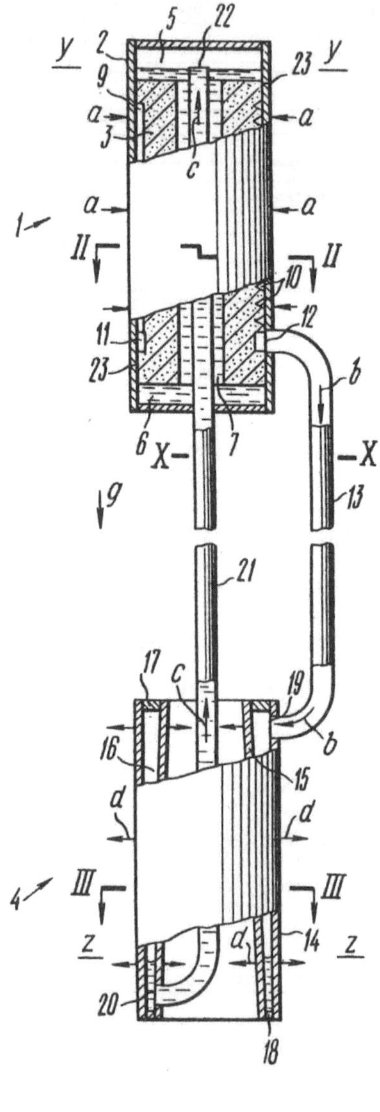
5
22
y
2
y
23
9
a
a
3
c
a
a
1
II
II
10
11
12
23
b
6
7
X
X
g
13
21
17
c
19
16
b
15
d
d
4
III
III
z
z
14
20
18

Fig. 124

The evaporator 3 is provided with vapour evacuation channels, which are in connection with a vapour collector 11, and with a longitudinal axial channel 7, which communicates with spaces 5 and 6, limited by an end face of the evaporator 3 and by the walls of chamber 1. The evacuation, channels are executed in the form of longitudinal grooves 9 and annular grooves 10, arranged at the outer surface of the evaporator 3 between annular shoulders 23. The chamber 4 has the form of a ring 14, in the interior of which another ring is disposed in coaxial arrangement so as to form a space 16, which is isolated from the surrounding medium and the transversal section of which decreases in the direction of the vapour current across the space. The effectiveness of the proposed heat exchanger is intensified by increasing the evaporation surface of the evaporator; by reducing the hydraulic resistance of the condensation chamber and by lowering the vapour temperature in the upper space of the evaporation chamber.

Patent (173.3) of Stein Surface S.A., relates to a heat exchanger, essentially comprising: two super-imposed chambers, destined for the circulation of a hot and a cold fluid current, furthermore, pipes, introduced partly into the said chambers but not communicating with them and containing a heat carrier fluid and an inert gas. The quantities of fluid and gas are so established, that at the operative temperature a certain amount of non-evaporated fluid remains in the pipe, while the volume occupied by the inert gas is smaller than the volume of the pipe disposed in the upper space, traversed by the cold fluid.

A feature of this construction is that the upper space is formed by a chamber, surrounding the upper sector of the pipe, which is in connection with the feed and discharge ducts of the fluid.

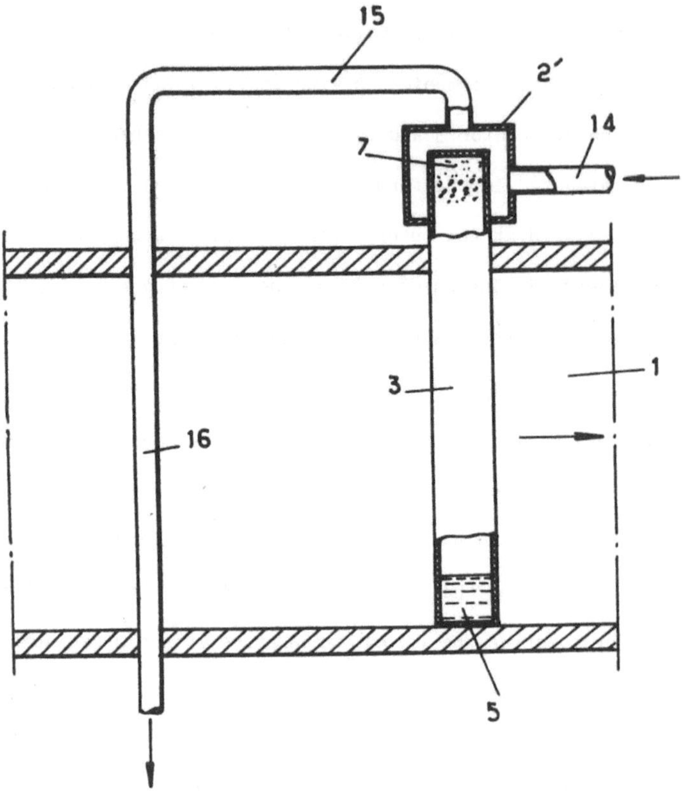

Fig. 125

This chamber may be connected to several tubular elements, (pref. vertical), which pass through the lower space, above the pipe, the said tubular elements intensifying the heat exchange between the hot and cold fluid. See Fig. 125.

Patent (174) of S. Michalak, (L. & C. Steinmüller GmbH.), describes a regenerative heat exchanger in which the heat transmitting elements 1 comprise hollow balls 2 filled partially with a heat conducting liquid 3 and partially with vapour 4 from the liquid. The heat transmitting elements may be used either as a fluidized bed or as a static bed. The elements may be carried in the rotor of a

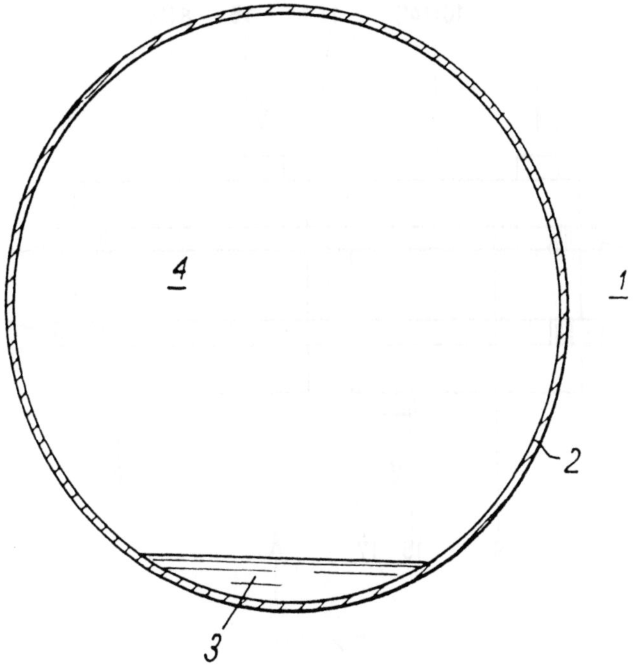

Fig. 126

rotary heat exchanger or circulated through the stationary column of a column heat exchanger. In Fig. 126, 3 - indicates the fluid and 4 - the vapour, formed from the fluid. The fluid may consist of water, freon, alcohol. The wall of the ball may be made of a metal or a non-metallic material.

M. Hage et al, (Wieland Werke AG.), provide in Patent (211.1) a heat exchanger, operating under especially safe conditions, with counter-current flow of the heat transfer media. The proposed exchanger, according to Fig. 127 comprises a cooling agent container 8 and a fluid container 10, into which water-filled heat pipes 9 are inserted. The heat pipes 9 traverse a double wall 11. The fluid container is formed by part 14', the cooling agent container 8 by part 14" of a casing 14, surrounding the heat pipe 9. The

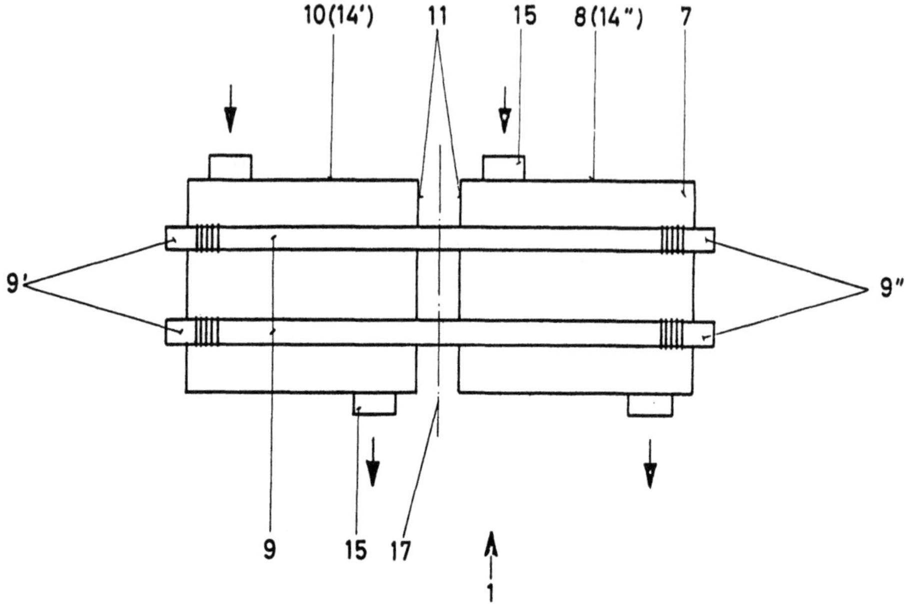

Fig. 127

double wall 11, disposed in the middle 17, is created by the connection of parts 14' and 14" with the heat pipes 9. Beside the double wall 11 apertures 15 are provided for the cooling agent and the fluid.

5.4 Solar heat exchangers using heat pipes

Patent (4.1) of M.J. Duncker, (Akzo N.V.), concerns a solar collector provided with a liquid-absorbing layer. More particularly, the solar collector comprises a closed housing 1, 2, a collector thermally connected to the housing for converting solar rays into heat, a working fluid, an evaporation space 3 within the housing, a condensation space 4 within the housing, an internal connection between the said spaces, and a heat exchanger. In order to obtain a uniform distribution of working fluid over the internal surfaces

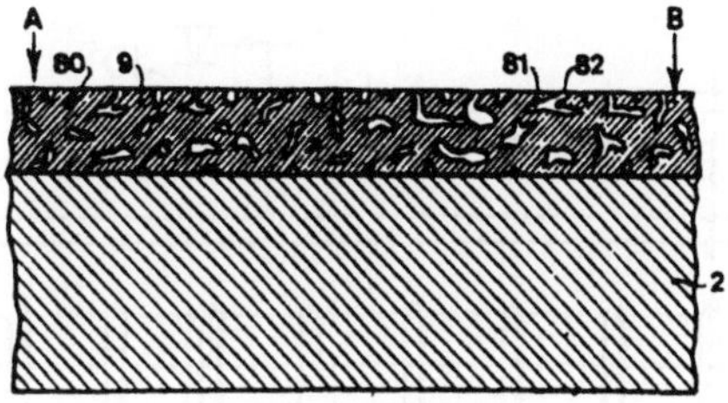

Fig. 128

of the housing, the collector is covered over at least part of its evaporation space with an optionally profiled material, which contains hollow spaces, capable of receiving condensed working fluid, the spaces being provided throughout the covering material and mainly disposed in spaced relationship. Preferably the covering material is thermally conductive, (for example carbon powder, mica powder or metal powder, furthermore polyimide resins and polyether resins and acrylate-based metal glues).

Fig. 128 shows the essential arrangement of the covering layer 9 and Fig. 129 a diagrammatic view of the collector.

A heat exchanger-comprising solar collector, revealed in another Patent (4.2) of M.J. Duncker, (Akzo N.V.), contains a closed housing of mainly rectangular cross-section, a collector thermally connected to the housing for converting solar rays into heat, a working fluid which, during operation, is partly liquid and partly gaseous, an evaporation space 3 withing the housing, a condensation space 4 within the housing, an internal connection between the spaces, and a heat exchanger 5. See Fig. 130.

An intensive thermal contact between the heat exchanger(s) and the working fluid is obtained by providing the condensation space 4 with a wall profiled inwardly of the housing and forming a space substantially shut off from the sur-

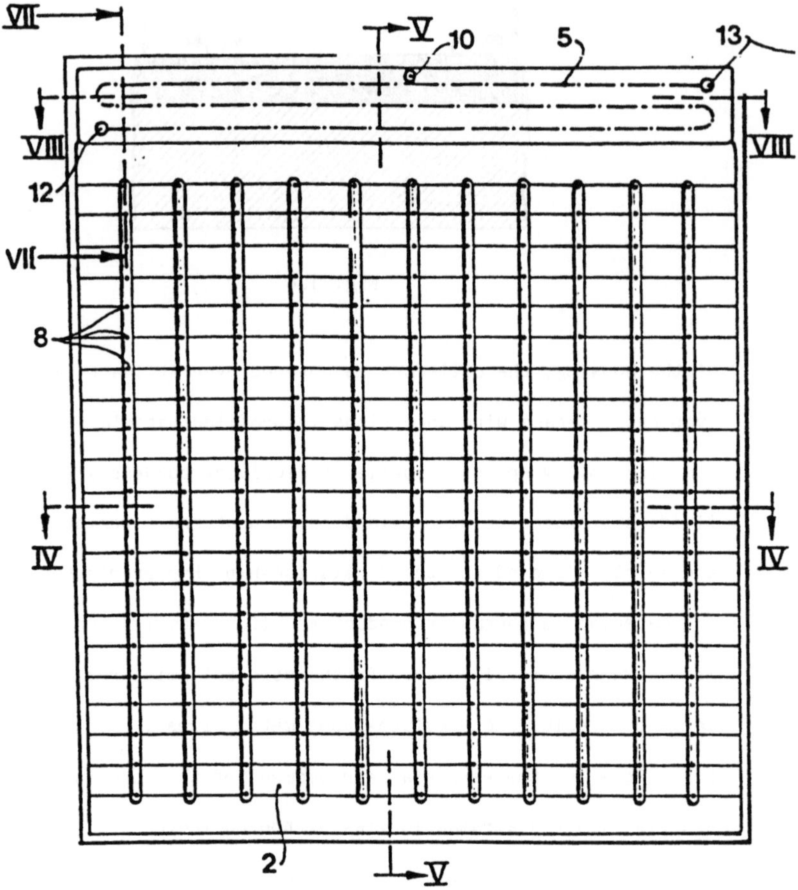

Fig. 129

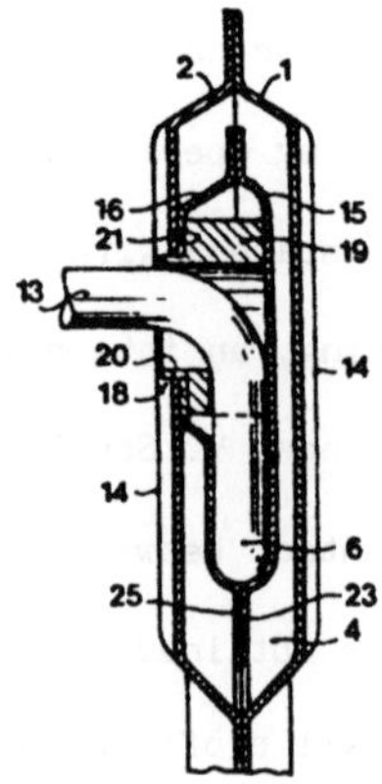

Fig. 130

rounding atmosphere, the said space being positioned within the outer boundary of the housing but outside the internal portion thereof, and by disposing the heat exchanger(s) within the said space and in thermal contact with the said wall.

Patent (8) of E. Bachli refers to a heat transfer system based on heat exchange and heat pipes, wherein the heat carrier fluid is transported between a solar collector and a heat accumulator, without having to use external power, the pump effecting the circulation of the heat pipe condensate being driven by a device, activated by the solar energy collected proper.

In Patent (11) of H.E. Barger, a solar heating unit is provided, including an upwardly opening generally hemispherical double wall boiler with spaced inner and outer hemispherical walls defining a closed chamber therebetween in which to receive a liquid to be heated.

An upwardly convex spherical lens structure is supported over the boiler and comprises a closure for the area enclosed within the boundaries of the inner hemispherical wall of the boiler and the lens assembly or structure functions to direct light rays incident thereon onto the inner wall of the boiler. The solar heater additionally includes a hollow downwardly opening transparent hemispherical cover of generally the same radius of curvature as the boiler and which is secured over the lens assembly in outwardly spaced position relative thereto with the center of curvature of the cover substantially coinciding with the center of curvature of the lens assembly. Water or other liquid inlet and outlet structure is provided on a second form of the invention for admitting liquid to be heated into the interior

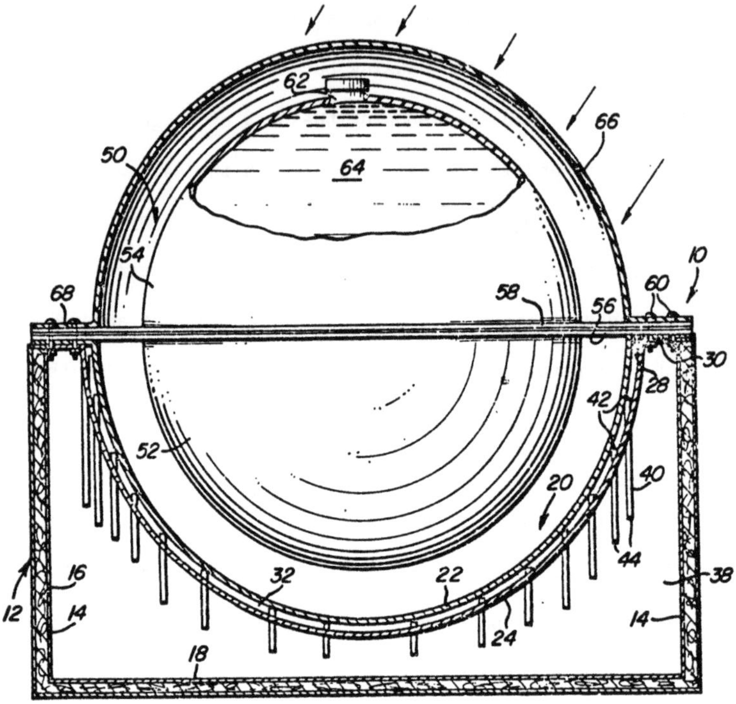

Fig. 131

of the boiler between the inner and outer plates thereof and for withdrawing heated liquid from within the boiler.

Fig. 131 is a vertical sectional view substantially upon a plane passing through the center of a first form of solar heater constructed in accordance with the present invention; (40 - heat pipes).

Patent (13) of C. Bennett refers to a fluid heating system by solar energy, according to which a heat transfer fluid is heated to boiling point by circulating it in heat exchange relationship with a solar plate, having a sun ray collecting (black) surface, the vapours emitted from the boiling fluid being circulated in heat exchange relationship with the fluid to be heated, whereafter the condensation fluid of the vapours

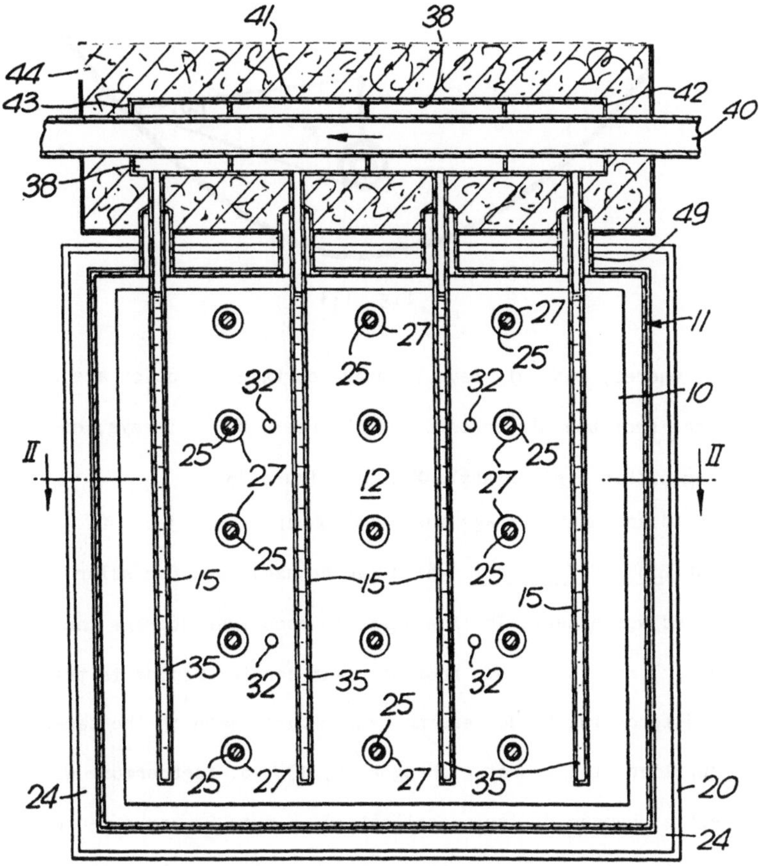

Fig. 132

is recycled to the solar plate for the next heat transmission cycle. Thus, the fluid, (water, toluol, methanol) is heated by a reflux condenser system. Fig. 132 illustrates a vertical section of the proposed solar energy system.

In Patent (15), A. Boettcher, discloses an approximately parabolic mirror dish, in the symmetry plane of which a heat pipe is arranged, which acts as an absorber and is provided with several external ribs. The external rib(s) and the heat pipe are disposed at either side of the focal line of the mirror

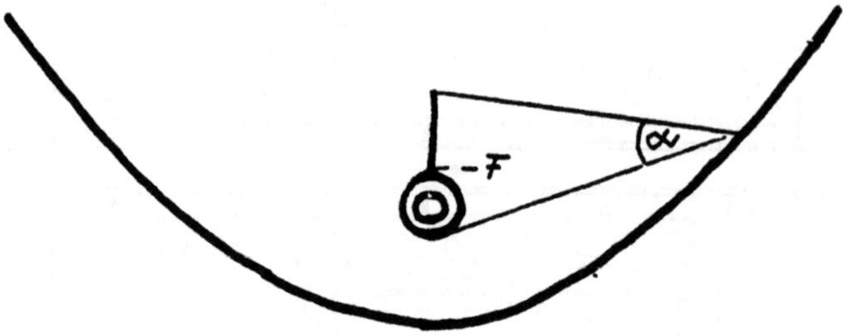

Fig. 133

dish, whereby not only rays from one certain direction are reflected from the mirror onto the absorber, but all rays, entering within a given angle α. See Figure 133.

A method and heat storage installation are the object of Patent (18) of V. Bossel, which can be used both for solar heat and waste heat. The installation comprises storage tanks 1, 2, 3 into which heat from the heat carrier medium is transferred according to decreasing temperature. Each of the tanks is supplied with a heat exchanger 16, 17, 18, furthermore with a heat exchanger 8, (solar collector), a pump 22 and a central device for charging, discharging, ventilating the installation. (See Fig. 134.)

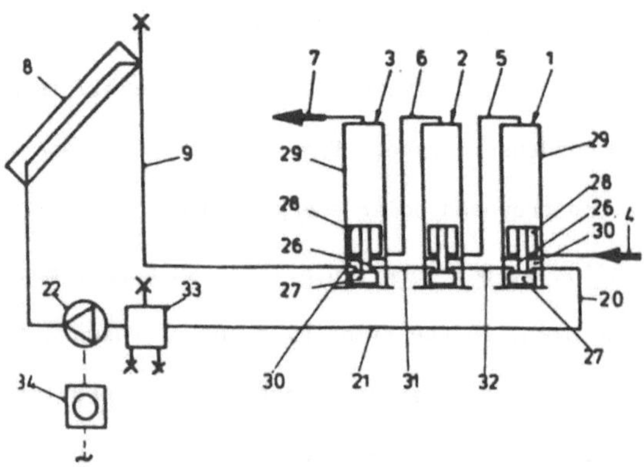

Fig. 134.

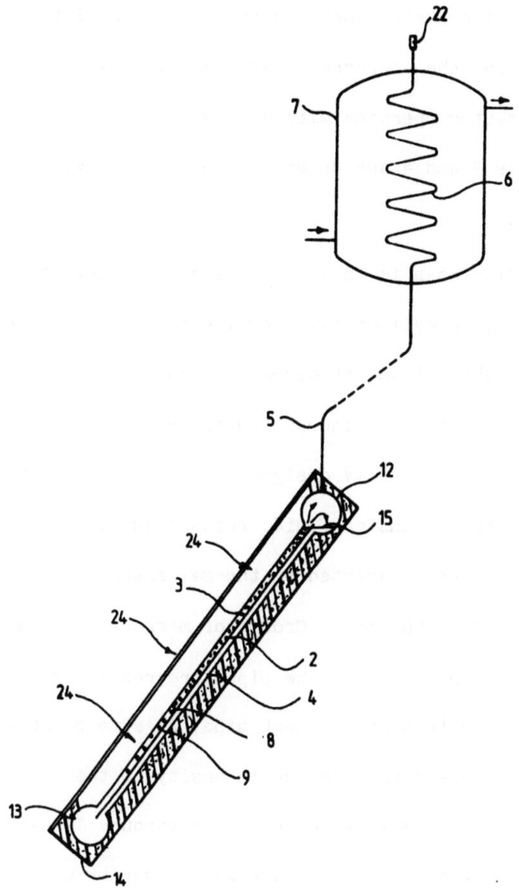

Fig. 135

The solar energy collection system, revealed in Patent (<u>20</u>) of B.R. Boyle, (<u>R. Boyle & Associates</u>), has a collector panel (Fig. 135) with relatively thin sheet metal outer walls 3, 4 and a relatively rigid, corrugated inner wall 2. The walls 2, 3, 4 define an evaporation space 8 and a condensate return passage 9, both of which are in communication with a vapour collection vessel 12 at the top of the panel. A single pipe 5 connecting the vapour collection vessel 12 with a coiled tube 6 in a preheat vessel 7 provides a path for upward transmission of vapour and downward return of condensate. The system operates by transfer of latent heat energy of vapourisation, the preferred heat transfer being de-ionised water. Normally the system operates with a negative internal pressure so that the outer walls 3 and 4 are drawn into engagement with the corrugations of the wall 2.

J.H. Carroll discloses in Patent (27) a solar energy focussing apparatus employing cylindrical focusing parabolic mirrors (parabolic troughs) of an optimized configuration capable of being manufactured by techniques of mass production to focus sunlight onto a specifically designed heat pipe which is to carry water or other heat transfer fluid. Focused sunlight energy is absorbed by the pipe, converted to thermal energy and transferred to heat the fluid in the pipe. Groups of mirrors are sequentially arranged so that the fluid in the pipes progressively absorbs more heat as it travels through the heat pipe. At some point in the fluid movement, depending upon the intensity of the specular sunlight the fluid, if liquid, may be converted to vapour. Transfer liquids other than water with higher boiling points may be used within the heat pipe and exit from the heat pipe in the liquid state. In either case, the steam, hot gas or super-heated liquid is transmitted to a collector system which is then transported by standard steam pipes enclosed within a silvered glass envelope to a central thermal-electrical generating station.

Collector mirror arrays are mounted to permit a single tracking motor to steer a large number of mirrors. All components are optimized for low maintenance and low cost to manufacture. (See Fig. 136)

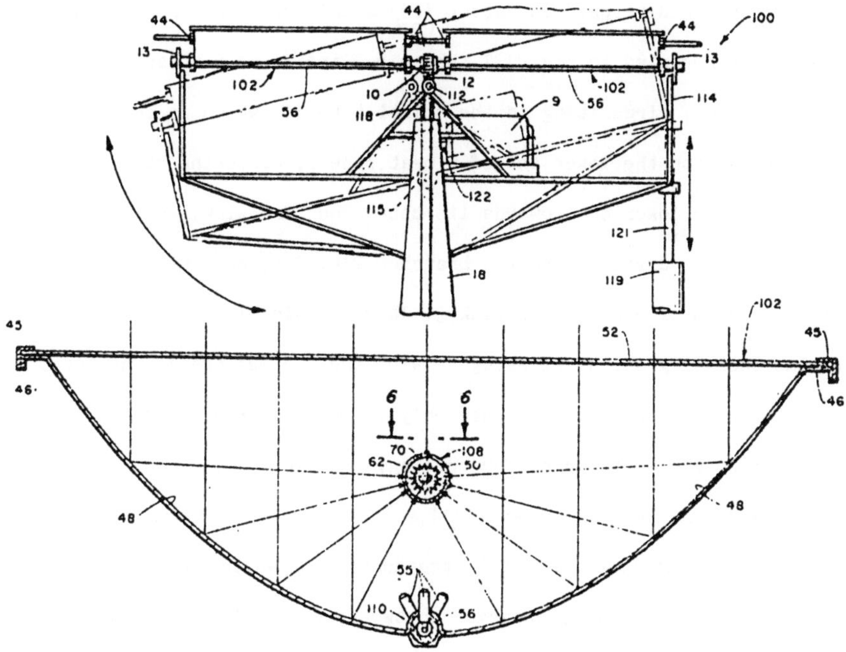

Fig. 136

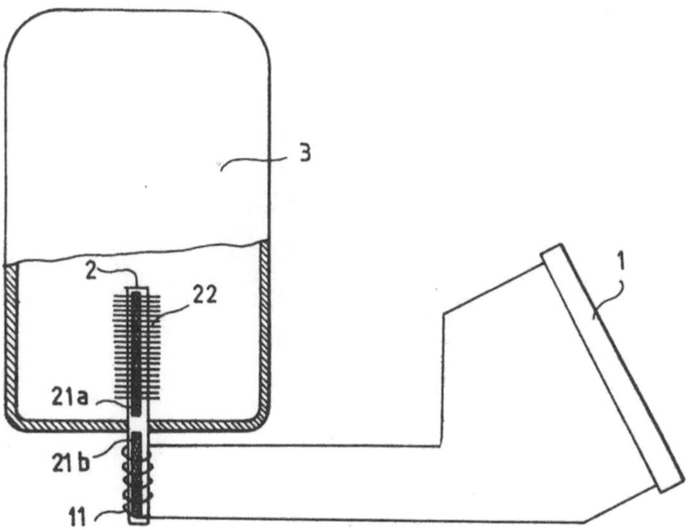

Fig. 137

R. Charbonnier developed in Patent (29) a unidirectional heat transfer system, represented in Fig. 137, wherein 1 indicates a solar collector, heating water, circulating in the spiral tube 11, which is in thermal connection with the lower end of a heat pipe 2, effecting heat transfer only from its lower end towards its upper end. The wick, applied to the heat pipe 2 is divided into two branches 21a, 21b, one inside and the other outside of a hot water tank 3. Thus the wick is cut at the place of the thermal gradient existing between the two sectors of the tank. In this way it can either be used only in the sector of entry of calories or the sector used for cooling. See Fig. 137.

P. Clement discloses in Patent (31) a solar heat collector, featuring a heat absorbing wall, the surface of which is provided with truncated pyramids, arranged in two perpendicular directions, between which heat pipes are arranged. (See Fig. 138.)

According to the inventor, such surface structure makes the solar heat collector independent of its position with regard to the direction of rays and increases the heat absorbing surface by at least 60%.

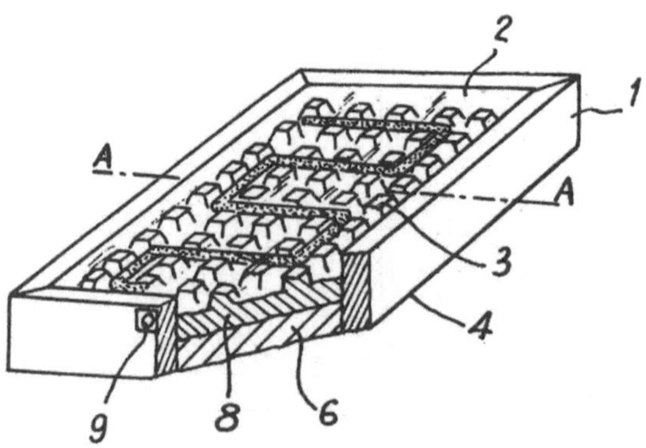

Fig. 138

The solar radiation collector, described in Patent (32) of N.G.A. Cobham, et al., comprises a cylindrical heat pipe, having a tubular transparent sector of for example glass, which is integrated with another sector of the heat pipe, this sector being made of a material of high thermal conductivity, like copper. The pipe comprises a wick of heat conductive material, in form, for example of a copper net, which is in direct (or nearly direct) contact with the transparent sector. The wick is coated with for example black chrome or other black dye. Figures 139 and 140 illustrate the proposed solar collector.

The solar heat collector, developed by T. Alleau et al. of Commissariat à l'Energie Atomique, (Patent 34.2), for heating water, comprises at least one heat pipe for heating water contained in a storage tank and is characterised in that the cold end of the heat pipe, disposed in the interior of the storage tank, is immersed into the water mass to be heated.

Solar energy is transferred through the heat pipe action towards the water to be heated, to which it transfers the entire latent heat of

Fig. 139

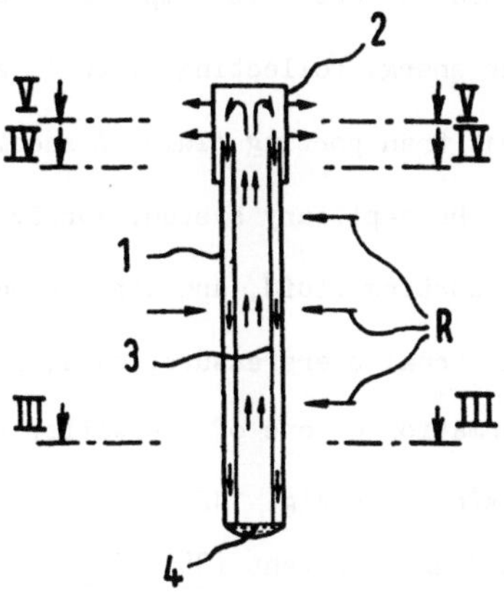

Fig. 140

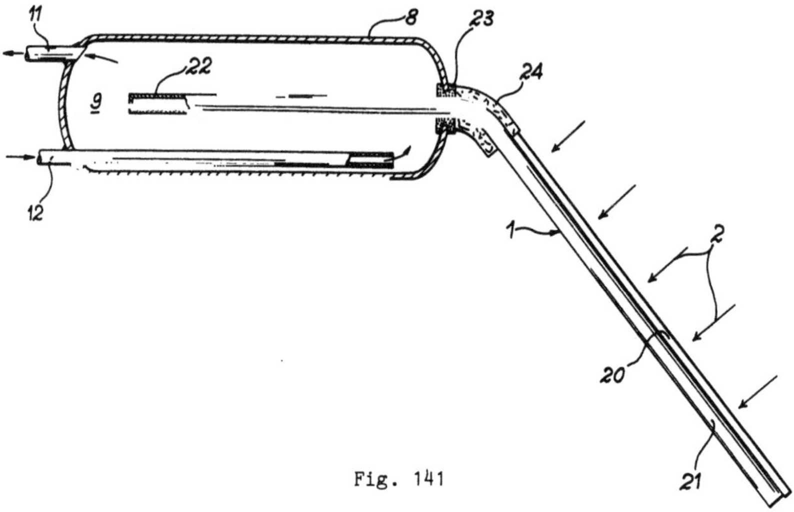

Fig. 141

condensation without any intermediary process or device, (auxiliary energy source, heat exchanger, heat siphon, circulation pump, etc.). See Figure 141.

Another Patent (34.4) assigned to the Commissariat à l'Energie Atomique, (inventor: B. Devin, et al.), reveals a solar collector, consisting of a casing, enclosing a solar energy absorber 8, an insulating element 6, placed between the absorber and the bottom of the casing; a glass plate 10.

The heat absorber is composed of an isolated case, having an upper solar energy collecting plate 1, a base 12, a capillary system and a heat transporting fluid. A special feature of the invention is that the capillary system, supplying the upper plate 1 with heat transporting fluid, and consisting of two parts 2a, 2b, is, in case of internal overpressure, capable of being disconnected due to the deformation of one of the walls, thereby interrupting the heat transfer. See Fig. 142.

According to Patent (36) of Compagnie Vendome, a solar heat collector comprises the conventional elements of such devices, with a heat

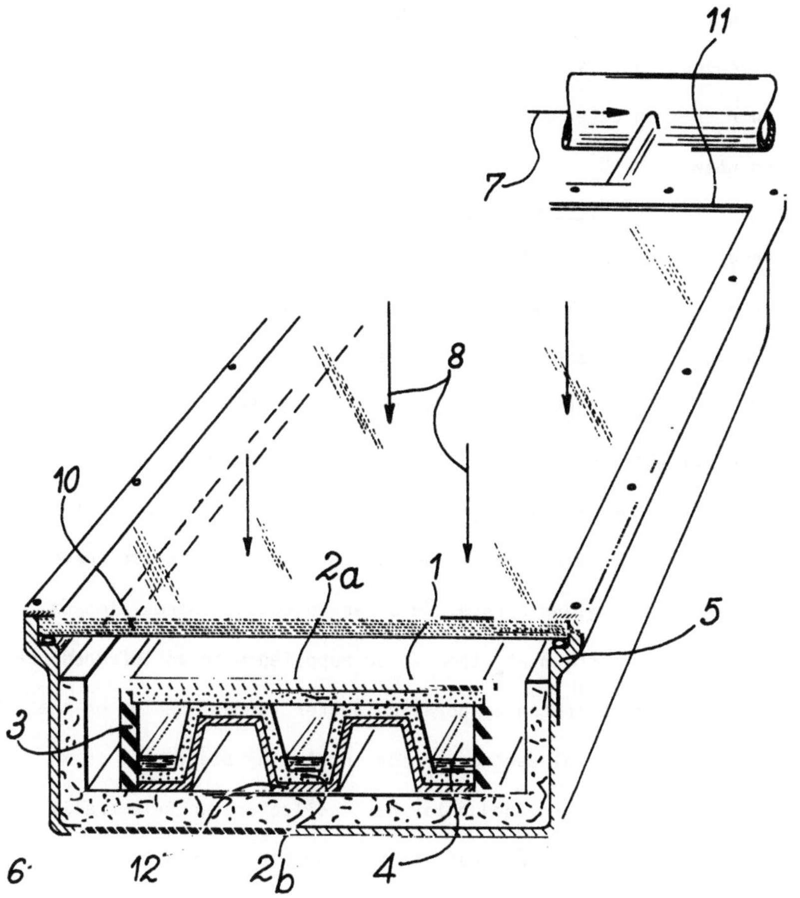

Fig. 142

pipe, which, in an embodiment of the invention, can be replaced by a heat pump, i.e. by a compressor, placed in vapour phase sector and a condensing element, placed in the sector of the fluid phase. The face of the collector, consists of metal of concertina structure, i.e. formed of a multiplicity of dihedral elements with vertical edges.

The other (non-collecting) faces can be made of concrete and consequently, can be used as structural (supporting) or decorative elements. As heat carrier fluid difluor-dichloro-methane is recommended. The device is also provided with means for supplying power from an external source, in case the solar energy becomes too weak.

Patent (37) of Corning Glass Works proposes a heat pipe type solar energy collector essentially consisting of a hermetically insulated casing, with a working fluid therein, furthermore with a capillary wick and a condensation sector. The solar rays are transferred through an evaporation sector of the casing, being directly absorbed by the wick arranged therein. The wick is made of transparent glass, which permits a complete transfer of the solar radiation. The evaporation sector of the casing is isolated from its surrounding by an evacuated glass sleeve. In order to reduce thermal losses, the exterior of the heat pipe can be supplied with an infrared reflecting coating in the evaporation sector thereof and/or on the internal or external surface of the insulating sleeve. It is claimed, that the proposed solar heat collector can be serially produced at low cost and that it considerably reduces thermal losses through convection, conduction or radiation.

See Fig. 143, illustrating a lateral (partly cutaway) view of the proposed heat pipe.

Patent (39) of J.M. Cutchaw concerns a solar energy collector panel, including a housing, having a solar window therein and which contains a loop of tubing through which a fluid is pumped for heating of the fluid. A heat absorbing porous wicking membrane containing a vaporizable working liquid is positioned within the housing for collecting solar heat and transferring that heat to the fluid within the tubing. The working liquid is vaporized by solar energy and the vapour which contains latent heat will move toward and condense proximate the

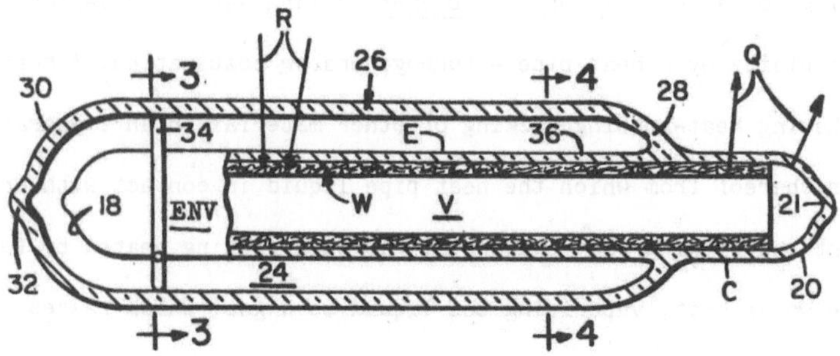

Fig. 143

ENV - tubular casing; W - wick, composed of glass fibres or glass tissue, pulverised material, ceramics; V - vapour conduit; R - rays; E - evaporation sector; C - cold zone of condensation.

tubing to release the latent heat and thus transfer that heat to the fluid in the tubing. The condensed vapour, or working liquid is reabsorbed by the wicking membrane and is returned by wicking action to the vicinity of the solar window so that a continuous cycle of vaporization, condensation and liquid pumping takes place to efficiently collect and transfer heat.

Fig. 144 is an isometric view of the solar energy collector panel of the invention partially broken away to show a portion of its internal structure.

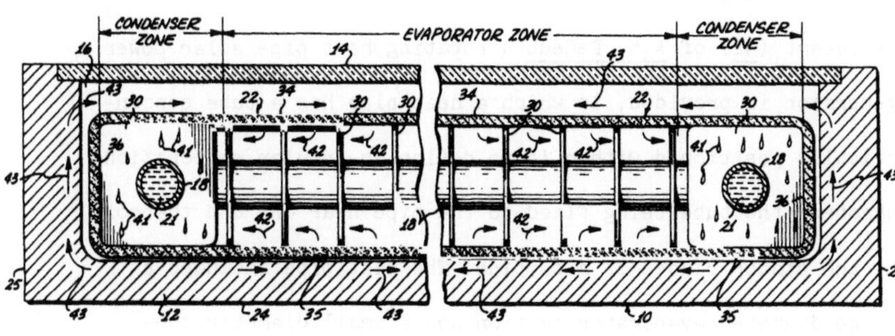

Fig. 144

According to Patent (41) of B.J. Decker, solar heat is converted to electricity by a heat pipe - turbogenerator combination. A heat pipe, having heat-sorbing wicking or other material in an evaporation section thereof from which the heat pipe liquid in contact with or adjacent the wicking may evaporate, has such wicking heated by for example solar heat, vaporizing the liquid to a gas, which passes through a turbogenerator to a condensation section wherein it is condensed to the liquid, which is returned to the evaporation section. The heat pipe - turbogenerator assembly is externally insulated, for example by a vacuum shield, to prevent heat losses and heat is recovered from the condenser portion of the heat pipe and returned to the evaporator portion. In an application of the invention it is employed in a building, where it is utilized on wall and roof portions and serves as at least a partial supporting structure for these.
In another application the solar heat pipe feedback turbogenerator may be used with reflective means, such as reflective sheet material of large area positioned to direct solar radiation onto the evaporator section of the heat pipe, which reflective means may be supported by gas balloons or other useful supports and may be changed in position to "follow the sun" to produce maximum power during operation.
Figure 145 is a partially cutaway sectional elevational view of a solar heat pipe panel feedback turbogenerator.
In Patent (42) of A.M. DeGeus a rotating heat pipe solar power generator is provided, in which a heat pipe has a tube concentrically positioned within it to define an annular evaporation chamber, the tube being fixed to the pipe near one end thereof by the stator blades of a turbine. The heat pipe includes a large diameter evaporator section and a small diameter condenser section remote from the turbine. The tube extends into

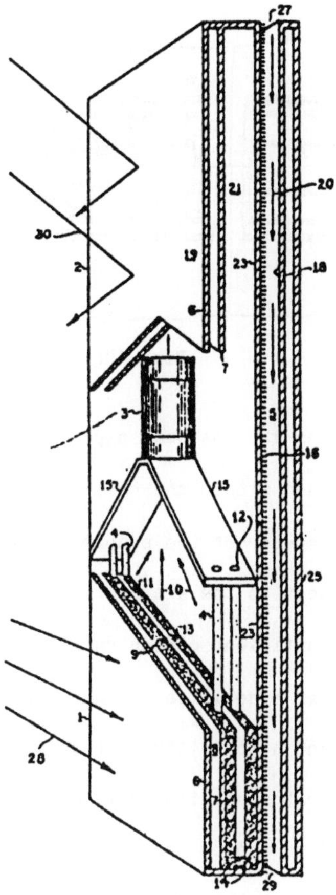

Fig. 145

the entrance portion of the condenser section, and liquid is pumped into the evaporation chamber through an annular throat with a tapered annulus constituting the juncture between the two sections of the heat pipe, this throat serving as a centrifugal pump. The turbine is geared to the heat pipe so that both rotate.

Fig. 146 is a plan view showing the heat pipe and its relationship to a reflector used to concentrate solar radiation on the evaporator section of the heat pipe.

In the figure 10 indicates a reflector, having a heat pipe 11

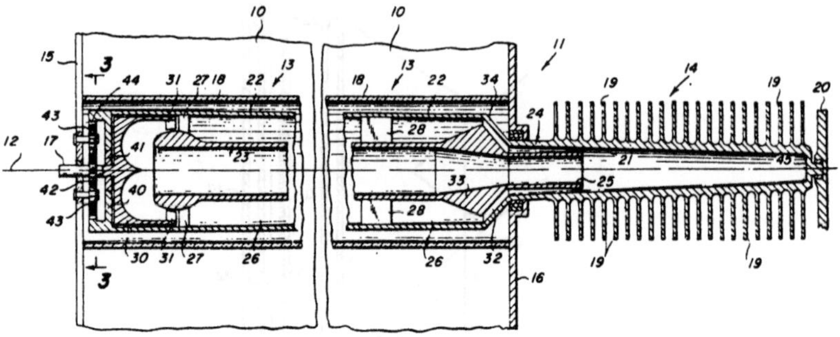

Fig. 146

mounted at its focus 12, the heat pipe 11 having an evaporator section 13 positioned within the reflector 10, and a condenser section 14. The diameter of the evaporator section 13 is relatively small with respect to the arc over which the reflector 10 extends so as to achieve sufficient concentration of the solar radiation. The reflector 10 has side walls 15 and 16 which rotatably support the heat pipe 11 and a power take-off shaft 17 extends through wall 15. The evaporator section 13 is disposed within a glass tubular housing 18 to reduce convection losses from the hot pipe, and the coolant airflow through the fins 19 carried by the condenser section 14 is shown by arrows A. The free end of the heat pipe is rotatably mounted in wall 20 which helps to confine the air flow.

In Patent (45) of M.A. Tolchard, (Denco Air Limited) provides an arrangement for thermally coupling a solar energy collector 12. (see Fig. 147), including a heat pipe 5, to a part of a heat-extraction system 3, 4. The arrangement consists of a first metal bearing member 6 provided with a male bearing surface and a second metal bearing member 2 provided with a corresponding female bearing surface adapted to receive the first member to form a rotatable coupling having at least one degree of freedom.

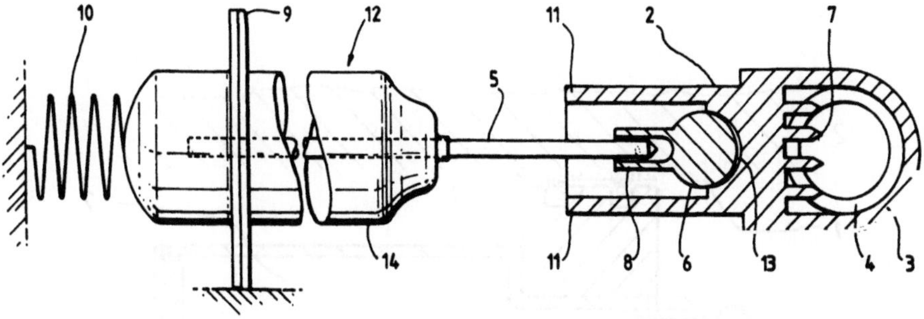

Fig. 147

The bearing surfaces may be spherical or part spherical, cylindrical or part cylindrical, or frusto-conical or part frusto-conical, and may be urged against each other by spring means 10.

Patent (46) of W. Frank, (Deutschmeister Bauelemente GmbH.) concerns a ventilation system, essentially comprising a fresh air conduit, a spent air conduit and a heat pipe, used for extracting heat from the spent air current and introducing it into the fresh air current. According to the invention the heat pipe is connected to a solar collector, under the effect of which the fresh air current is not only heated by the heat separated from the spent air current, but also by solar heat. In this way the effectiveness of the heat pipe is considerably increased.

Fig. 148 illustrates the cross-section of the proposed ventilation system, as applied to a window.

I. Lorschiedter, (Dornier System GmbH.), reveals in Patent (47.3) a solar collector, which can be mounted on flat roofs.

The collector comprises a hermetically sealed heat pipe, provided with a flat absorbing plane, which is inclined under an angle of 45°-60° with regard to the horizontal line, towards the sun, while the axis of the heat pipe is adjusted under an angle of 5° with regard to the

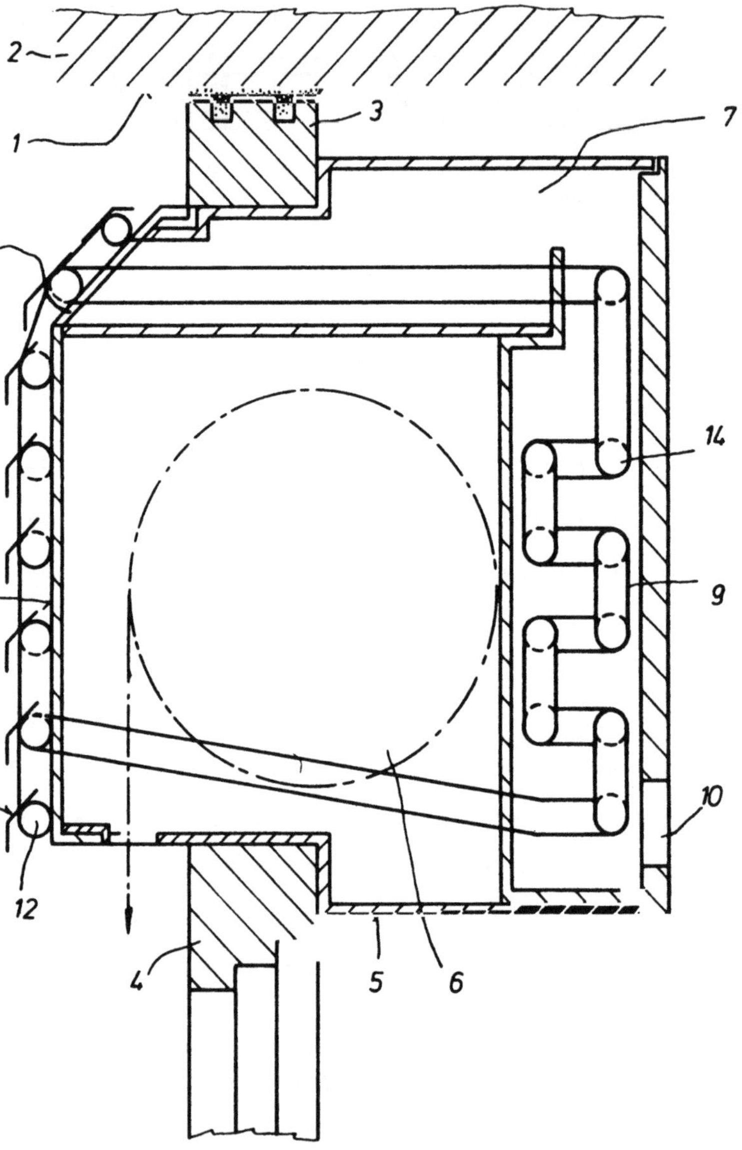

Fig. 148

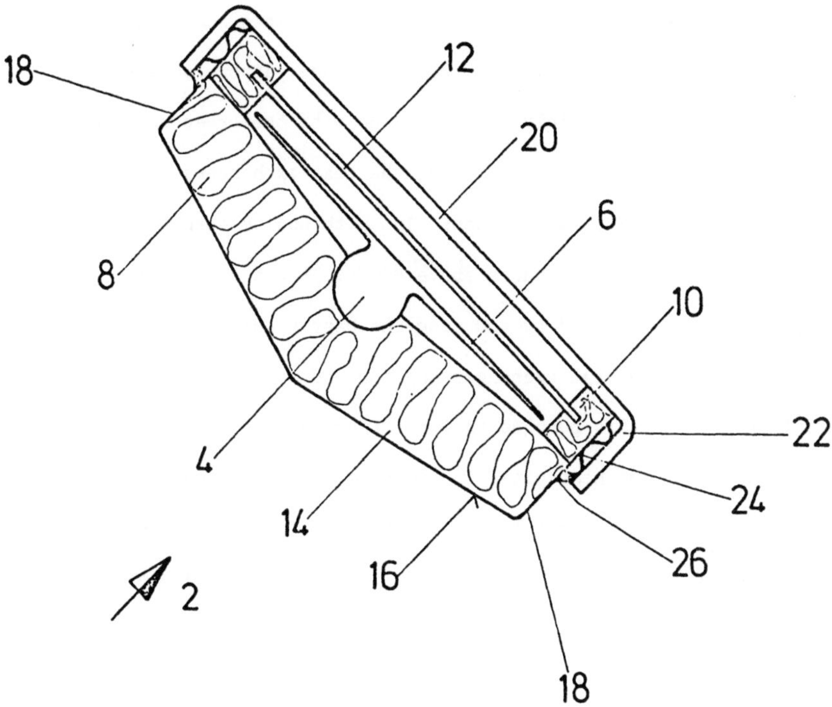

Fig. 149

horizontal line. Such arrangement is realised by placing the heat pipe in a casing of thin sheet metal, the edges of which are in connection with the edges of a V-shaped glass element (of at least 5 mm thickness) which covers the collector plane in a radiation-permeable manner. See Fig. 149.

In another Patent (47.4) of Dornier System GmbH., (inventor: H. Kreeb, et al.), there is revealed a heat exchanger for solar energy plants, wherein solar heat is transferred at temperatures over 800°C, under high pressure through fluid-metal heat pipes (of high-alloy steel, or tungsten, tantalum, molybdenum), the evaporation sectors of which project from a plane heat exchanger wall, wherein they are fastened, while their condensation sectors are arranged in an intermediary space, through which a gas current is passing. The ends of the heat pipes are fixed in a second heat exchanger wall. However, high temperatures cause, after long-term use, various problems, mainly in connection with the creep behaviour of the high alloy metals, the walls of the heat exchangers are made of, while no problems arise in connection with sodium-operated heat pipes, even at temperatures above 800°C. It was found that this problem can be solved by reducing the heat transit between the rear wall of the heat exchanger and the ends of the heat pipes fixed therein. For this purpose an insulating gas cushion had to be created within the heat pipe, in the range of the second heat exchanger wall, (for example of N_2, inert gas). If small amounts of inert gas are fed into the heat pipe, it will collect at the heat-releasing rear end of the heat pipe, thereby blocking the end part of the heat pipe and reducing the heat transit between heat pipe and heat exchanger wall.

See Fig. 150 wherein: 16 - evaporation zone; 24 - condensation zone; 42 - gas cushion; Wärmesenke - heat sink; Behälter- tank; Rohr - heat pipe; Dampfraum - vapour space; 38 - vapour current; 40 - condensate current; 36 - capillary structure; X - solar radiation.

A solar radiation collector, which can be rolled-up and unrolled, is disclosed in yet another Patent (47.5) of Dornier Systems GmbH., the collector consisting of two thin-walled foils or webs, in superimposed arrangement and tightly connected at their edges, thereby forming a intermediary space, filled with fluid, under high pressure. The collector can be unrolled to such an extent, that its absorption and/or emitting faces are adjusted to the length, which, under given conditions, is necessary for the heat amount to be absorbed or emitted.

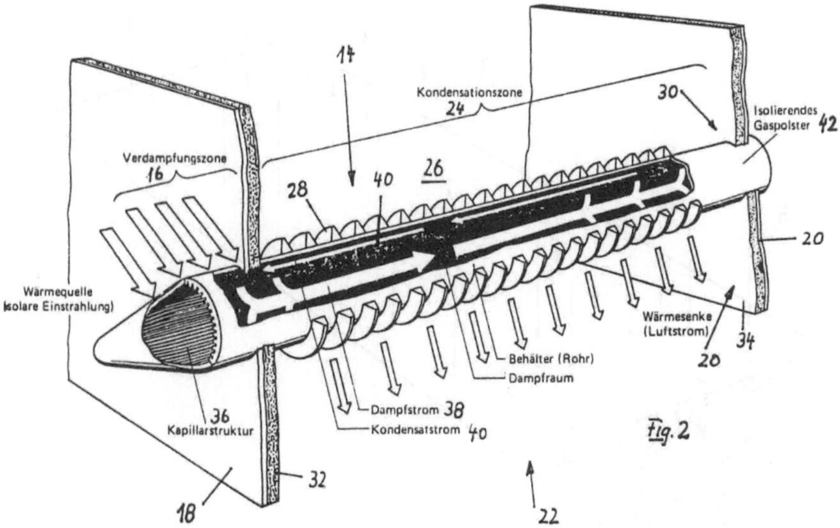

Fig. 150

Figs. 151, 152 illustrate the proposed, mat-like collector 1 in partly and completely unrolled state.

6 - absorption/emission face; KZ - cooling zone; HZ - heating zone.

In Patent (47.10) of H. Koch, (Dornier), a solar heat collector is revealed, wherein the heat conductor and the heat absorbing and heat emitting faces thereof form an integrated, structural unit, while the profile of the block and the disposition of the faces with regard to the heat conductor can be arranged in various

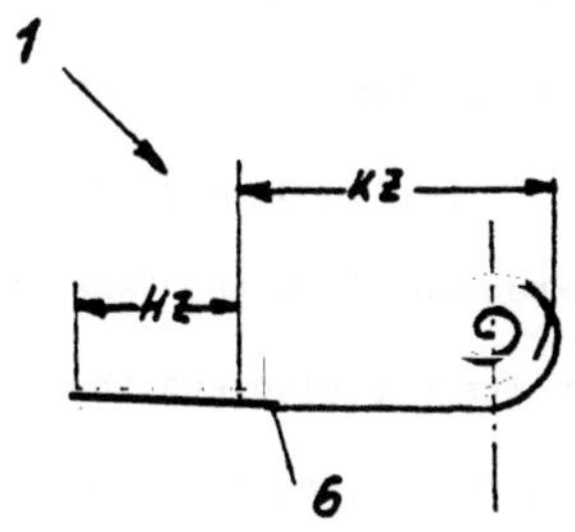

Fig. 151

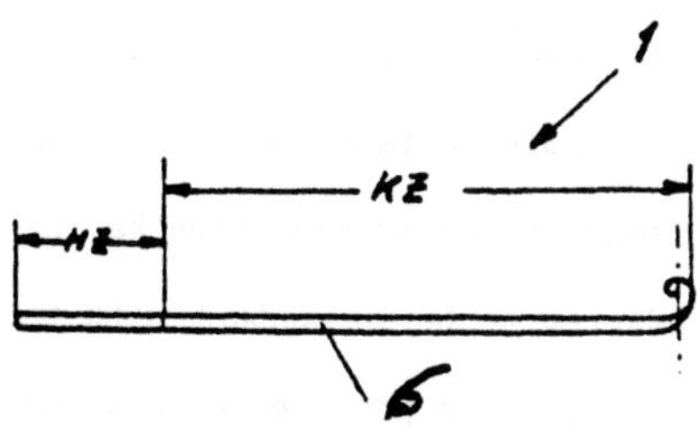

Fig. 152

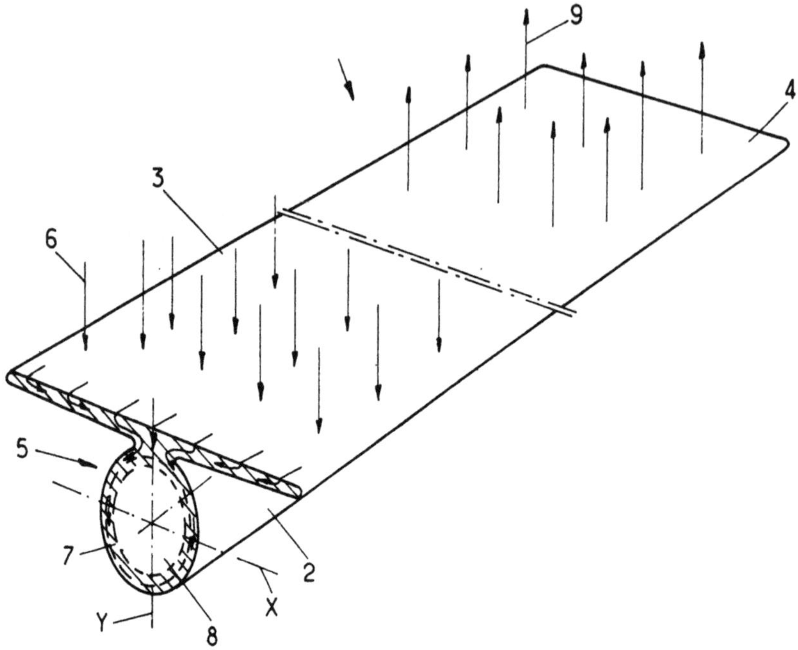

Fig. 153

forms, (T, H, V, X or Y). It is also possible to arrange the absorbing and emitting faces radially to the axes of the profile, in a symmetrical or asymmetrical form.

See Fig. 153, wherein: 2 - heat conductor; 3, 4 - absorbing and emitting faces, 5 - profile in T-form; 6 - heat supply; 7 - capillary system.

Patent (51.1) of D.S. Ward, (Entec Products Corp.), concerns a passive solar assistance device for use with a heat pump having a heat transfer fluid inlet and outlet, with means for circulating the fluid, a tank for containing an amount of the fluid, and at least one solar heat pipe having one end extending into the tank and adapted to be in contact with the fluid. At its opposite end the heat pipe is provided with a solar receiving device, such as an enclosed box having one or more transparent walls facing the sun

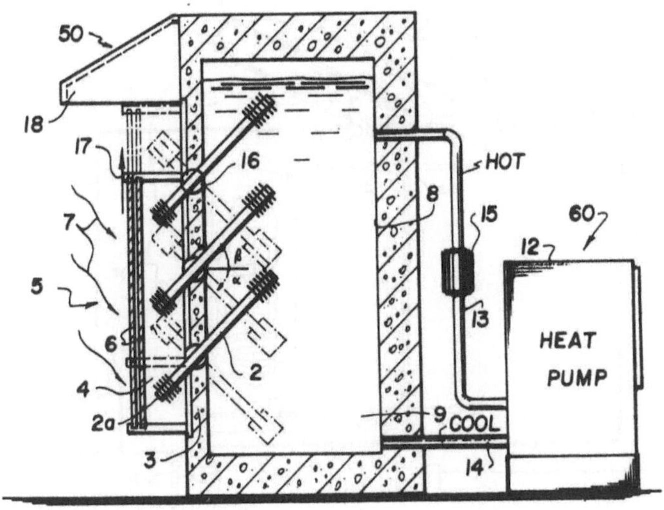

Fig. 154

or a plate construction comprising an extension of the heat pipe itself. In its heating mode, the heat pipe is mounted at an angle of at least 30° , with the end associated with the solar receiving device in the lowermost position so that the heat pipe acts as a one way heat valve conducting heat from the solar receiving device to the fluid in the tank but not in the opposite direction. The heat pipe may be pivotally mounted to provide for either a heating mode or cooling mode.

Fig. 154 is a side elevation cross-sectional view of an embodiment of a solar collector embodying the invention, showing its association with a heat pump.

Patent (52) of M.W. Frank, et al., (Entropy Ltd.), refers to an apparatus for collecting and transporting solar energy as vaporized fluid. The vapour will flow from a heat absorbing chamber C and through a conduit V to a point of use where it is condensed. The heat absorption chamber C includes a sump in its lower portion to

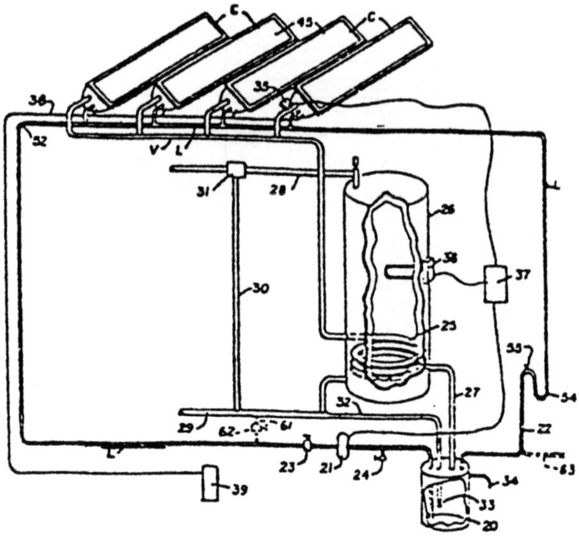

Fig. 155

hold a puddle of fluid. A heat source directed against the chamber C vaporizes the fluid therewithin and a replacement fluid supply L connects with the chamber to maintain the level of the puddle during vaporization. The chamber C of the preferred embodiment of the invention is an elongated heat absorption pipe mounted horizontally to maintain the puddle at the lower portion of the pipe.

The heat absorption pipe is mounted in a solar collector C to receive concentrated solar energy as from a reflector system.

Fig. 155 is a diagrammatic view of a complete solar domestic water heating system incorporating a sequence of unit collector modules, a water supply line and a vapour conduit.

The water heating system, for example for domestic use, designed by J.J. Evans, (Patent 56), has an evacuated sealed vapour transfer duct 1 through which vapour is transferred from an evaporator section, which may for example be included in a solar heat collecting panel 3, to a water heating vessel 7, in which the vapour condenses,

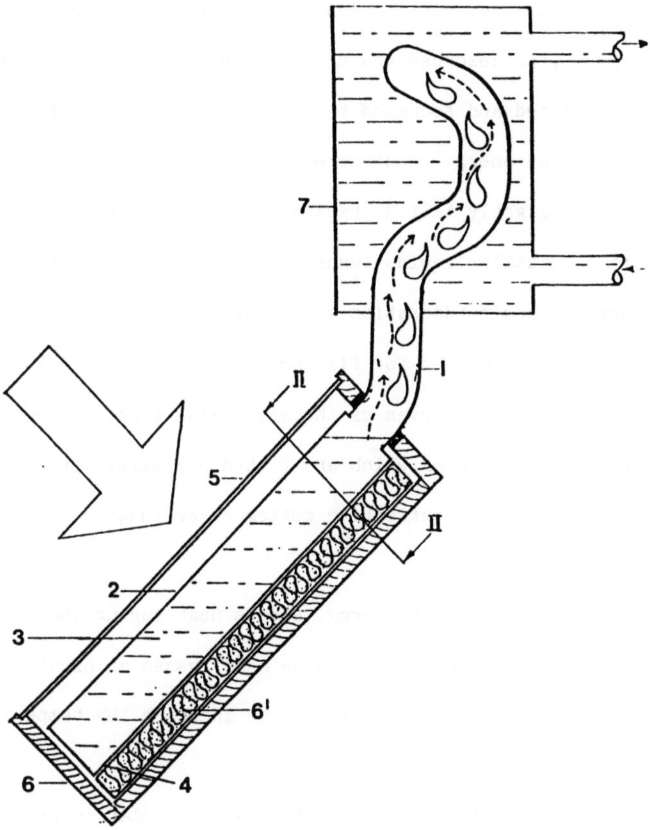

Fig. 156

the condensate flowing back to the evaporator section. An automatically controlled pump may be employed to convey the working liquid to the evaporator via a liquid reservoir, if the solar collector is located above the level of the heating vessel. The evaporator section may be integral with an incinerator.

Fig. 156 is a schematic cross-sectional view of a simple solar hot water heating system according to one embodiment of the invention.

Patent (58) of K.T. Feldman relates to a method and a passive apparatus for absorbing, transferring and storing solar energy as heat, without pumps, solar tracking devices or electric power.

The apparatus comprises an improved trough-type concentrating collector, a heat pipe absorber and an insulated storage tank. Solar energy is reflected and focussed by the concentrator onto the absorber where the energy is absorbed as heat. The absorber, made of one or more slightly tilted gravity-assisted heat pipes partially filled with a volatile liquid, transfers the heat by evaporation, vapour transport and condensation into a slightly elevated heat storage reservoir. A method for filling the heat pipes is also disclosed. The absorber serves as the main axial support for internal structural ribs over which are fitted a flexible transparent top cover and a flexible reflective bottom cover that comprise the concentrator.

The apparatus collects solar energy, stores heat during the day and automatically shuts off to minimise heat losses at night. The apparatus may be refocused periodically to increase the temperature of the heat collected.

See Fig. 157, illustrating the apparatus, with a solar energy

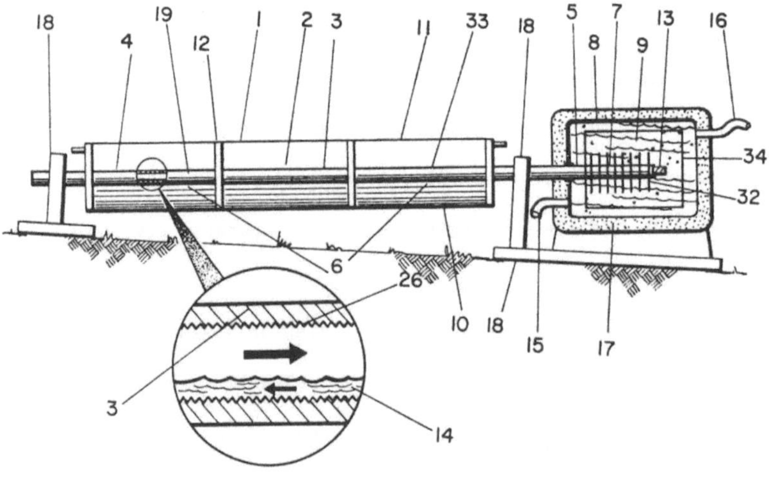

Fig. 157

concentrator 1, structures to reflect and focus solar energy into a linear region of small volume 2 where the temperature is increased. A linear heat absorber 3 having a first end 4 and a second end 5 is located in this high temperature region of small volume 2 to absorb the focused solar energy as heat. The heat absorber is composed of at least one heat pipe 33 which absorbs the heat along the first end 4 of the absorber 3 in a section known as the heat absorber section 6. Heat is transferred by the heat pipe(s) 33 to the second end 5 of the absorber where the heat is removed in a heat removal section 7. The heat is then transferred to and stored in a heat absorbing fluid 9 contained in a heat storage reservoir 8 fitted over the second ends of the absorber 3.

M.W. Frank discloses in Patent (67) a solar thermal energy collection system, having a collector of solar energy which reflects solar energy to a heat pipe, transferring vapour produced therein to a condenser to release and utilise the thermal energy carried by the vapour either to perform heating functions or to be dissipated as the vapour is condensed.

Fig. 158 is a diagrammatic, three-dimensional view showing a solar energy collection system using channel-type solar collectors with

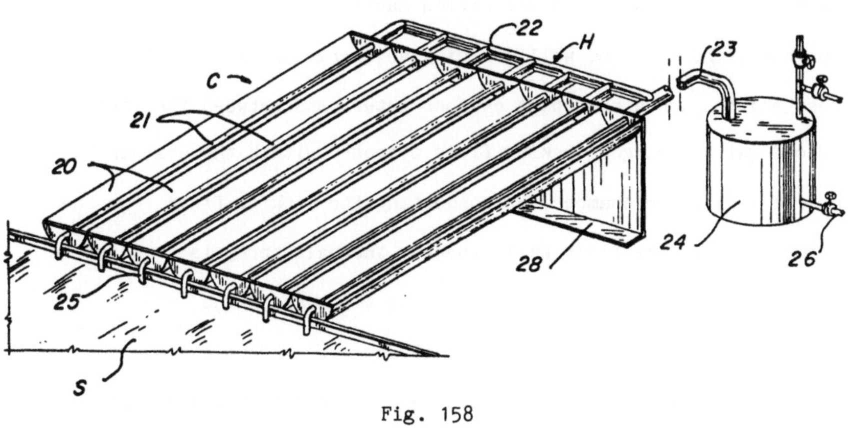

Fig. 158

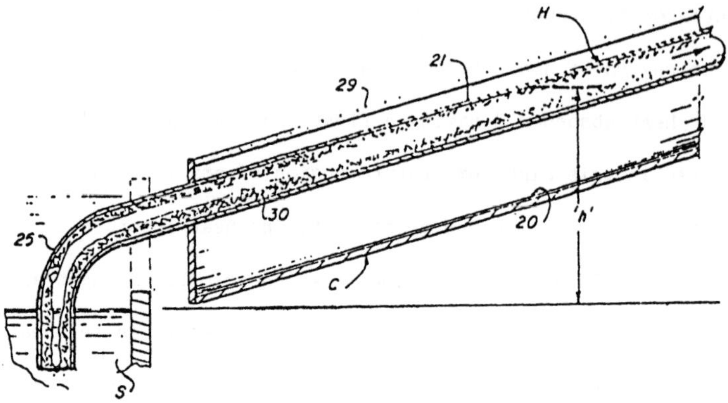

Fig. 159

open heat pipes extending through the channels to pick up and evaporate water from a source, pond or lake and to convey the vapour to a condenser as an open system.

Fig. 159 is a longitudinal sectional elevation through the lower portion of a single heat pipe and collector, indicating the arrangement of a wick within the heat pipe.

R.A. Frosch, et al., provide in Patent (68) a thermal energy transformer for use in combination with a heat engine, the flux receiver having a first wall defining therein a radiation absorption cavity for converting solar flux to thermal energy, characterised by a first wall defining a radiation absorption cavity having a solar flux entry aperture, and a second wall defining an energy transfer wall for the heat engine, and a heat pipe chamber interposed between the first and second walls having a working fluid disposed within the chamber and a wick lining the chamber for conducting the working fluid from the second wall to the first wall, whereby thermal energy is transferred from the radiation absorption cavity to the heat engine.

According to Fig. 160 the system includes a transformer 10 which includes a flux receiver 12 coupled with a heat operated engine 14.

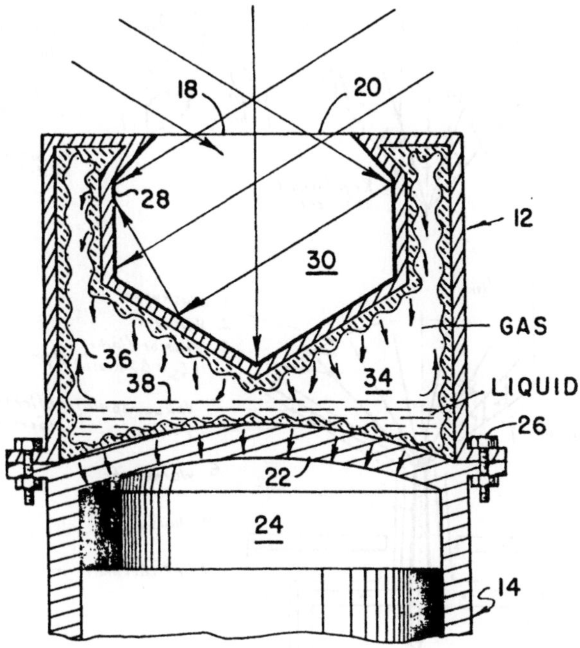

Fig. 160

The transformer 10 is disposed along the axis of a parabolic reflector, generally designated 16. The reflector 16 serves to direct a highly concentrated beam of solar flux through an aperture 18 defined within the receiver 12, having an entry plane 20.

In Patent (77) of Annemarie Genswein-Schmitt discloses a solar heat collector, the absorption capacity of the proposed collector being considerably increased by a special construction thereof, realising multiple reflections of the oncoming solar rays on the reflecting collector surface and thereby a nearly complete absorption of radiation. These effects are achieved, according to the inventor, by a collector, composed of several conical collector faces, symmetrically arranged within one another around a common cone axis. As shown in Fig.161, depending on the altitude of the sun, the rays (S_1 - S_3) hit the collector faces under various angles and are re-

Fig. 161

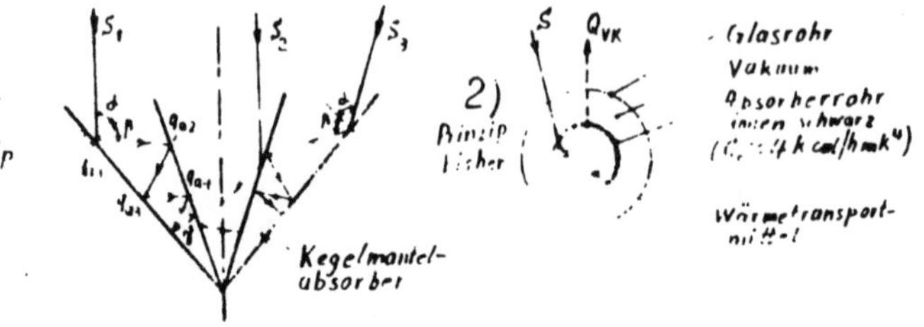

Fig. 162

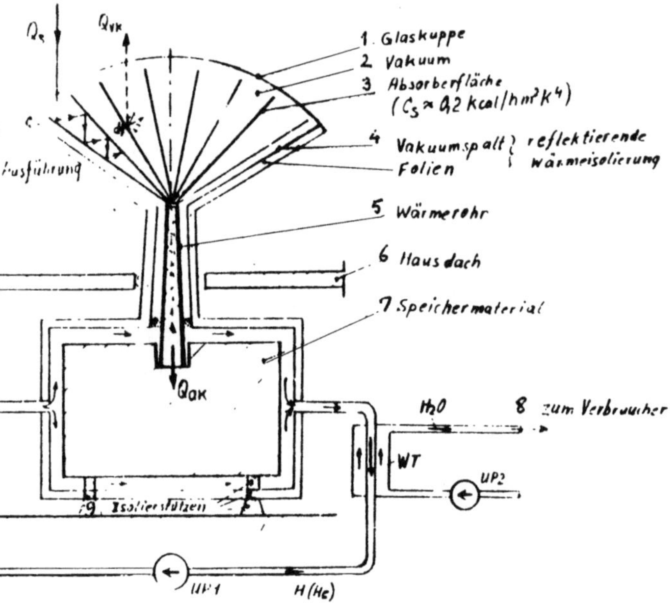

flected several times between the collector faces, which are in an acute - angle arrangement. In this way the collector achieves an absorption capacity, which surpasses its emission capacity and consequently displays a higher energy collecting capacity than that of prior art collectors.

Fig. 161 is a scheme of the invention's basic principle, while Fig. 162 illustrates the proposed solar heat collector, wherein: 1 - glass cupola; 2 - vacuum; 3 - absorption face; 4 - vacuum gap (reflecting heat insulation, foils); 5 - heat pipe; 6 - roof; 7 - storage material; 8 - towards the consumer; 9 - insulation support.

In the passive solar heating and cooling panels, devised by J.A. Cogliano, (W.R. Grace & Co.),in Patent (79), a heat pipe is used to transfer heat from an outside solar panel into a building where the heat can be stored in a heat absorber. Various heat pipe structures are disclosed where the pipe passes through the wall of the building and where it has an end portion thermally contacting the solar panel. Heat transfer can be stopped by rotating the heat pipe, by the use of plungers and trap chambers in the pipe or by specially designed heat pipes that can be easily removed. The heat pipe can also be used to remove heat from the inside to the outside. The inside end portion of the heat pipe is in thermal contact with a heat absorber inside the building which contains a phase change material. Modular wall units can also be made having the solar panel and heat absorber integral on each side of the wall with the heat pipe in between.

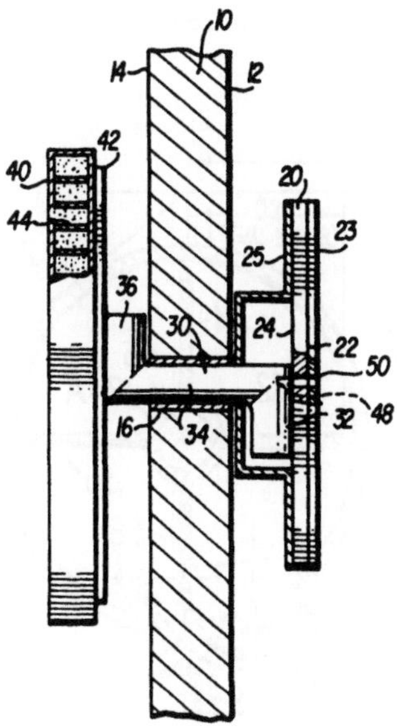

Fig. 163

Fig. 163 is a cross-sectional view through a wall showing the heat pipe positioned for heating in the winter.

Patent (82.7) and essentially Patent (82.9) of G.W. Knowles, (Grumman Aerospace Corp.), relate to a solar energy collector for collecting, concentrating, and utilising solar energy, including a target for transferring solar energy into another useable energy form and a reflector positioned to increase the amount of solar energy reaching the target and prevent solar energy from escaping around the target. The target includes in its preferred form a transparent envelope and a heat pipe containing a heat transfer fluid, the heat pipe having an evaporator portion disposed within the transparent envelope and an emergent condenser portion with a flange forming a dry thermal interface with a manifold for conducting heat energy directly from the heat pipe to the manifold.

See Fig. 164.

In Patent (83) of F.H. Theakston, (The Guelph Manufacturing Group Ltd.), a solar-powdered refrigeration apparatus is disclosed, in

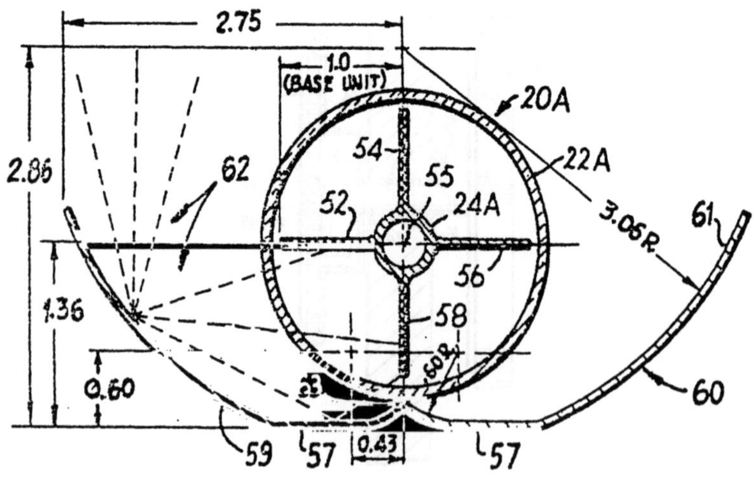

Fig. 164

which an absorption refrigeration system 16 is operated directly by solar energy. One end of a heat pipe 24 is thermally connected to the boiler 18 of the absorption refrigeration system by a curved plate 30, and a solar collector 26 is thermally coupled to the other end of the heat pipe 24. The heat pipe 24 is a sealed, evacuated metal tube partially filled with water. The solar collector 26 is a double walled glass vacuum tube with a central axial opening For accommodating the remote end of the heat pipe. Heat energy collected by the solar collector boils the water in the heat pipe to subsequently condense in the area of the boiler thus transferring heat energy along the heat pipe to the boiler. The heat pipe is installed sloping downwardly away from the boiler to permit the return of condensate down the pipe to the solar collector area thus permitting continuous operation.

As shown, the refrigerator 10 also includes a part of auxiliary heating elements 46 which are electrically powered to operate refrigeration system 16 in the event that there is no solar energy available for that purpose.

Fig. 165 is a perspective view, partly broken away and exploded, showing a preferred embodiment of the solar powered refrigeration apparatus of the invention as applied to an absorption type refrigerator.

A solar heat collector, designed by H. Harrison, (Halm Instrument Co. Inc.), (Patent 84), features a thin evacuated tank which has upper and lower parallel sides, the upper side having a black exterior surface, the tank being mounted at an angle to the horizontal. A small amount of vapourisable fluid is contained in the tank. A fluid conduit is connected in heat exchanging relation with the upper end of the tank. Solar rays striking the upper tank

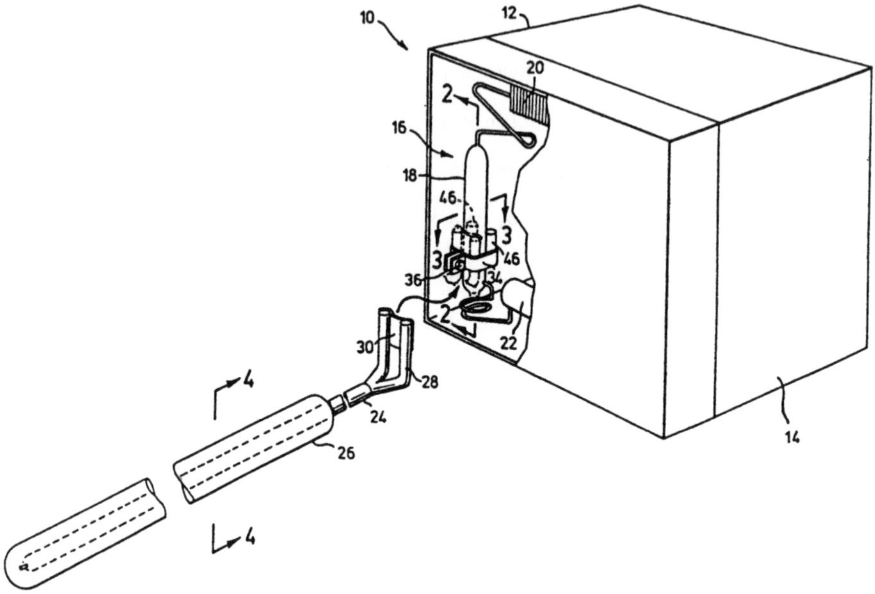

Fig. 165

surface will vapourise the fluid in the tank and the vapour will flow to the upper portion of the tank where it will be condensed and transfer its heat to fluid in the conduit.

See Fig. 166, wherein 1 - tank; 2 - black upper side of the tank; 3 - lower side of the tank; 4, 5, 6 etc. - curved ridges on side 3; 8 - clamp; 11 - transparent covers; 12, 13 etc. - spacing members; 14, 15 - elastomer end pieces; 14', 15' - nailing flanges.

Patent (85) of R.A. Hamilton refers to a wickless heat pipe assembly, comprising a pair of conduits, each having a vapourisation region at one end and a condensing region at the other. The vapourisation regions are configured to prevent the heat-transferring fluid, while in liquid form, from flowing downwardly into their respective conduits. A pair of liquid return passages connect each condensing region with the vapourisation region of the opposite conduit through a check valve. Cyclically alternating changes in pressure in the

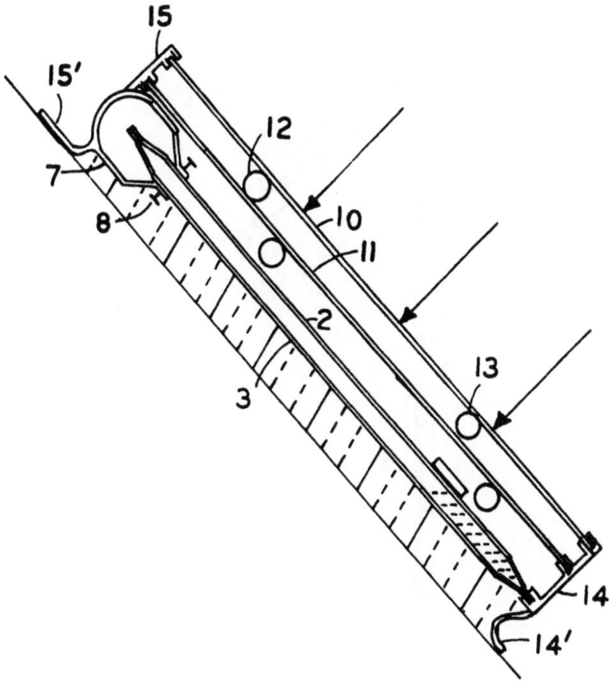

Fig. 166

two conduits are produced by alternately applying heat to the two vapourisation regions. The pressure differentials cause any liquid collected at the condensing region to be returned to the vapourisation region. This positive pumping action of the heat transfer fluid permits the heat pipe assembly to operate in almost any attitude and is particularly suitable for the downward transfer of heat. See Fig. 167.

The heat source required to supply thermal energy to the input end of the system, comprises a solar energy collector 15 indicated at 14. Collector 14 comprises a rectangular Fresnel lens 15 for concentrating incoming radiant energy from the sun onto the uninsulated upper portions 3 and 4 of the system.

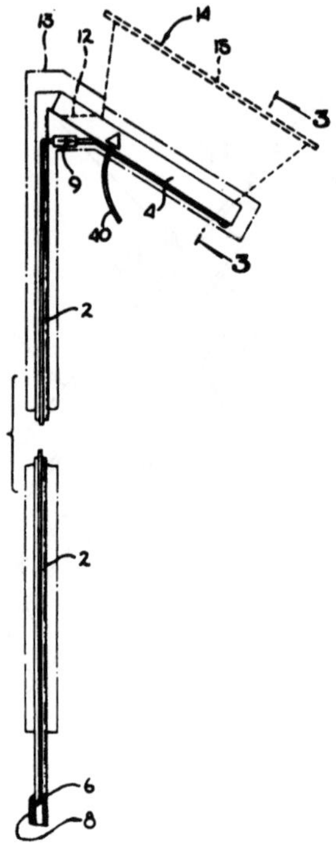

Fig. 167

Patent (87) of W.A. Harper concerns the utilisation of solar energy to heat a domestic hot water supply, wherein an apparatus employs fluid pressure generated by solar radiation as both the motive force of pumping and controlling agent for maximising heat gain and timing of required transfer of the heat absorbing medium within the system. As the temperature of a simple solar collector rises when exposed to solar radiation, pressure is developed at a predetermined temperature by the vapourisation of a fluid whereby a positive fluid pressure directly transfers a heat absorbing medium such as water into a second flat plate collector to absorb available solar generated heat developed by the collector. Upon removal of the solar radiation and concomitant collapse of the energy emission/absorption equilibrium in both collectors, the fluid pressure becomes negative with the condensation of the vapour and transports the water and its absorbed heat energy from the solar collector. The hot transport medium is transferred to a heat exchanger to heat a domestic water supply and transferred to a heat exchanger to heat a domestic water supply and to contribute to space heating needs. Upon the return of the next cycle of solar radiation the system responds by transferring fluids under pressure to collect available energy in the collector. Such a system is protected from freeze damage and controls the undesired loss of heat energy during times when solar radiation is inadequate for a heat buildup.

Fig. 168 is a simplified perspective view showing the two flat plate solar collectors, heat exchange tank, vapour bag, connecting means, fluid transport medium, domestic water lines and expansion bag as an example of the invention.

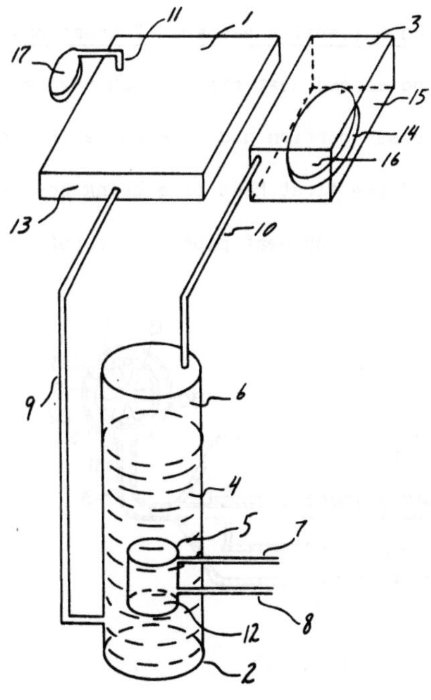

Fig. 168

A solar heat collector, designed by Sumida Isao, et al., (Hitachi Ltd.), described in Patent (89.8) and shown in Fig. 169, comprises an evacuated transparent pipe 1; a solar heat collection plate 7, disposed in the transparent pipe 1, a heat pipe 3 disposed in the transparent pipe 1 so as to contact with the solar heat collection plate 7, and containing an evaporable working liquid; a heat medium pipe 11 containing a heat medium to be heated; a heat releasing fin portion 131 extending from the axis to the inner surface of the heat medium pipe 11; and a cylindrical casing 9 surrounding coaxially the heat medium pipe 11 to provide an annular space 11 communicating with the heat pipe 3. The evaporable working liquid evaporates, receiving solar heat collected by the heat collection plate 7. The resultant vapour heats the heat medium through the heat medium pipe 11 and the heat releasing member 130. See Fig. 169.

A self-contained passive solar heating system, provided by J.E. Woods, (Iowa States University Research Foundation) in Patent (98)., includes first and second heat pipes, each having a refrigerant medium therein, a condenser portion and an evaporator portion, with the condenser portion of the first heat pipe being coupled to the evaporator portion of the second heat pipe for transferring heat

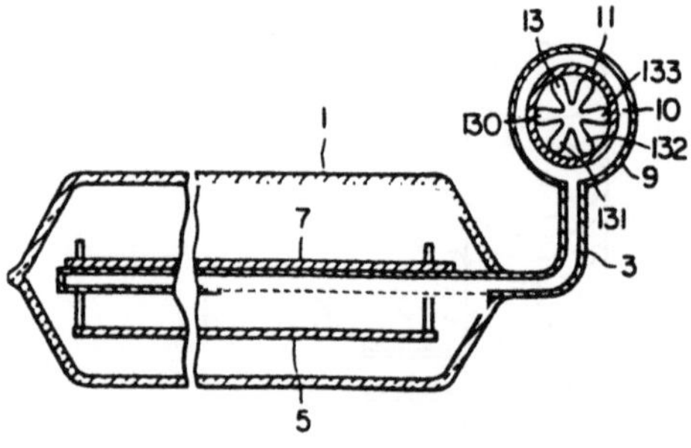

Fig. 169

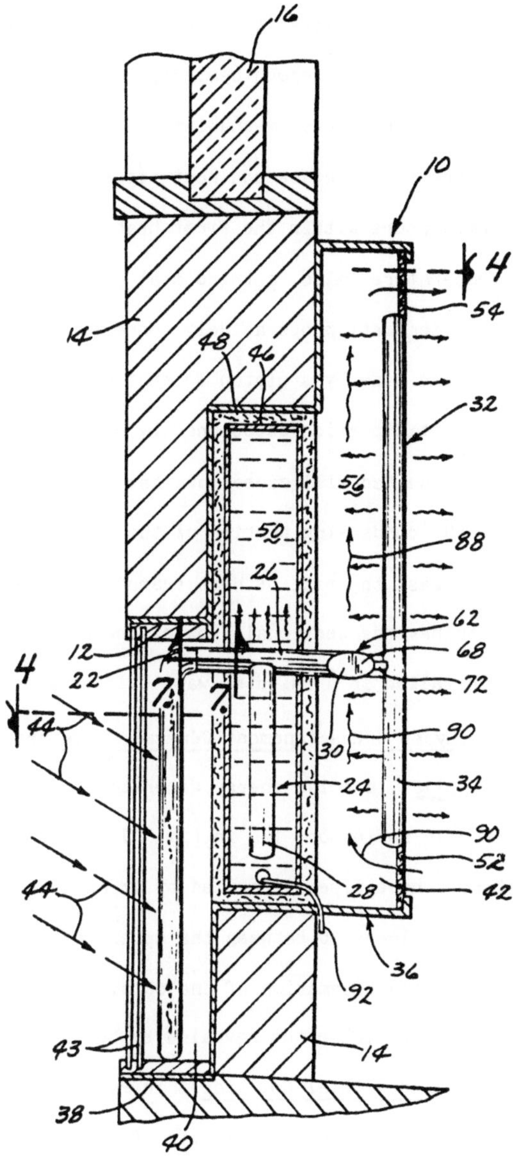
16
10
4
4
14
54
48
46
32
56
50
88
26
62
12
68
22
72
7
7
44
90
30
24
34
90
52
28
42
44
36
92
43
14
38
40

Fig. 170

thereto when the pressure within the first heat pipe is greater than the pressure within the second heat pipe. The evaporator portion of the first heat pipe is adapted to be exposed to a heat source and the condenser portion of the second heat pipe contacts a medium to be heated. A temperature control mechanism may be installed as the coupling between the first and second heat pipes for uncoupling the same when the temperature within the first heat pipe falls below a predetermined temperature. Also, a third heat pipe may be provided having a thermostatic portion operatively connected to the condenser portion of the second pipe by a piston so that changes in pressure within the thermostatic portion occasioned by changes in temperature of the medium to be heated will cause movement of the piston to vary the size of the condensing portion of the second heat pipe to increase or decrease the rate of heat transfer to the medium. See Fig. 170, a transverse sectional view of the proposed solar heating module installed in a building wall below a window.

G. Lety, et al., (Lycée d'Enseignement Général et Technologique), propose in Patent (112), a solar energy generator comprising a cylindrical-parabolic reflector, (2, in Fig. 171), a support 1 for the reflector; a heat pipe 3, carried by 1; an adjusting and thermally conductive sleeve 16 between the upper end of the heat pipe and a device for the transformation of the obtained thermal energy.

The heat-pipe type solar collector revealed by Soichi Kitajima, (Matsushita Electric Industrial Co. Ltd.), in Patent (119), comprises as its principal portion a plurality of heat pipes 1 arranged parallel with each other and having vapourising portions 7 and condensing portions 9 and a plurality of heat collector plates 2 traversing the heat pipes 1 and arranged parallel to each other. The vapourising portion 7 of each heat pipe penetrates and is fixedly

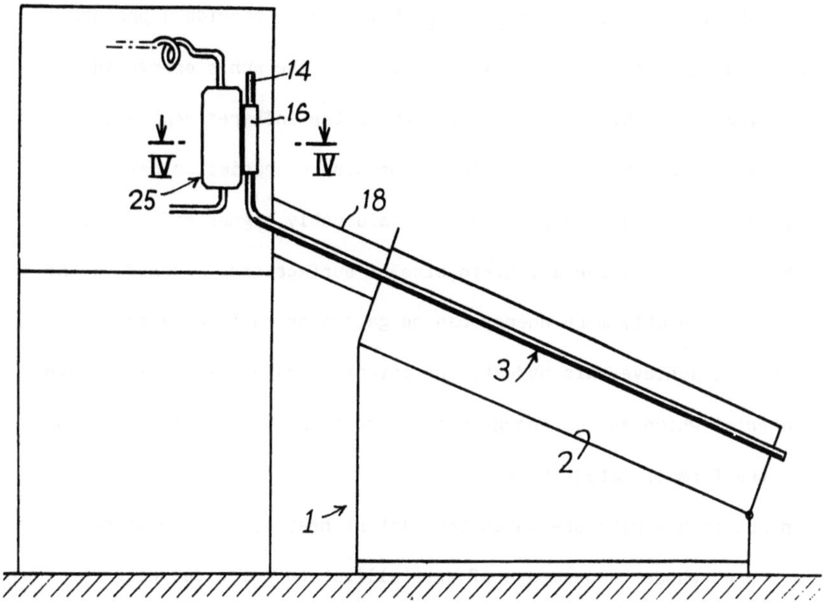

Fig. 171

mounted to the collector plates in intimate heat conductive relation therewith, and each collector plate has a lower portion 2' bent substantially perpendicular to the remaining portion to cover a space between itself and an adjacent collector plate 2.

According to this arrangement, solar heat can be effectively collected by the collector plates 2 and transmitted to the heat pipes 1 with a negligible heat loss.

The numerals given for the components in the above description refer to Fig. 172.

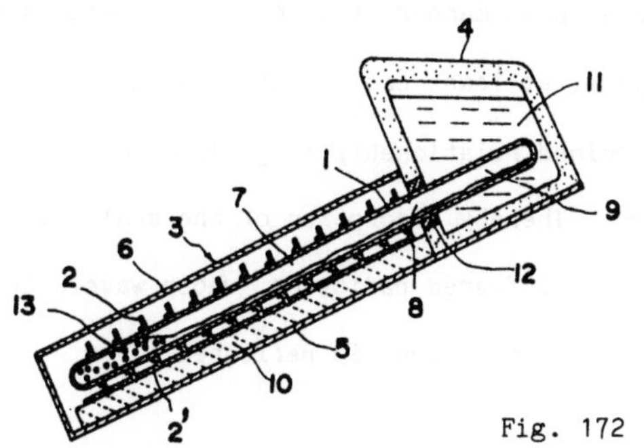

Fig. 172

The object of Patent (108.2) of N. Laing is to provide a process and a device for gaining heat by receiving the sun's energy and/or cooling spaces by emitting radiation in the infrared region of the spectrum, at which the heat absorption or radiation emission by the layer can take place at a considerably higher efficiency, which means that for any given size of surface area exposed to the sun considerably more energy can be gained or radiated away. In order to achieve this object, the invention makes use of a storage material which stores energy not in the form of sensible heat, but in the form of latent heat.

In accordance with the invention, latent heat storage substances are used whose temperature of crystallisation coincides with the desired process temperature, e.g. the temperature of the water in the case of water heaters. Thereby the temperature of the collector is considerably reduced and, in operation, maintained at a constant value which lies considerably below the hitherto conventional temperature. The invention further envisages providing the sun collector with a layer which, over the frequency range of the sunlight, behaves substantially like a black body, i.e. which exhibits as high a degree of absorption as possible, whilst, to a large extent, it does not emit the longwave radiation. Such layers may, for example be realised by providing, in a known manner, the collector surfaces with a number of layers, at least one of which has a thickness which has a predetermined relationship (e.g. 1/4) to the mean wavelength of the sunlight. Thereby absorption of the sun's radiation results, whilst the infrared radiation, whose wavelength is greater by more than a factor of ten, is reflected.

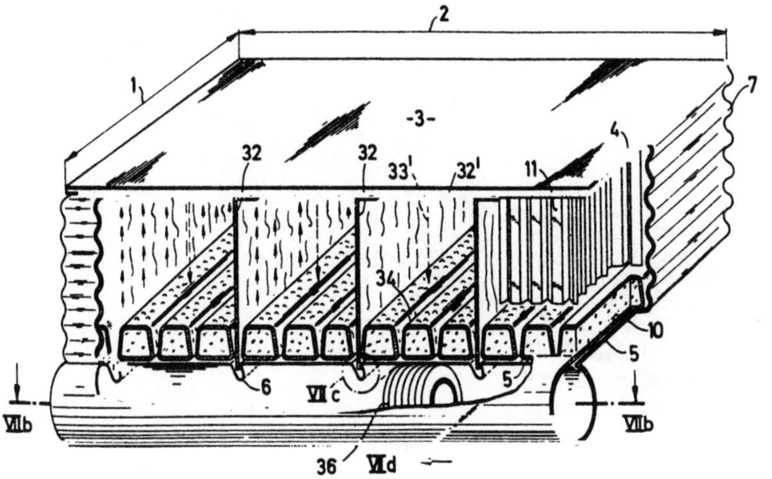

Fig. 173

Fig. 173 is a perspective view, (partly in section) of a device (roofplate) according to (108.2).

In Patent (122) of G. Meckler an infrared solar energy collector is disclosed. The collector comprises a heat absorber which in a first embodiment is a tube through which a heat transfer fluid is circulated. The heat absorber is disposed within a larger glass tube. In a modified embodiment, the heat absorber is in the form of a heat pipe which conducts heat to a heat transfer fluid circulated through a manifold. A wick carrying a volatile fluid may also surround the heat pipe. Absorbed heat evaporates the fluid which is in turn condensed on the said manifold. Either the entire interior of the glass tube surrounding the heat absorber is under reduced pressure, or an annular region between the surrounding glass tube and a second, larger diameter surrounding glass tube is under reduced pressure. An energy director such as a reflector within the

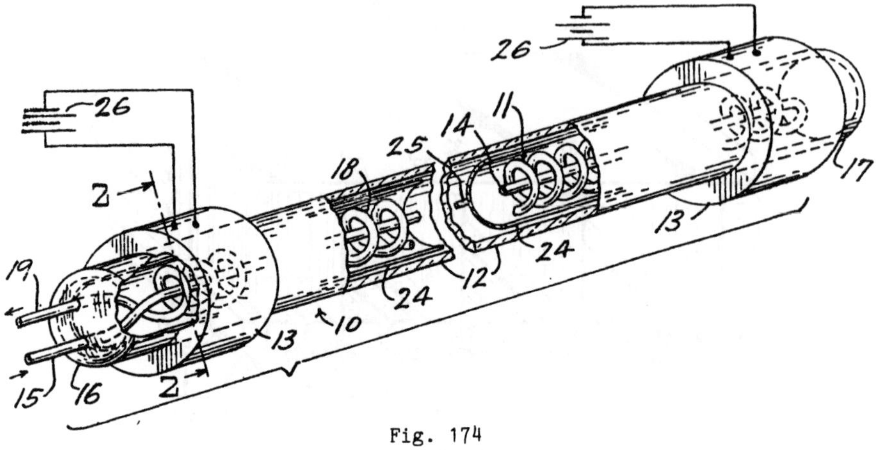

Fig. 174

enclosing glass tube directs solar energy on the heat absorber. The relative positions between the energy director and the heat absorber are changed progressively during the course of each day to enable the maximum utilisation of solar energy.

Fig. 174 is a fragmentary perspective view of an embodiment of the invention, while Fig. 175 is a cross-sectional view along the line 2 - 2 of Fig. 174.

The collector for solar energy, described in Patent (123.1) of J.H. Millar comprises an outer glass tube 2, having therein a compound parabolic reflector 4, extending substantially along

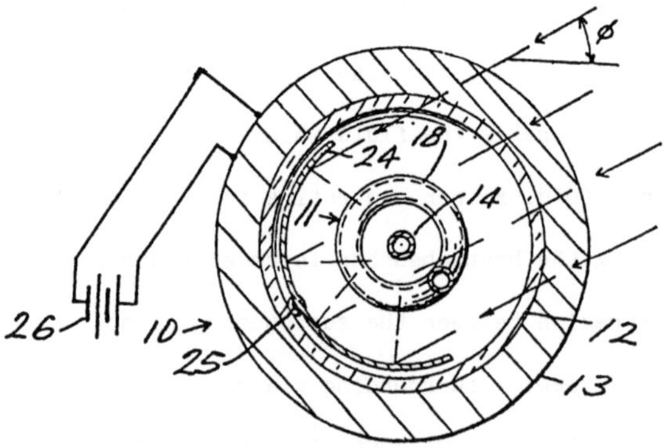

Fig. 175

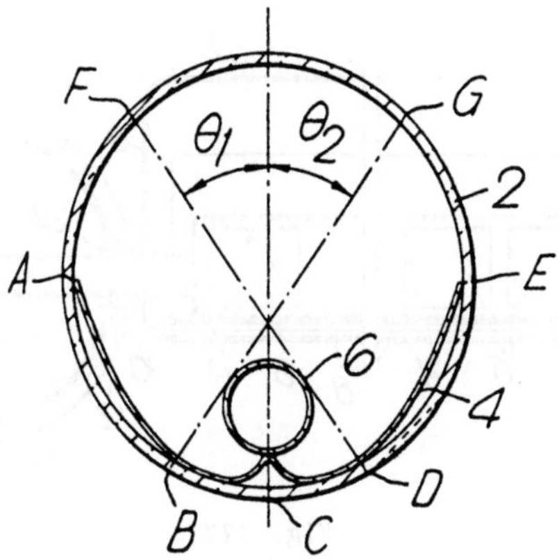

Fig. 176

the length of said tube and a straight conduit 6, which may take the form of a heat pipe or may contain a heat pipe, extending along the focus of said reflector, the ends of the outer glass tube being sealed and the space within the sealed outer glass tube being evacuated.

Fig. 176 represents a transverse cross-section through a collector tube in accordance with the invention.

Another Patent (123.2) of J.H. Millar refers to a solar energy collector, comprising a conduit 4, which may be in the form of a heat pipe or may contain a heat pipe, positioned within an outer glass tube 2, at least a portion of the outer surface of the conduit being constructed and arranged to support one or more solar cells 6. A compound parabolic reflector 8 within the outer glass tube is arranged to focus incident radiation upon the conduit and solar cells. The ends of the outer glass tube are sealed and the space between the outer glass tube of the conduit is evacuated. The collector may be used to obtain both electricity and heat, e.g. hot water, simultaneously by absorption of solar energy.

See Figures 177 and 178.

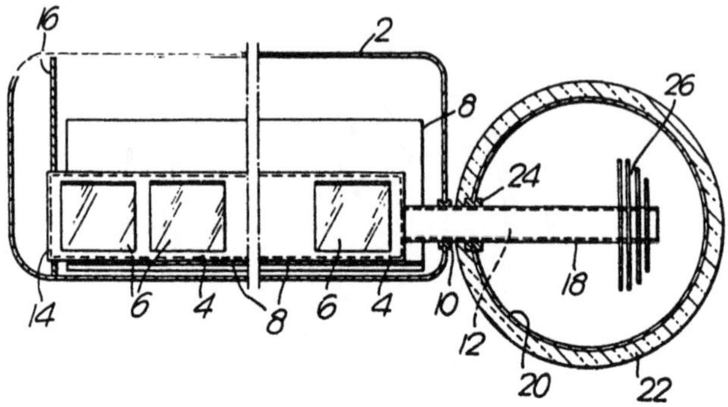

Fig. 177

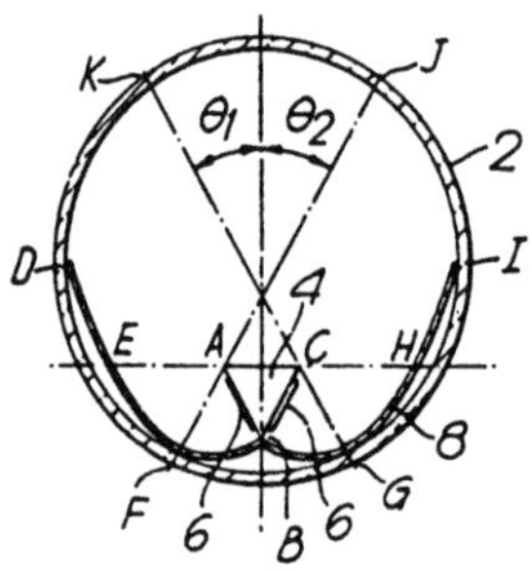

Fig. 178

In Patent (125) of A. Bertels, (Nagron Steel & Aluminium Co.), concerns a solar energy system comprising a collector consisting of a sloping space exposed to solar light for a heat transporting medium, performing a primary circulation, which is thermally coupled by the downward run with a working medium circuit.

An object of the inventor is to design a solar energy system in a manner, such that in contrast to the conventional system no control or safety apparatus is required, since solely natural, physical mechanisms are used so that a self-controlling operation is obtained, as a result of which the risk of disturbances and standstill is considerably reduced and the intrinsic safety is enhanced.

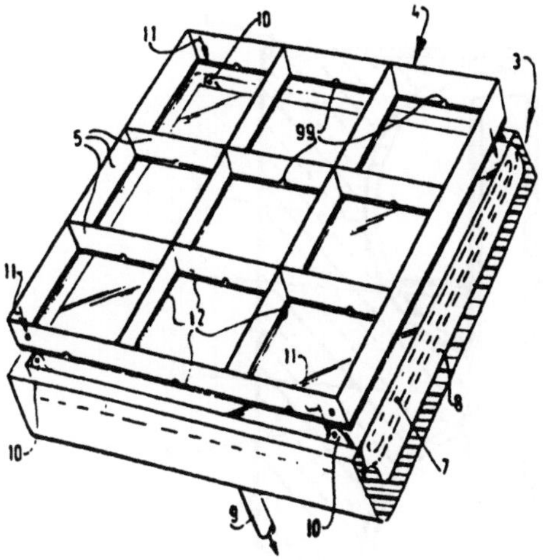

Fig. 179

The above mentioned purposes are achieved by means of a system, the primary circulation of which takes place in a vessel partly filled with fluid and partly with the vapour thereof and operating as a heat pipe.

The invention further provides a solar heat collector, the active surface of which is divided in cells bounded by upright ribs. Owing to the provision of such cells the dissipation of heat by convection due to wind brushing past can be reduced to negligible proportions. See Fig. 179.

In Patent (131) of E.J. O'Hanlon provides means and method for transferring heat downwardly to an out of sight area. This is particularly applicable to the reception, transport, and storage of solar heat in a manner that prevents its prompt loss back to the sky if the sky clouds over.

Likewise when incorporated in the roof of any shelter structure it can provide internal solar warmth to the structure and at a minimum construction and maintenance cost.

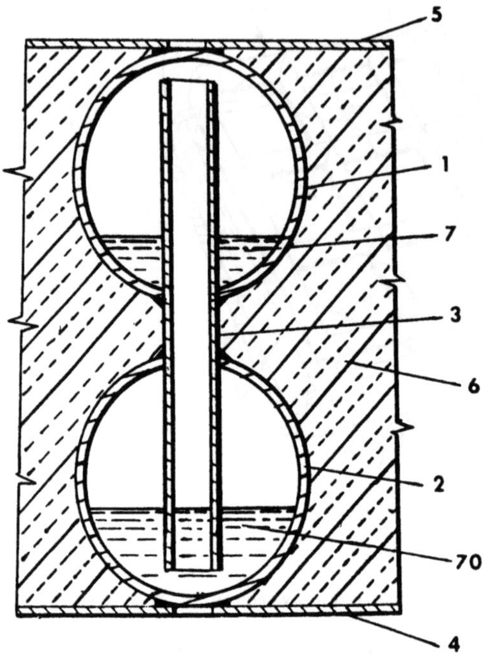

70 - the condensate

Fig. 180

In Fig. 180, 1 is an upper sheet metal container, 2 is a lower sheet metal container positioned under it, 3 is a metallic passage way joining container 1 and container 2 together and reaching almost to the top of container 1 and almost to the inner bottom surface of container 2. All tubular and container contacts and connections are permanently sealed and made leak-proof so no liquid sulphur dioxide or sulphur dioxide fumes can possibly escape and become noxious. Numeral 4 represents a sheet metal surface fastened by welding or brazing to the bottom of container 2, numeral 5 being a similar sheet of metal fastened by welding or brazing to the top of container 1. Around the outer surfaces of containers 1 and 2 and tubular passage way 3 is packed heat insulation material 6, in this case rigid polyurethane foam. Within the upper container 1 there is a supply of liquid sulphur dioxide 7. It needs not fill the entire container, but could occupy at least the bottom one-third thereof.

A solar energy conversion device, forming the object of Patent (133) of G.M. Edge, et al., (Pa Management Consultants Ltd.), includes a sealed container, the container comprising a first part forming an energy collector and a second part forming an energy transmitter, the said collector part having a first wall portion through which solar energy falling thereon can pass to a second wall portion which is adapted to absorb solar energy, the container including a fluid which, when the device is in operation, is vaporised by energy absorbed by the second wall portion, the vapour passing to the second part where it condenses and gives up its latent heat of vaporisation, the condensed fluid returning to the first part.

Fig. 181 shows a radiant collector C, which provides a large flat area and which forms the heated end of a heat pipe. The working fluid in the heat pipe system, which includes a small bore tube T, coupled to the collector C, is vaporised at the collector C during operation and conveyed along the small bore tube T, to an emitter end E, of

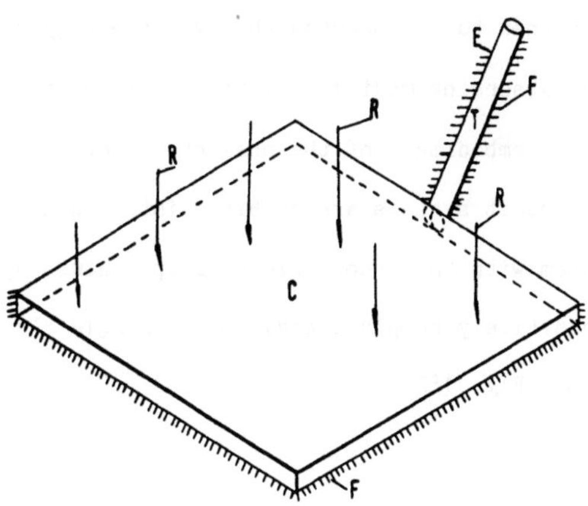

Fig. 181

the tube T, where it recondenses and gives up its latent heat of vaporisation. The liquid then returns to the collector area by gravity or through a wick, incorporated in the walls of the heat pipe. The collector end C of the device has an upper surface U, which is transparent to solar radiation.

Patent (134) of F.J. Pasarelli et al., refers to an energy collector and transferring apparatus which in its first embodiment utilises three connected closed loop conduit systems.

The first closed loop conduit system is to include an energy collector which is to collect not only solar energy but ambient energy as well. The energy is absorbed within a black liquid flowing through the first conduit system, which includes a first energy exchanger wherein the absorbed energy is transferred from black liquid to a second energy absorbing medium.

The second energy absorbing medium is cycled within a second closed loop conduit system. A compressor is included within the second closed loop conduit system to condense this medium from a gaseous state to a higher density fluid, thereby substantially raising the temperature. The energy from the second energy absorbing medium is then to be absorbed through an energy exchanger to a third energy absorbing medium of a third closed loop conduit system. In a second embodiment of the present invention, the first and third loop conduit systems are combined into a single closed loop conduit system with the second closed loop conduit system functioning to additively supply energy to the single closed loop conduit system. See Fig. 182.

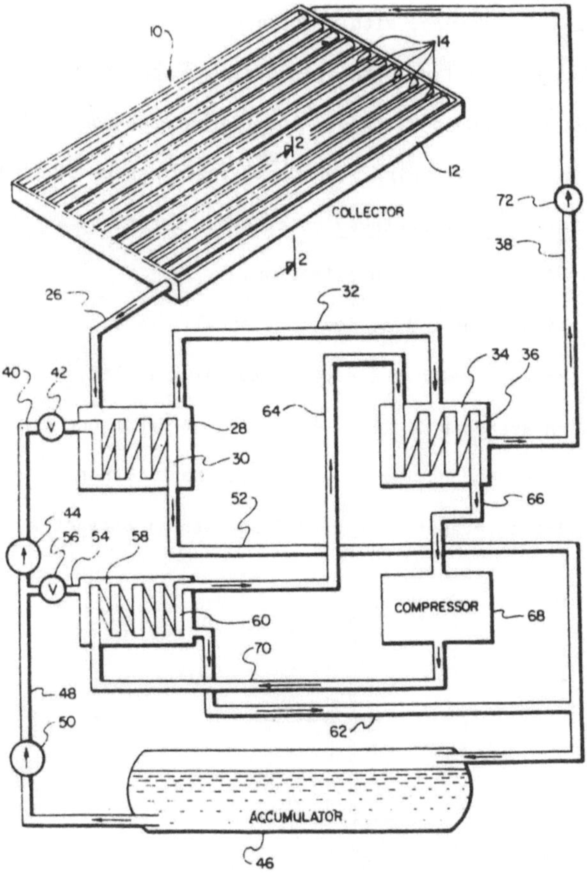

Fig. 182

In Patent (136.1), H. Scholz, reveals a heating system, for heat supply to consumers. The system consists of a closed space, containing several closed containers, filled with a heat accumulating medium obtained from a first heat source, the said medium being in heat exchanging relation through the container walls, with a heat transferring medium, present in the closed space, heat being transported through an evaporation/condensation circuit from the containers to the consumers. The temperature in the closed space can be established as high as 1500°C. A feature of the invention is, that each container contains identically evaporable heat transport media, for adjusting the pressure, within the containers at each temperature to the pressure, prevailing outside the closed space. In this way the complications, caused by pressure differences exposing the container walls to extremely high mechanical loads can be eliminated.

N.V. Philips Gloeilampenfabrieken carries out intensive activity in the field of solar energy utilisation.

Thus, Patent (136.2) of G.A.A. Asselman et al., (assignors to Philips) refers to a solar collector comprising a flat, metal plate solar radiation absorber 1, (Fig. 183), and at least one elongated sealed tube 2, which contains an evaporable heat transport medium 5, The duct having an evaporation zone 3 in thermal contact with the

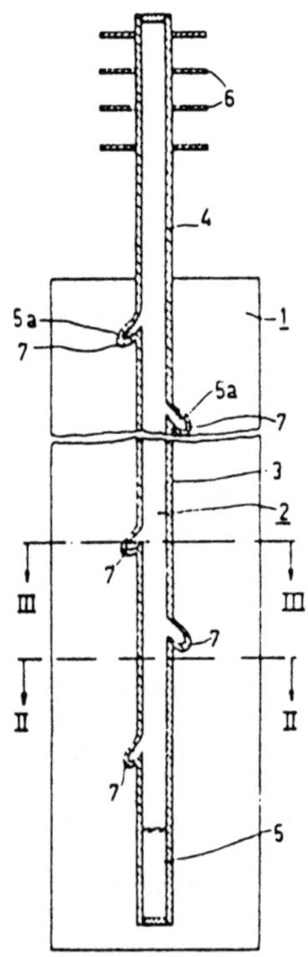

Fig. 183

absorber, a condensation zone 4 and a number of protusions 7 distributed along the length of the evaporation zone 3 for collecting condensed heat transport 5a medium on being returned to the evaporation zone 3 by gravity. See Fig. 183.

Another Patent (136.3) of G.A.A. Asselman et al., also assigned to Philips, concerns a solar energy absorber, comprising a sealed tube 2 of metal or glass containing an evaporation heat transport medium 5 and comprising an evaporation zone 3 in which a number of pieces of liquid-absorbing material 7 are provided at a distance from each other in the axial direction in order to accumulate any returning condensed heat transport medium therein. A flat plate 1 may be secured to the tube adjacent the evaporation zone. The portion of the tube comprising the evaporation zone may be enclosed within an evacuated glass tube. The solar energy absorber is shown in Fig. 184.

According to Patent (136.4) and essentially Patent (136.19) of H. Bloem, (Philips), there is provided a solar collector, comprising a plate-like absorber 13 for collecting solar radiation which is thermally conductively connected to an evaporator section 3 of a heat transport system which is constructed as a heat pipe 22 and which, in use, can exchange heat with a further heat transport system 5 by way of a condenser section 4. The heat pipe containing a working medium has a critical maximum temperature in the further heat transport system. As the working medium isobutane is recommended, which

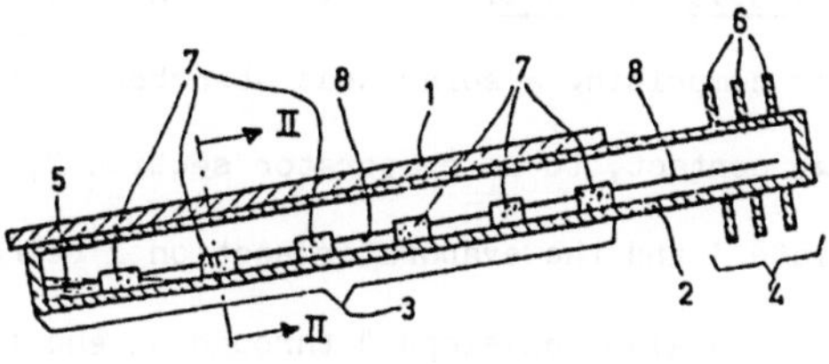

Fig. 184

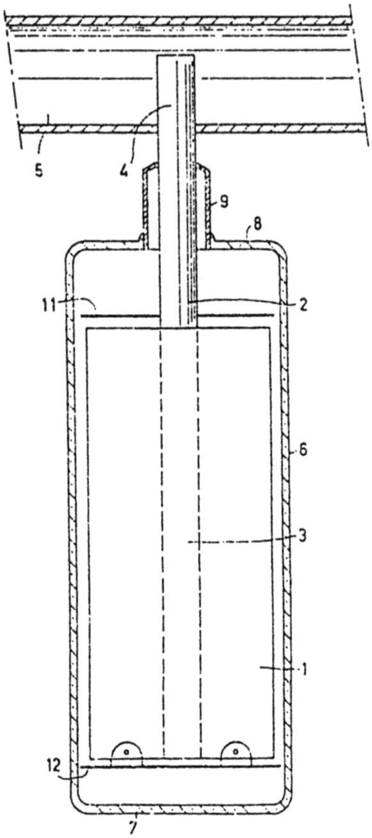

Fig. 185

has the advantage that even at higher temperature it is a stable compound which does not decompose. Furthermore, isobutane has a critical temperature of 135°C which is an acceptable maximum temperature for the further heat transport system 5. (See Fig. 185.) Another Patent (136.5) of Philips, (inventor: J.C. de Grijs), discloses a solar collector comprising a solar heat absorber plate 1 connected, in heat exchange contact, to an evaporator section 2, of a heat pipe 3, the absorber plate 1 and the evaporator section 2 being surrounded by an evacuated, sealed glass envelope 4 through an end face 6 of which the heat pipe 3 passes to a condenser section 11 situated outside

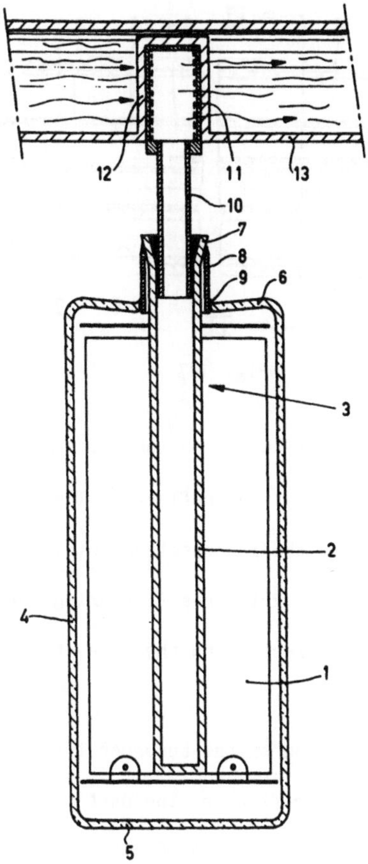

Fig. 186

the envelope 4 for exchanging heat with a heat transport medium of a heat transport system 13. The part of the heat pipe 3 between the condenser section 11 and the location where the heat pipe 3 passes through the end wall 6 of the envelope 4 is of a flexible construction. See Fig. 186.

Patent (136.8) assigned to Philips concerns a solar heat collector of symmetrical tube structure, essentially of U-form, the absorber element being arranged in the central part of the tube, while the reflector adjoins the absorber and extends in transversal section, under an angle of about 120°, cutting the tube at points situated

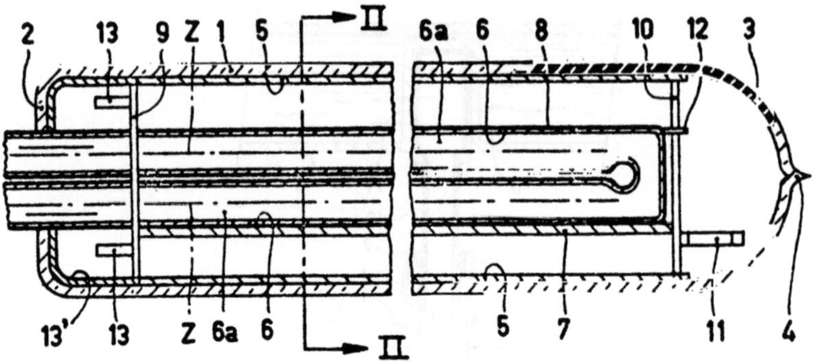

Fig. 187

on the X-X diameter of the tube's periphery, X-X being perpendicular to the symmetry axis Y-Y- of the reflector. The center of the tube section M is disposed at the intersection point of the Y-Y axis of the reflector and the straight lines R, passing through the intersection points of the reflector and the tube, tangentially to the absorber.

Due to the central position of the absorber in the plane of symmetry of the interior reflector, the part of radiation directly arriving at the absorber, is increased. The end of the tube has a cupola-like form, the inner surface of which is coated with a reflecting material of AC, Cu, PG, Ag or Au.

Fig. 187 shows a longitudinal section, while Fig. 188 is a section through points II - II of Fig. 187.

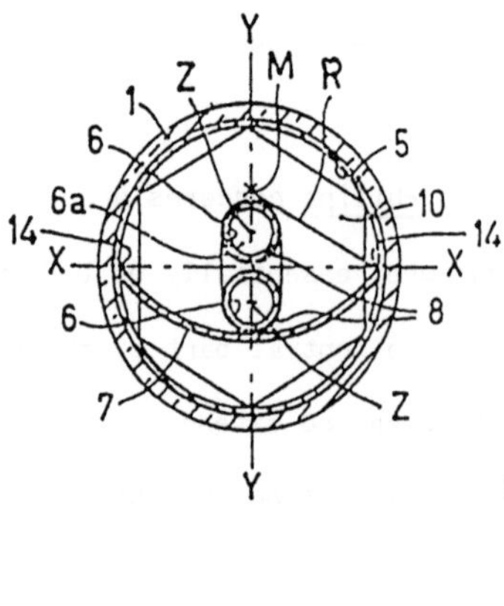

Fig. 188

In order to prevent the increase of pressure, which inevitably takes place after long-term operation of solar collectors, in spite of using substances like Ba, Zr, Ti, etc. for bonding the forming gases, Patent (136.11) of Philips (inventor: R. Doevenspeck, et al.), provides in the collector casing a heating body operating in the 500-2000°C temperature range for cracking the residual gases. The invention is based on the idea that gaseous impurities released from the absorber and casing, (CO, H_2O), react with the bonding agents, forming hydrocarbons (methane). The applied heating body permits the cracking of hydrocarbons, whereby the carbon forming can be absorbed by the heating body, while the liberated hydrogen can be bonded by the bonding agents.

A.A. Slaats, (Philips), reveals in Patent (136.13) a heat pipe solar collector, comprising an evacuated transparent envelope 22 which surrounds the evaporation zone 20a and a jacket 23 which surrounds the condensation zone 20b and which is provided with circumferential flow passages. The jacket 23 is connected to the transparent envelope 22 and protects the joint 8 between the transparent envelope 22 and the heat pipe. See fig. 189.

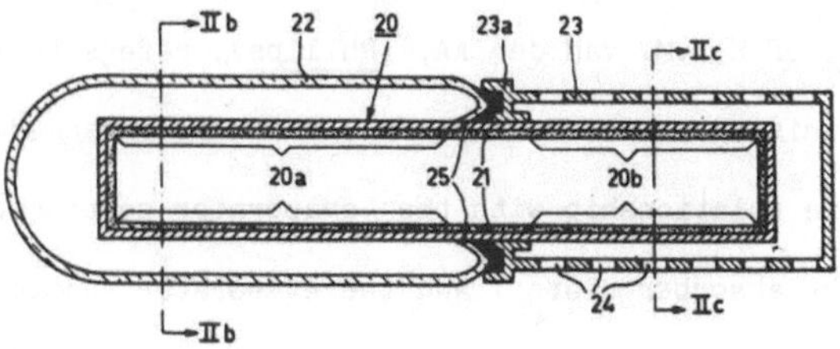

Fig. 189

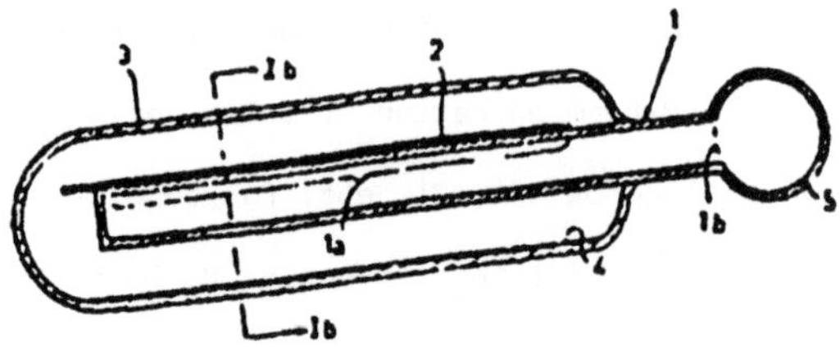

Fig. 190

In Fig. 190 illustrates a Patent (136.14) of H. Bloem, (Philips), wherein the condenser end portion 16 of the heat pipe 1 is in open communication with a duct 5 wherethrough a heat absorbing liquid flows for a direct exchange of heat between the vapour formed in the heat pipe and the liquid.

In yet another Patent (136.18), H. Bloem of Philips, claims a heat-transport device 1 comprising a sealed and evacuated tube 2 which contains water, which in the operation of the device 1 evaporates in a heated section of the tube and condenses in a cooled section, the condensed water collecting at the bottom of the tube 2, in the form of a liquid column, and a fine-meshed gauze 11 being attached to the inner surface of the tube 2 at the area of the liquid level 10 in the bottom of the tube when the device is not in operation.

The numerals indicate the components in Fig. 191.

Patent (136.15) of H.H.M. van der AA, (Philips), refers to a solar collector, comprising an absorber plate 7 which is arranged in thermally conductive relationship with the evaporator section 8 of a heat pipe 9, the absorber plate 7 and the evaporator section 8 being surrounded by an evacuated transparent envelope 1, the condenser section 10 of the heat pipe projecting from the envelope 1 to exchange heat with a medium to be heated, the evaporator section 8 of the heat pipe being enclosed by a tube 5, one end of which is closed while

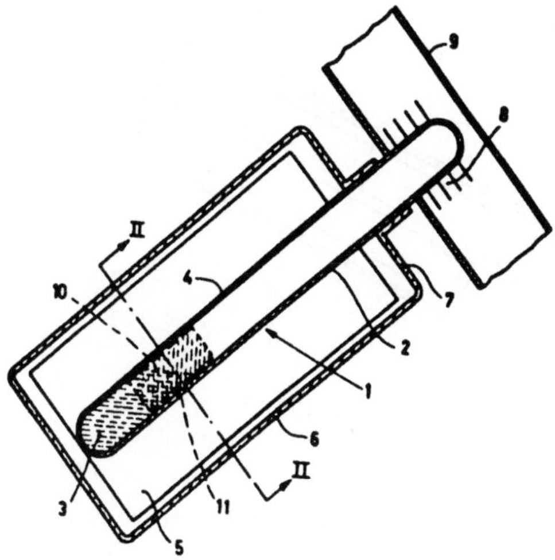

Fig. 191

its other end is connected to the end wall of the envelope 1 through which the heat pipe passes, the absorber plate 7 being connected to the tube and a suitable thermally conductive medium 11 being present between the tube 5 and the evaporator section 8. See Fig. 192.

Another Patent (136.16) of H.H.M. van der Aa, (Philips), concerns a solar collector, comprising a plate-shaped absorber 4, one side of which is provided with a black layer for selective absorption of incident solar heat, and which exchanges heat with the evaporator section 3 of a heat pipe 1 whose condenser section 2 exchanges heat with a medium to be heated, the evaporator section 3 and the absorber

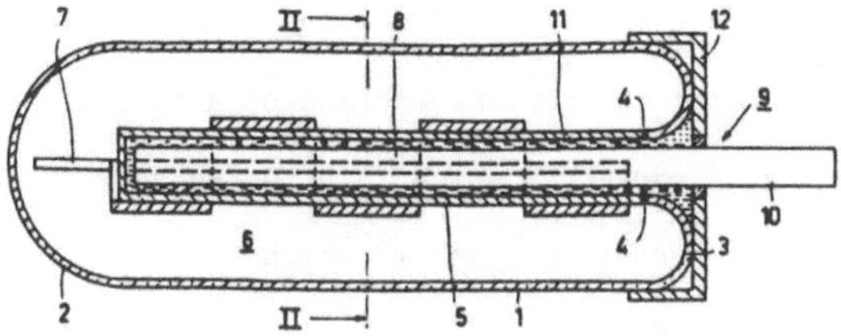

Fig. 192

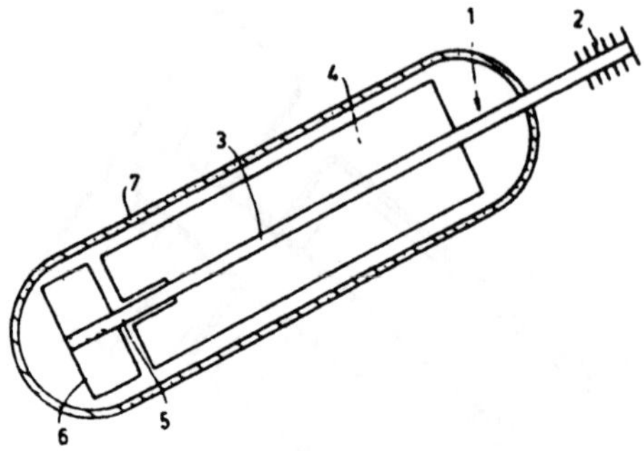

Fig. 193

plate 4 being enclosed in an evacuated transparent envelope 7, a part 5 of the heat pipe 1 which is situated within the envelope 7 being free of the absorber plate 4, said part 5 exchanging heat with one or more layers which has or have the same absorption coefficient as and a higher radiation loss than the layer on the absorber plate 4 per unit of length of the heat pipe 1. See Fig. 193.

R.L.C. de Vaan of Philips, describes in Patent (136.17) a solar collector comprising one or more elongate, plate-shaped absorbers 7 which are or each of which is in heat-exchanging contact with a pipe 9 which contains a heat transport medium which discharges the heat incident on the absorber 7 which is enveloped by an evacuated, transparent envelope 5 having a circular cross-section. On one side of the absorber a reflector 15, 17 outside the envelope 5 is arranged, which consists of two parts which contact the envelopes 5 along a common generatrix, and each of which has the form of a part of the cylindrical surface of a circular cylinder and has a centre of curvature which is situated substantially on the circumference of the envelope, and in which

$0.8\ b \leqslant R \leqslant b$ and $0.7\ b \leqslant d \leqslant 1.2\ b$, wherein=

R = the radius of curvature of the reflector parts 15, 17,

d = the distance between the centre line of the or each envelope 5 and the edge of each of the associated reflector parts 15, 17,

b = the diameter of the or each envelope 5.

It is claimed that the small depth of the solar collector ensures that all intercepted solar radiation is incident on the absorber plate either directly or via only one reflection, and which enables optimum interception of solar radiation with a comparatively cheap construction. See Fig. 194.

Another solar collector unit is disclosed in Patent (136.20) of W. Lucassen, (Philips), which comprises a collector or a plurality of collectors which or each of which comprises an absorber plate arranged in thermally conductive contact with the evaporator part of a heat pipe which contains a heat-transporting medium, the absorber plate and the evaporator part being surrounded by an evacuated transparent envelope and the condenser part of the heat pipe projecting beyond the envelope and being provided with means which enlarge its heat-transferring area.

The unit further comprises a duct section which comprises an inlet duct and an outlet duct for the medium to be heated, which ducts are separated from one another by a wall.

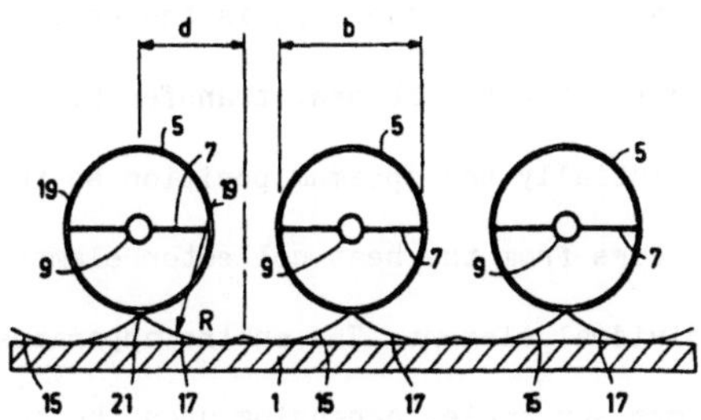

Fig. 194

This separating wall is provided with an aperture or apertures which or each of which is aligned with an associated aperture of the same or larger diameter in an outer wall of the duct section to form therewith a pair of aligned apertures, the aperture or apertures in said outer wall having a diameter corresponding to that of the collector envelope or envelopes and a collector being arranged in the or each pair of aligned apertures so that the envelope of the collector engages the outer wall of the duct section in a sealing manner, the means which enlarge the heat-transferring area of the condenser part of the heat pipe of the or each collector being surrounded by a wall which adjoins the boundary of the aperture or respective aperture in the separating wall.

In Fig. 195, 1 - the number of collectors; 2 - absorber plate; 3 - evaporator part of heat pipe 4; 5 - evacuated transparent envelope.

A solar heat collector, designed by M.E. Phillips, (Patent 138), incorporates a hexagonal shaped collector element, which achieves maximum heat utilisation in conjunction with a heat pipe for transfer of the heat collected by the element. Various minor modifications can be made to the heat collector element and/or the heat transfer pipe without departing from the spirit or the scope of the invention. Basically, however, it is the hexagonal shaped heat collector element with the heat transfer pipe passing therethrough at substantially the optimum position so that maximum heat transfer occurs from the heat collector element to the pipe within each individual element. The entire apparatus can be adjusted for optimum sun angle, depending upon the hemisphere or

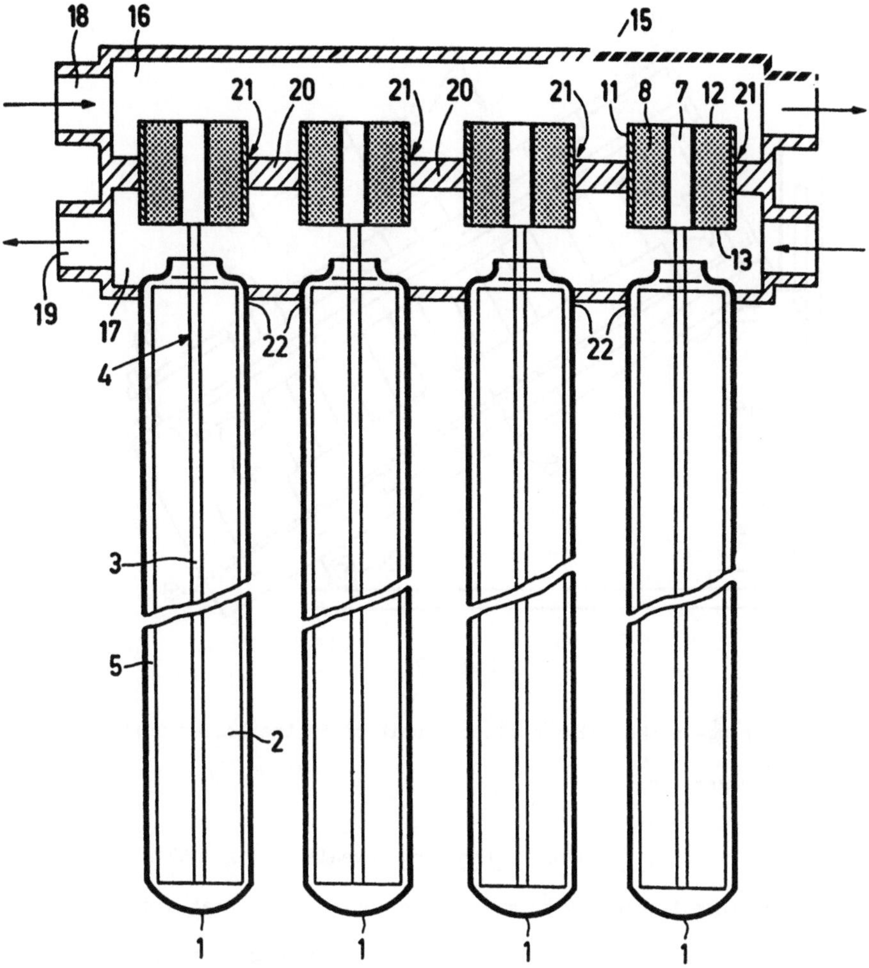

Fig. 195

latitude of use, and extremely efficient heat transfer and use of solar energy is possible. See Fig. 196.

The absorption-type refrigerator, described in Patent (139.1) of E. Pöhlmann, comprises a heat transport system, composed of a heat pipe-operated closed cooling agent circuit, which can be connected to a solar heat collector. According to the inventor the proposed heat pipe unit can have a modular structure, which can be

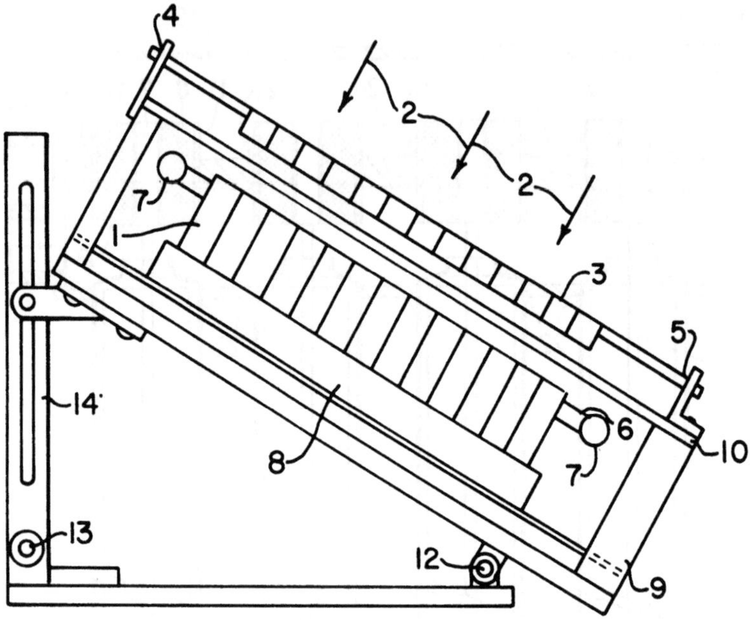

Fig. 196

simply plug-in the solar collector unit or in the boiler unit of the refrigerator.

Another Patent (139.2) of E. Pöhlmann reveals in Fig. 197 a heating/ cooking installation, with a solar collector 2 for heat absorption; a heat-storage block 2 and a heat pipe 4, which connects the solar collector 2 with the heat-storage block 1, furthermore with a heating and/or cooking element 7. During heat absorption and heat

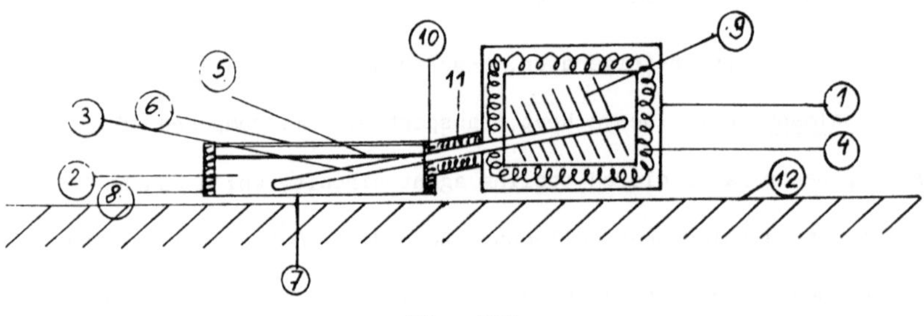

Fig. 197

storage the heat pipe 3 in rising arrangement from the solar collector 2 towards the storage block 1, while during heating and/or cooking periods the heat pipe 3 rises from the storage block 1 towards the heating/cooking element 7.

Patent (142) of K. Raetz, concerns a solar heat collector, which can be integrated into central heating systems. Pure, degasified water is passed through the collector and heated therein by solar heat. In periods of high radiation intensity water flows through a three-way valve into a hot water storage tank and thence in the heating system. In case of weak solar radiation, (cloudy weather), water flows through a low-temperature storage tank, from which heat (at low temperature) is collected by a heat pump and tranferred to the hot water storage tank to assist in heating.

A solar energy collector, disclosed by G.L. Fitton, (Patent 143.3), (Redpoint Associates Ltd.), includes a thermally conductive plate, one surface of which is coated with a layer of material which acts to absorb and convert into heat electro-magnetic radiation of a wavelength of two micrometers or less and which is transparent to electromagnetic radiation of a wavelength of five micrometers or more. The other surface of the plate is thermally coupled to a heat pipe which acts, in use, to carry away energy collected by the collector. Fig. 198 is a sectional end view of the solar energy collector described, wherein: 10 - an aluminium or copper plate; 11 - a layer of hydrated aluminium or black nickel coating; 12 - glass pane (of greenhouse glass); 13 - a heat pipe; 14 - C-shaped groove.

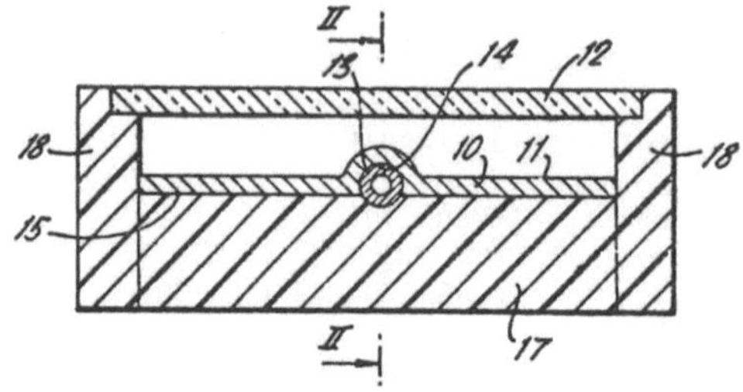

Fig. 198

In Patent (150.1) of F.M. Sabet there is provided a solar collector for the exploitation of the thermal energy of sunlight, in which the absorber has a transparent pipe containing an absorbing plate connected with a channel containing a transfer-fluid having a critical temperature of the same order of magnitude of the highest temperature of the whole system so as to dispense with difficulties under no-load conditions or with a surplus of solar irradiation.

In Fig. 199 : 1 - indicates a pipe; 2 - an intermediate piece of flint glass; 3 - a hood; 4 - a metal pipe of low heat conductivity, (iron); 5 - an absorber; 7 - reflectors; 8 - a gettering pad; 10 - a channel; 12 - a connecting piece (copper pipe, metal bellows); 34 - the connection place of 4 and 3 in vacuum-tight manner.

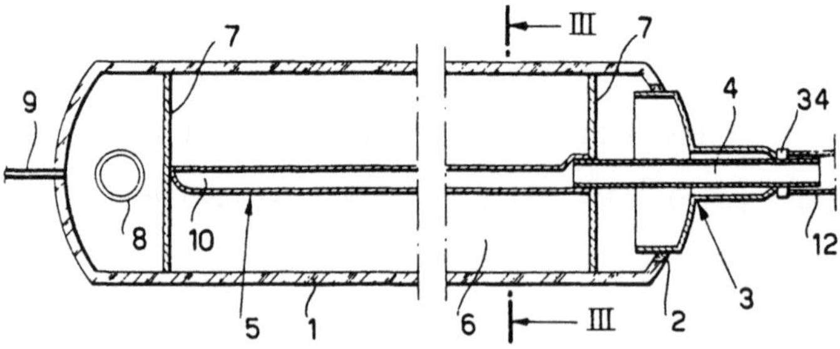

Fig. 199

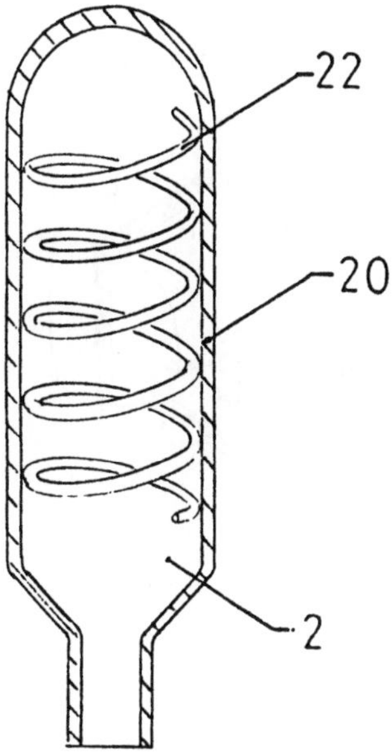

Fig. 200

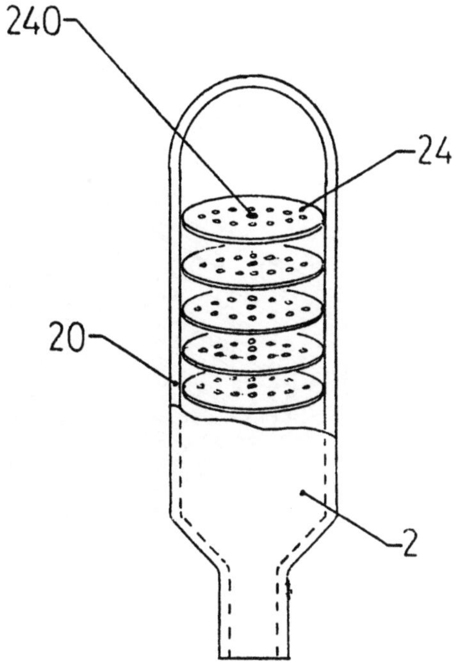

Fig. 201

The solar collector of F.M. Sabet revealed in Patent (150.2) comprises a heat-pipe having an absorber plate in heat transferring contact to its evaporation compartment. The condenser compartment 2 is constructed to be secured in a water/air manifold. The interior surface of the condenser compartment is increased by inserting a metal heat conducting member or members in thermal contact with the condenser body. The heat conducting member may be a helical body 22. Alternatively, the condenser compartment may contain perforated discs 24, a wavy metal sheet or metal mesh.
See 22 and 24 in Figures 200 and 201 resp. 2 - condenser; 240 - holes.
Another Patent (150.3) granted to F.M. Sabet concerns a solar heating system comprising a water storage tank 7 and an array of solar collectors 1, each comprising an elongate, evacuated, substantially cylindrical, transparent envelope and an absorber

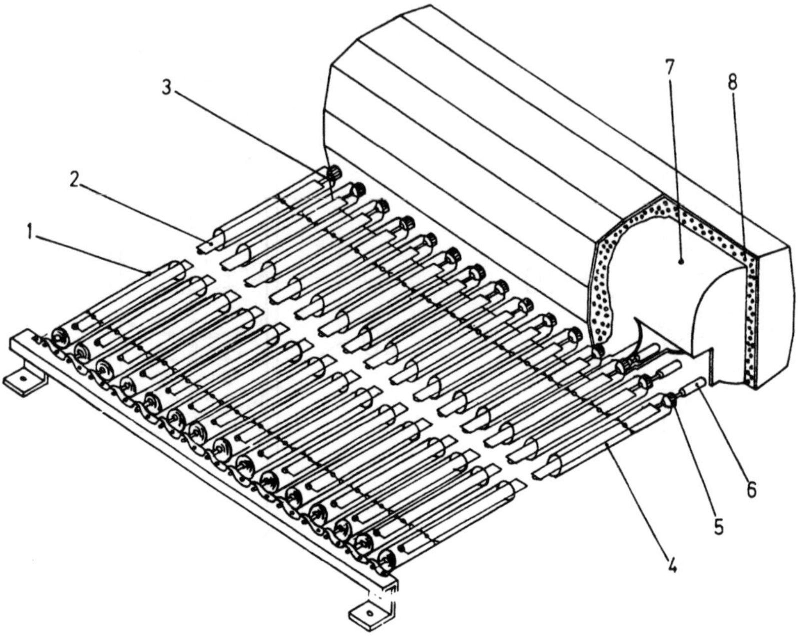

Fig. 202

plate 2. Each absorber plate is thermally conductively connected to the evaporation zone of a heat pipe 3 which extends through the end of the transparent envelope so that its condense zone is situated in the tank.

The heat transfer medium of the heat pipe can be freon or propane. The proposed system does not require a recirculating pump, control unit, heat exchanger in the storage tank and will not be subject to frost in the winter and no danger of overheating in hot summer days, it is claimed.

See Fig. 202.

Another Patent (150.5) of F.M. Sabet improves the operation of heat-pipe-operated solar collectors. It is known, that if a cloudy weather is followed by a sudden hot sunny period, the temperature of the vapour in the heat pipe steeply rises to such a degree, that the condenser is not capable of operating, until the vapours' temperature

is not lowered to the combustion degree. In order to eliminate such delay in operation, the proposed evaporation sector comprises a number of receptacles; in spaced-off relationship, along the sector, and having an aperture area, which is smaller than the cross section of the evaporator sector, permitting to collect the heat transfer medium at the rate at which it is condensed.
The total volume of the heat transfer medium surpasses the sum of volumes of the receptables, destined to collect this medium.

In the solar heat collector, described in Patent (154) of H. Schreiner, solar radiation is collected by a flat lens (operating as a cylinder collector lens) on a focal line, which moves over the focal plane according to the changing position of the sun. In the range of the focal plane several longitudinal heat absorber elements are arranged, parallel to the focal line, of which one or two are always within the range of the focal radius and are heated. In order to establish a high temperature, only the heated absorbers are connected to the heat transfer system, while the other ones are inactivated.
In cloudy weather, all absorbers receive the same amount of radiation and can be operated as in a flat collector, for generating heat at low temperatures.
See Figures 203, 204.
E. Schwarz proposes in Patent (155.1) and essentially in Patent (155.2) solar collector columns, for the mounting of flat collectors on places, which are not suitable for such purpose (roofs, inclined northward) or where no suitable mounting place exists at all (swimming pools, etc.).
The columns proposed can be carried out in various forms and be provided with flexible, black or transparent hoses, through which

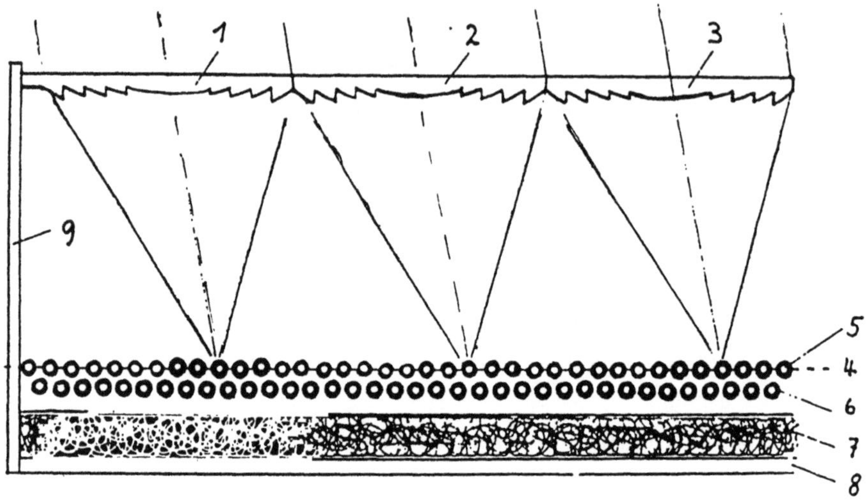

Fig. 203

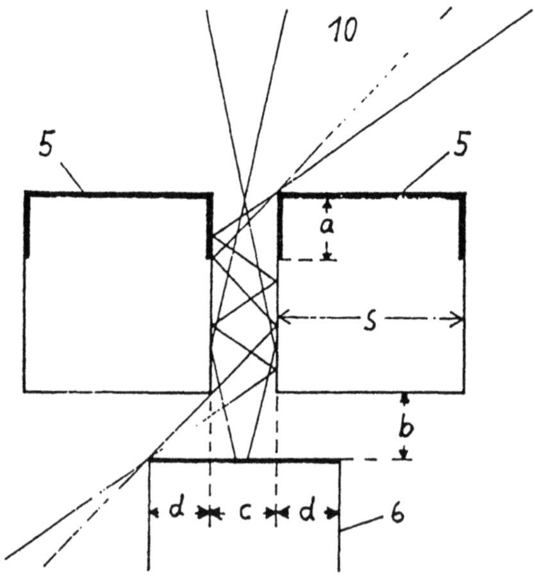

Fig. 204

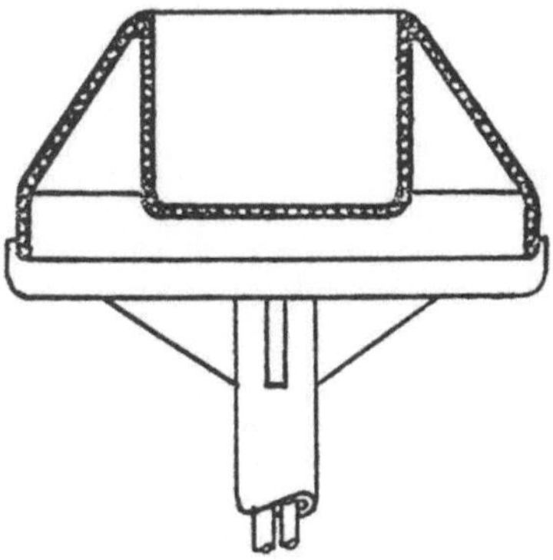

Fig. 205

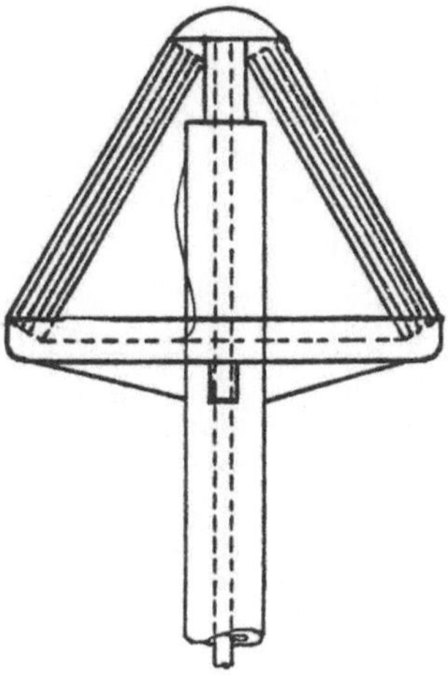

Fig. 206

a working fluid is flowing. They are mainly used for water heating. The columns represented in the figures have, as a rule, a radiated surface of 4-5 m^2.

See Figures 205, 206, 207, 208, 209, 210, showing the column structures and the arrangement of hoses therein.

Patent (158.7) assigned to the U.K. Secretary of State for Defence, (inventor: M.J. Graveney) provides a heat pipe assembly including a vacuum pump for maintaining a low pressure regime within the assembly. The vacuum pump is preferably a jet pump. The assembly includes a heat pipe in sheet form, and may be rigid as in a solar heat collector panel, or conformable as in a conditioning garment.

In an embodiment of the invention the heat pipe assembly comprises a panel for employing solar heat to heat water.

A suitable working fluid is water, though other liquids, for example methanol or one of the freon refrigerants, may be employed, dependent on the temperature to which it is required to

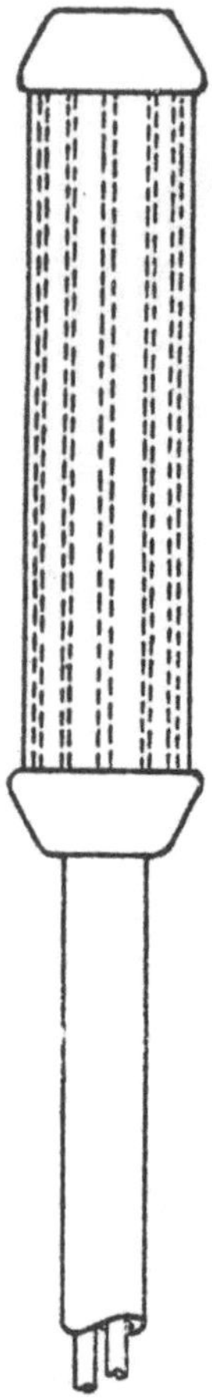

Fig. 207

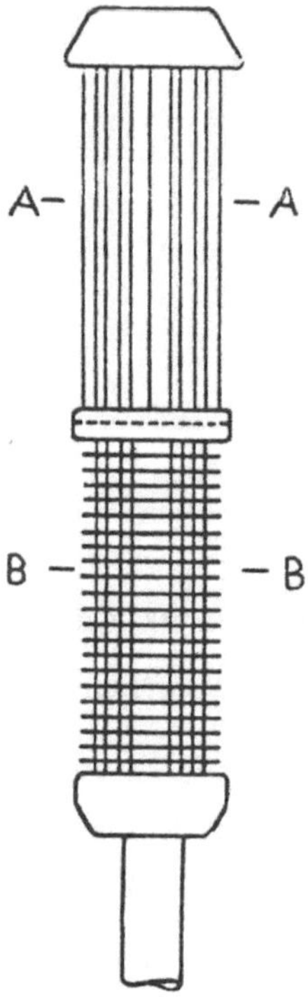

Fig. 208

Fig. 209

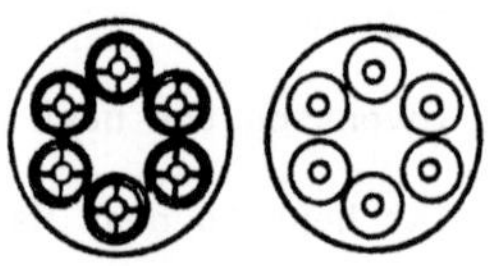

Fig. 210

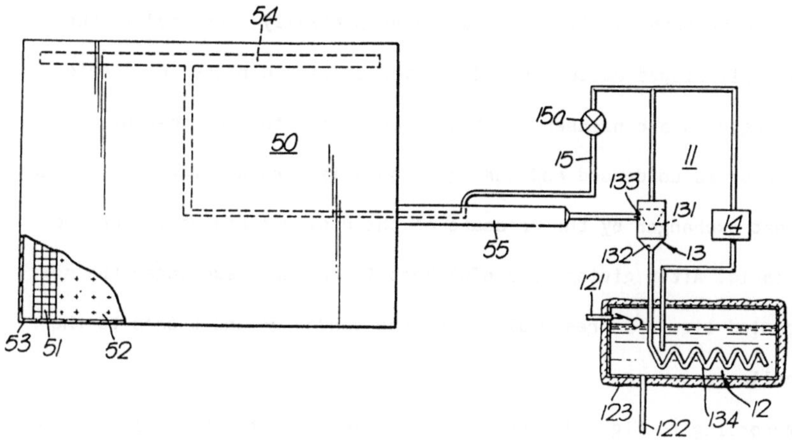

Fig. 211

cool. Fig. 211 shows a heat pipe assembly comprising a solar panel 50 of rigid heat pipe construction associated via pump apparatus 11 with a domestic water reservoir 12. The panel 50 comprises a reticulated structure 51 in sheet form, supporting an outer wall thereof a perforated wick sheet 52 and enveloped in a rigid metallised plastics sheet 53. It has a liquid distributor 54 which, together with a vapour outlet duct, communicates with the pump apparatus 11 via an umbilical 55. The pump apparatus 11 comprises a jet pump 13, and a pump 14. The jet pump 13 comprises a chamber having a liquid inlet nozzle 131 and an outlet nozzle 132 and a vapour inlet controlled by a non-return valve 133.
See Fig. 211.

Patent (171) of W. Sorensen concerns an apparatus and method for circulating a heat transfer liquid from a heat collector, which is generally but not essentially a solar collector, to a heat exchanger or heat storage means which is located at a level below that of the heat collector without the use of any externally powered pumps.

The heat transfer liquid is at least partially vaporised in the steeply sloped collector and the vapour/liquid rises in a series of "slugs" to a condenser located adjacent the top end thereof. The vapour is condensed and the hot liquid is forced downwardly to the heat exchanger by the pressure of the rising slugs of vapour and liquid. After giving up useful heat in the heat exchanger the now cooled liquid is recirculated to the condenser and thence to the collector.

According to Fig. 212 tube 1 gathers solar heat both incident upon and conducted to it from plate 2, causing the liquid 3 to rise in temperature and boil in turn causing gas bubbles 4 to form and move

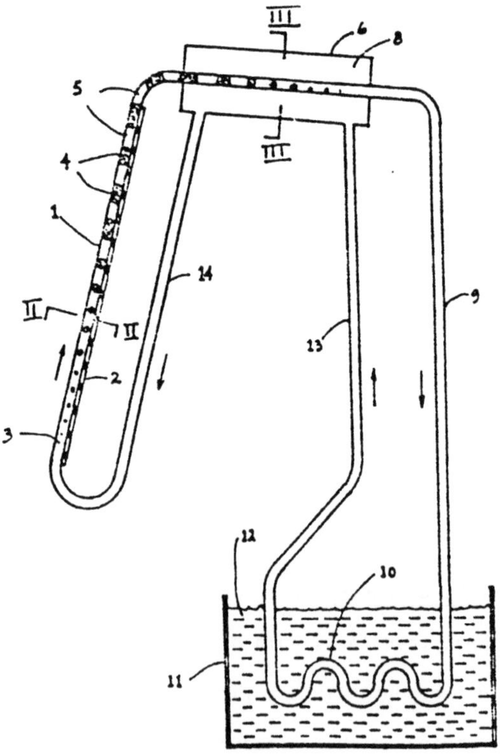

Fig. 212

upward in tube 1. With proper sizing of the tube to suit the viscosity of the liquid, the gas bubbles will enlarge to the inside diameter of tube 1 producing cylinders of upward moving liquid 5.

J.M. Stewart claims in Patent (175.1) a versatile heat exchange apparatus, which may be used in a stream of solar heated air for heating flowing water, wherein an array of heat pipes is utilised in heat exchange relation with a manifold, wherein vapour medium is exchanged therebetween for conducting heat from one to the other.

Fig. 213 shows the proposed apparatus positioned within the air flow of a duct, carrying air from a solar collector; A - open frame; B - elongated manifold; C - pipe; D - heat pipes; E - collector for pipe; 10 - duct; 11, 12, 13 - parts of frame A; 14 - support; 15 - cap; 16 - fins; 17 - space around 16.

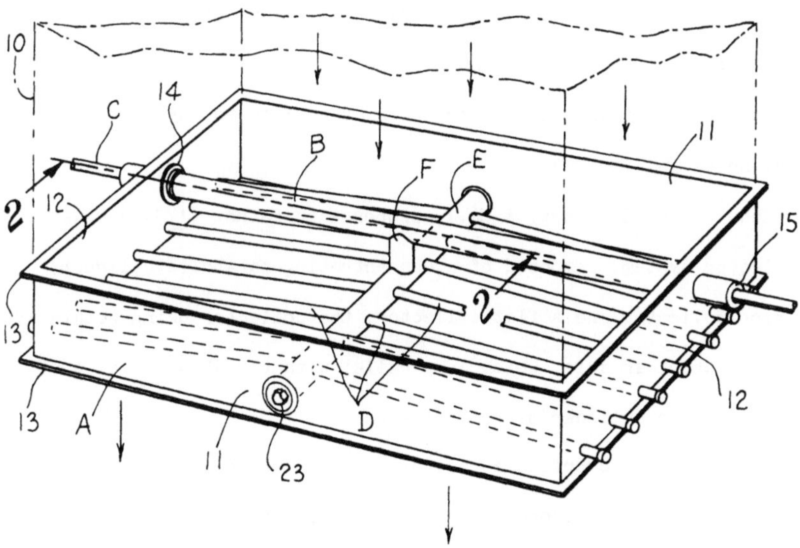

Fig. 213

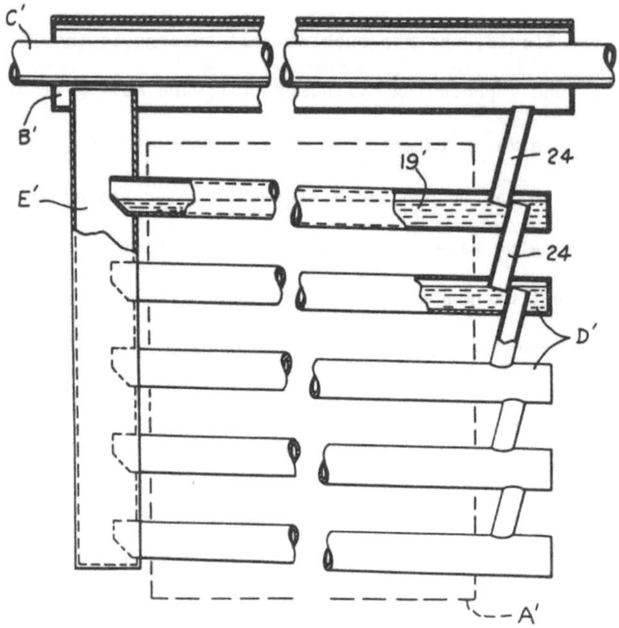

Fig. 214

In Patent (175.2), Fig. 214 illustrates a variant of J.M. Stewart's heat pipe manifold heat exchanger, wherein, the components are indicated essentially by the numerals as applied to the corresponding components in Fig. 213.

K. Schwab, (D. Swarowski & Co.), proposes in Patent (179) a solar collector system, composed of a concentrating parabolic cylinder mirror collector and a flat collector, the collector being rotatably mounted about an axis, which is parallel with the absorber pipe of the parabolic cylinder mirror collector.

The system also comprises a control element, which automatically rotates the system according to the position of the sun.

The absorber pipe of the parabolic cylinder mirror collector and the absorber elements of the flat collector are combined

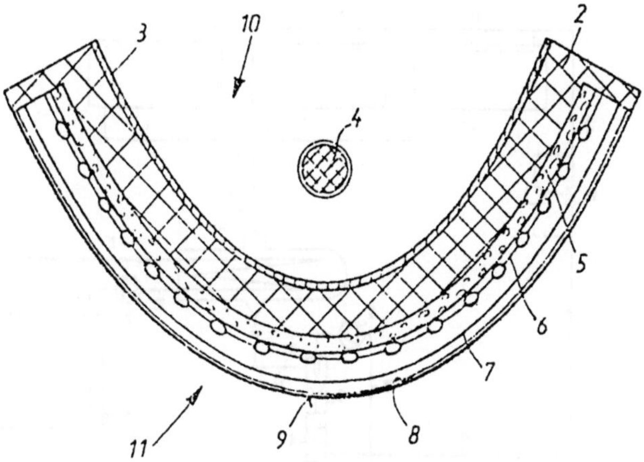

Fig. 215

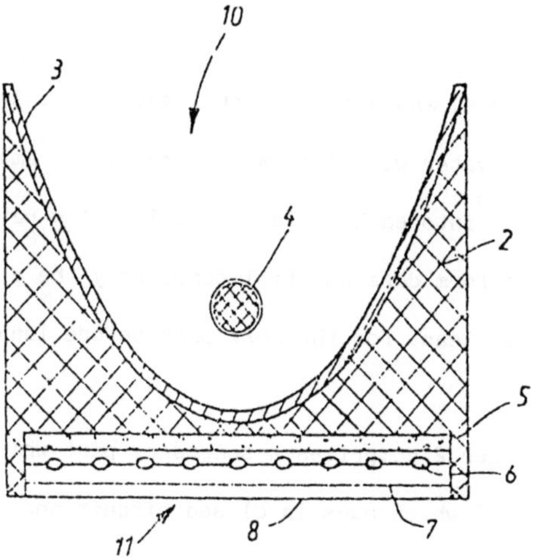

Fig. 216

in a common heat circuit. Figures 215 and 216 illustrate embodiments of the proposed systems.

In the sun heating equipment, assigned by P. Vandeplas, (Patent 176), to the Studiecentrum voor Kernenergie S.C.K., and illustrated in Fig. 217, the outlet of the collectors 1 is connected to the inlet thereof by means of a line 12 in which are mounted the primary part of a heat exchanger condenser 2, a vacuum pump 3, a

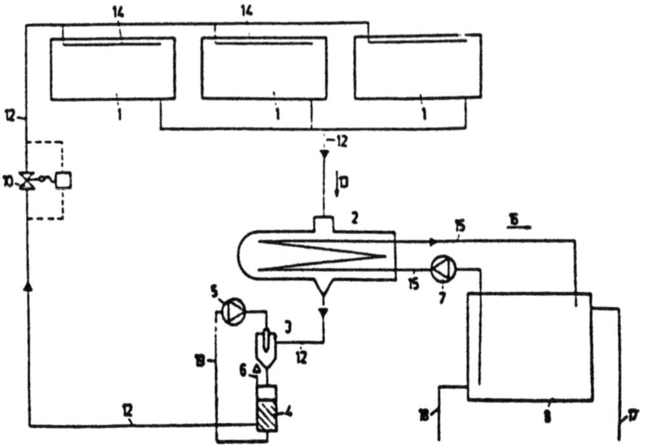

Fig. 217

venting cylinder 4, and a flow-control device 10. A distributing device 14 feeds water with a pressure drop and a small flow rate from the line 12 into each collector 1. The flow rate is such that water in liquid form does not fill completely the collectors 1 and vaporises at least for the most part before leaving the collectors 1.

The heat exchanger 2 comprises a secondary part which is part of a hot water line 15 which runs in closed circuit and wherein are mounted a circulating pump 7 and a storage tank 8.

See Fig. 217.

Tubular solar collectors, having an absorbing cylindrical surface wherein the heat carrier agent flows, are as a rule made of metal or glass, and are coated with an absorbing substance, which is applied thereto by physical or chemical means. In case of glass structure, such coating involves difficulties. The inventors A. Schrolf, et al., (Thomson-Brandt), propose in

Patent (183.1) a fluid heat carrier, displaying solar heat absorbing properties, in order to avoid the application of a separate absorbing layer at the level of the collector.

In another Patent (183.2) of A. Schrolf, et al., (Thomson-Brandt), propose for the same purpose a composite material, displaying solar heat absorbing properties, due to having a dendritic structure and being executed in the form of a grid of one or several layers. This composite material is placed in the interior of the collector.

Patent (184.1) of D. Grauleau, (Thomson - C.S.F.), concerns the evacuation of the evaporator of solar heat pipe heat collectors and more particularly modes, facilitating the evacuation of the evaporator, using for this purpose a glass shell, covering the evaporator part of the heat pipe of the collector which can be of glass or metal.

Another Patent (184.2) of Thomson - C.S.F., (inventor: A. Friedrich et al.), refers to a solar heat collector, consisting of a first glass shell 2 and a second glass shell 1 in such a form that a space 4 is forming between them. The first shell 2 is in combination with means 3 absorbing solar radiation, such means comprising a coating deposited on its external wall and containing a first layer 31 of a glass-metal composition and a second layer 32 consisting of an oxide of the said metal. A layer 30 of the glass-metal composition may also be arranged on the interior wall of the first shell 2. See Fig. 218.

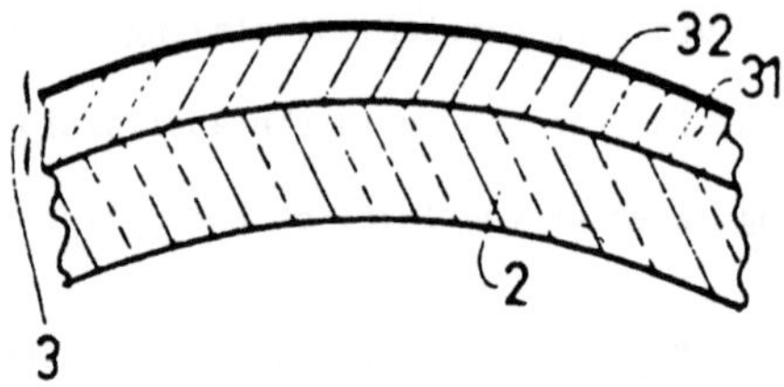

Fig. 218

In Patent (190.2) M.J. Davies, (UK Atomic Energy Authority), discloses a heat exchanger module 10, (see Fig. 219), which is provided by at least one thermal device 11 in the form of a heat pipe or thermal siphon sandwiched between layers of heat exchange fins 19, defining passageways transverse to the length of the device 11. A number of the modules 10 may be clustered together to form an assembly 30. Each module 10 in the assembly 30 has an elastomeric sealing strip 31, (adhesive rubber "SILASTIC"), around its midway along the module 10 so as to divide the assembly 30 into two sealingly separated portions 32 and 33. See Figures 219 and 220.

Patent (193.2) of J.M. Corliss, et al., (US Dept. of Energy), concerns a passive environmental temperature control system according to Fig. 221, showing a solar energy heating system 10 for utilising sunlight indicated by ray 12, to heat an indoor area or room 14 of a building. The system includes a solar heat absorber 16 which receives the sunlight after it passes through a window pane 18 that isolates the absorber from outdoor air which may be cooler than the absorber. Heat from the absorber 16 is transferred, by heat pipe devices, through a thermal insulation barrier 20 to a heat storage reservoir 22 which may be a thick concrete wall or a water tank, that can store heat and transfer it gradually to the room 14. The efficient transfer of heat from the

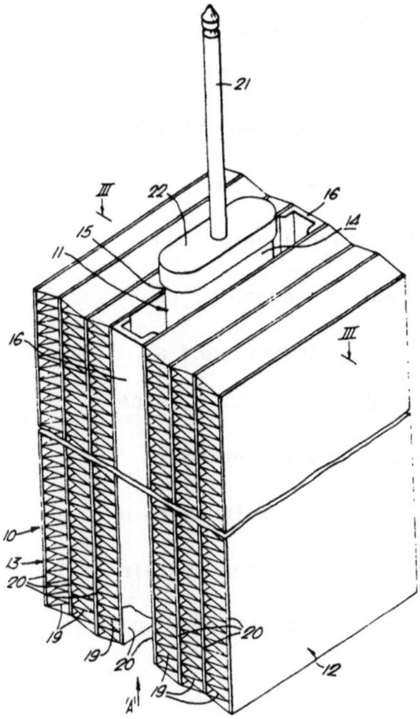

Fig. 219

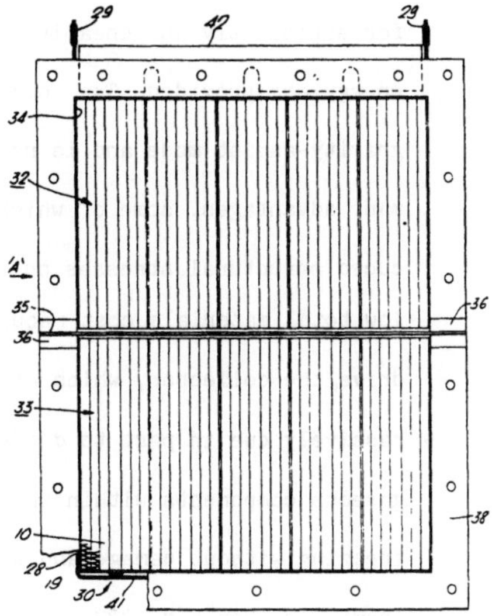

Fig. 220

absorber 16 to the reservoir 22 is accomplished by the use of heat pipe devices 24 that each can efficiently transfer heat between its opposite ends, and with the outer end of the heat pipe device connected to the absorber 16 to receive heat therefrom and the inner end connected to the reservoir 22 to deliver the heat thereto.

See Fig. 221.

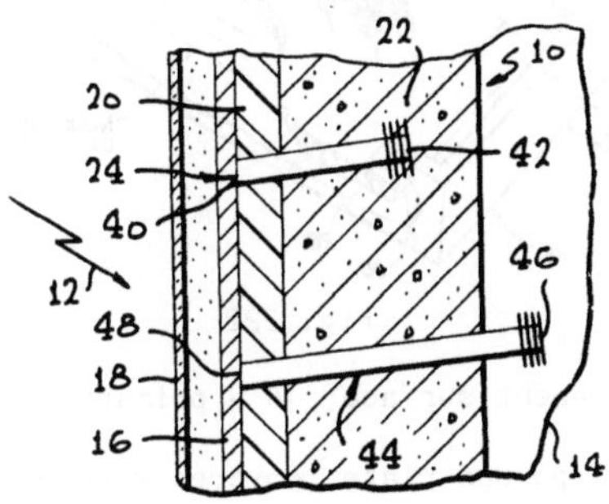

Fig. 221

In another Patent (193.7) assigned to the US Dept. of Energy, (inventor: M.A. Merrigan), a heat collector and method suitable for efficiently and cheaply collecting solar and other thermal energy are provided. The collector employs a heat pipe in a gravity-assist mode and is not evacuated. The collector has many advantages, some of which include ease of assembly, reduced structural stresses on the heat pipe enclosure, and a low total materials cost requirement. Natural convective forces drive the collector, which after start-up operates entirely passively due in part to differences in molecular weights of gaseous components within the collector. The collector sustains freezing without damage and without the use of potentially toxic chemicals. See Fig. 222.

Patent (194.3) of H. Hinterberger, (US Energy Research and Development Administration), refers to a radiant energy con-

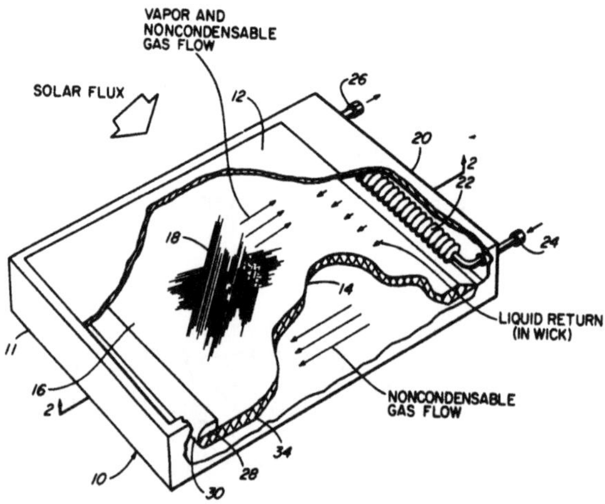

Fig. 222

12 - solar transparent surface; 14 - primary heat absorption surface; 16 - non porous plate, covered with porous material 18 acting as capillary wick.

centration and collection device, having a parabolic reflector directing incident energy to a focus, and a plurality of energy absorbers. The absorbers are positioned so that the focus, which moves as the sun moves, is incident on at least one, and ideally no more than two, absorbers at any one instant.

Each absorber is composed of a bulb and each bulb is coupled to a main header. The header is partially filled with fluid and communicates with the bulb so that the bulb is filled with fluid. In one embodiment, a tube from each bulb extends into the header above the level of the fluid. When energy is incident on a bulb, the fluid therein boils and the resultant gas is transported to the unfilled portion of the header via the tube where it is utilised. In another embodiment, a separate conduit is contained within the header and the boiled fluid from the bulb heats the conduit which then heats the fluid within the conduit. Those absorbers not heated by incident radiant energy

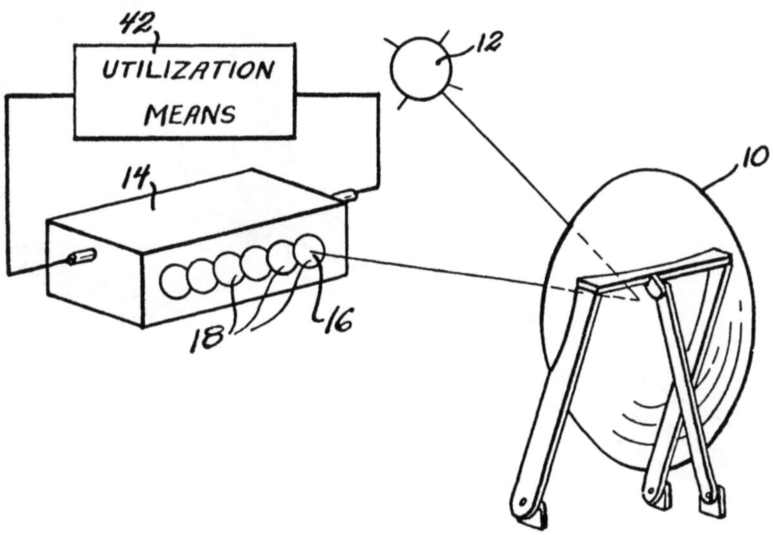

Fig. 223

are inhibited from dissipating heat since they are not directly heated by the heat absorbed by adjacent absorbers upon which the energy is actually directed.

See Fig. 223, wherein 10 - primary mirror; 12 - sun; 14 - receiver; 16 - focus; 18 - absorbers.

Patent (200) of K. Kühnke, (VAMA GmbH), concerns a solar heat installation, consisting of evacuated heat pipe collectors 1, provided with condenser heads 5, furthermore of a storage space 6, 6a; 6, 14; 6a, 14a, storing the entire amount of heat.

The said condenser heads 5 are arranged in the lower range of the storage space. The installation operates by gravity force.

The storage space may consist of one or two superimposed containers, in horizontal or vertical arrangement.

See Fig. 224.

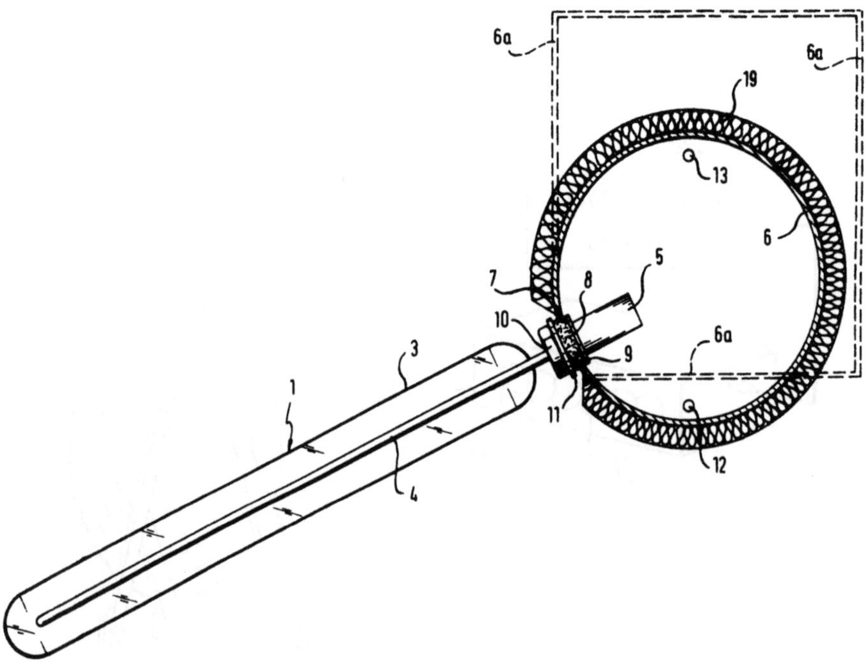

Fig. 224

In Patent (202) of O. Vasilina, et al., a solar photoelectric module is revealed, mainly applicable in power installations, which are not easily accessible. Such modules, as a rule comprise a photo-converter, a concentrator of solar energy and a cooling element for the photo-converter. In order to simplify and to improve the functioning of the photo-converter, according to the present invention, it is immersed in a fluid heat carrier agent, arranged in a sealed, hollow container, whereby the fluid heat carrier agent fills up only a part of the container's interior. This construction ensures an increased capacity of the photo-electric module, due to the proposed efficient cooling of the photo-converter, based on the principle of thermo-siphons, without losses of electric energy, while at the same time reducing the module's weight by 30-40%. Fig. 225 illustrates the proposed module, with parabolic mirror reflector and a counter-reflector outside of the sealed container.

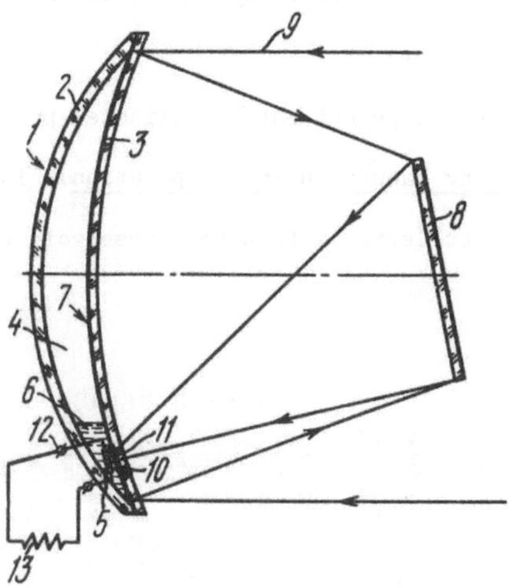

Fig. 225

An isothermal process solar collector panel, designed by R.E. Watt, (Patent 204), includes a collector plate for absorbing radiant heat and a plurality of isothermal process heat pipes in an array over a surface of the collector plate. Each heat pipe is closed at both ends and contains thermodynamic working fluid for transferring heat energy from the collector plate to a second fluid flowing through a manifold pipe for conducting the heat energy from the collector panel. The manifold pipe is coupled to the heat pipes. One end of each pipe is coupled to the collector plate and has an evaporator section wherein heat energy is transferred from the collector plate to the thermodynamic working fluid and the other end of each heat pipe is positioned within the manifold pipe and has a condenser section wherein heat energy is transferred from the thermodynamic working fluid to the second fluid flowing through the manifold pipe. Fig. 226 is a sectional view taken lengthwise of a heat pipe included in the panel.

A thermosyphon heat pipe hot water appliance, presented by Ying-Nien Yu, (Ying Manufacturing Corporation), in Patent (212), combines a solar collector with a heat reservoir wherein fluid

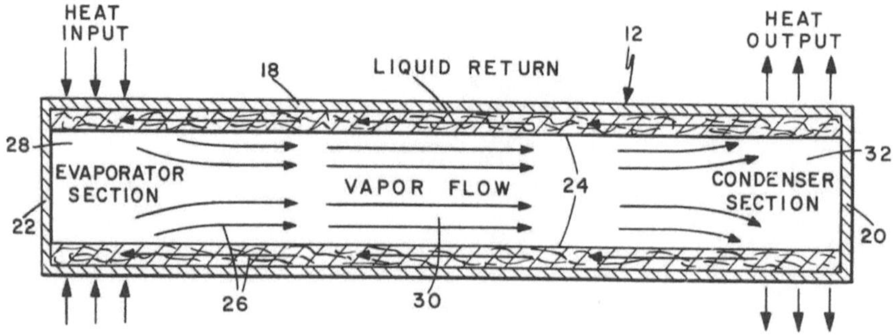

Fig. 226

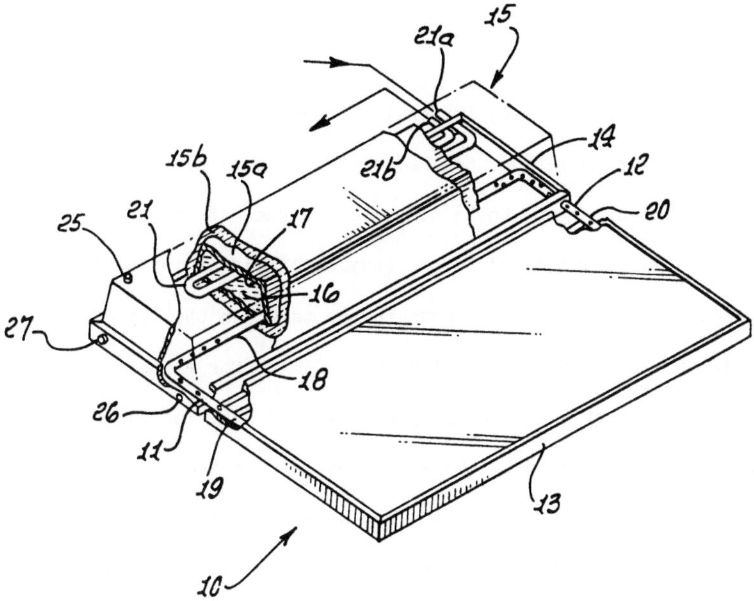

Fig. 227

vaporised in the collector transfer latent heat to reservoir fluid, for re-transfer to fluid such as water supplied by city reservoirs, or from municipal mains.

As shown in Fig. 227, a solar collector 10 has an inlet (defined by duct 11 for a first fluid to be heated and an outlet (defined by duct 12) for the solar heated fluid. The collector is carried by a first rectangular frame 13. A second and smaller rectangular frame 14 carries a tank 15 for containing a heat transfer second fluid 16. The tank includes a metallic liner 15a and insulation 15b. The insulation may for example consist of styrofoam, and the metal liner of heat conductive metal (copper or copper alloy).

First ducting in the tank interior 17 communicates with the solar collector to receive the heated first fluid and transfer heat therefrom to the second fluid body 16 in the tank interior.

That ducting is shown at 18 in endwise series communication with inlet 11 and outlet 12.
First ducting 18 may be integral with header pipes 19 and 20 in the solar collector, and may contain freon to be vaporised in the collector and to be condensed in duct 18, to release latent heat transferred to the body of liquid 16 in the tank. Second ducting 21 also extends in the tank, in the heated body of liquid 16, for passing third fluid (for example water) into and out of the tank via ports 21a and 21b. The second ducting 21 is heated by the water body 16 and transfers heat to the third fluid flowing in that second ducting 21. Accordingly, water to be used in the home, is heated by the hot liquid reservoir 16, which acts as a heat source during the night or cloudy weather when the solar collector itself is not being heated by the sun's rays; also, the heat storage capacity is enhanced due to latent heat of vaporisation release to the water body 16 upon condensation of freon in the duct 18.

Patent (206) of K. Weller refers to a solar heat collector system, comprising, in addition to the usual components of such systems, cooling air ribs at the absorber face of the collector, thermal coupling of the collector with a collector head or directly with the water tank; thermal insulations; transfer systems, capillary lining, shut-down and control elements, pressure reducers, and control means operating outside of the transfer system. The inventor claims, that the system can operate in case of current supply failures as well.

Patent (213.1) of Shao W. Yuan provides a system for transferring solar energy for instant utilisation and for long duration underground storage.

More particularly, the system includes solar collectors attached to liquid tubes or unidirectional heat pipes for transferring solar flux into heat energy for instant use and storing the heat energy into earth ground for auxiliary and long-term use. An expanding array of heat conductive pipes distributes the heat throughout a large volume of underground earth and extracts the same when it is required.

The system provides for long-duration earth storage of the heat energy which can later be used as independent source for both space heating and hot water heating in homes, multiple-unit housing, commercial and public buildings, swimming pools, greenhouses, industrial needs, etc.

See Fig. 228, wherein: 10 - the system; 12 - solar collector; 14 - heat absorber tube, filled with fluid transfer medium; 16 - pumps for circulating the medium; 18 - conduit; 20 - tank (heater); 22 - inlet; 14 - tube; 24 - tank exit; 26 - conduit; 28 - collector inlet; 17 - thermostat, turning off the pump when there is no sunshine.

According to another Patent (213.2) of S.W. Yuan, the heat collected by a solar heat collector system, consisting essentially of collector 12, heat pipe 12, evaporator sector 16 can be conducted through an adiabatic sector 18 to an underground condenser sector 20, which is placed in a depth of 1.5 - 9 m under the ground surface. In the range 22 within the earth, heat will be absorbed from the condenser sector

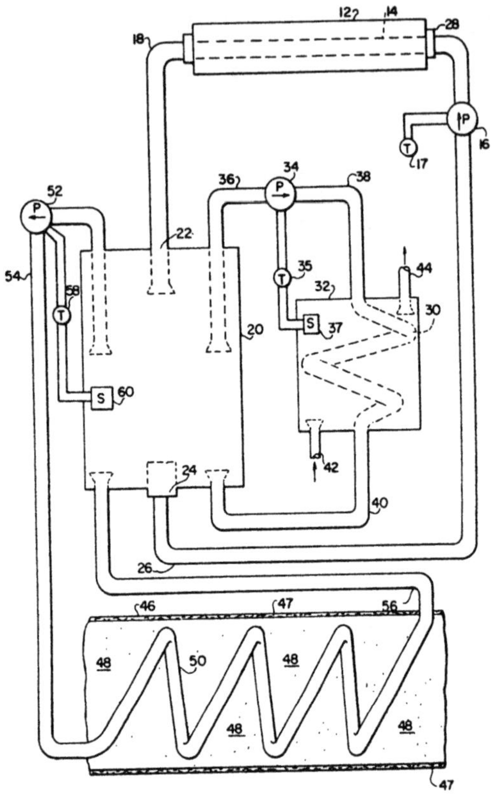

Fig. 228

20 and transferred from here in various directions, merely cooling the condenser and distributing heat in the surrounding zones. The earth has some, from the viewpoint of heat storage, significant characteristics: its huge mass displays on an area of 1 acre, in a depth of 4.6 m, at a temperature gradient of about 27.8°C an energy flow of $0.38.10^9$ Kcal, furthermore it has a very low heat conductivity.

The heat collected is conducted via the heat pipe through the evaporation sector to the condenser sector, with the lowest possible resistance, while the return flow of heat towards the

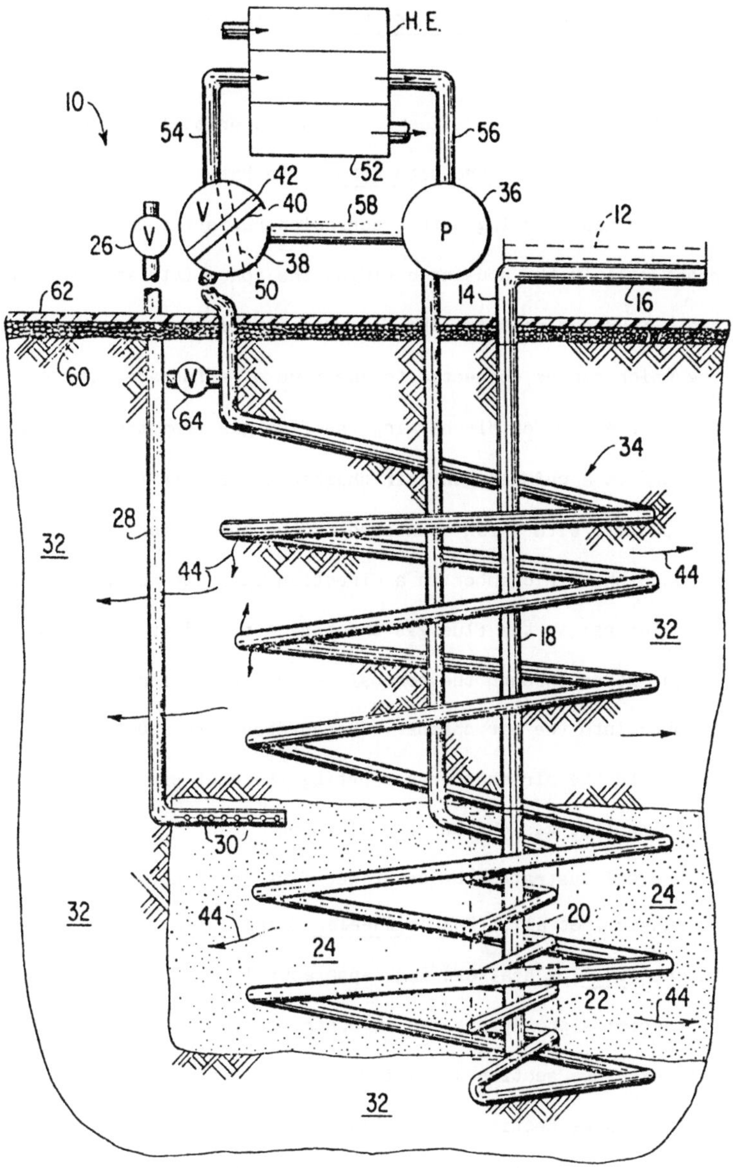

Fig. 229

ground surface is prevented if the surface temperature is lower than the temperature in the underground heat storage space. Fig. 229 is a schematic illustration of the proposed solar heat collector/underground storage system.

5.5 Heat pipe heat exchangers adapted to industrial applications

A flue heat recovery device for furnaces or boilers, revealed by E.R. Powell, (Entec Products Corp.), in Patent (51.2), comprises: a rectangular housing with a corrosion-resistant or corrosion-proof wall in the housing dividing the space into an air chamber and a flue gas chamber. The flue gas chamber has a gas inlet opening which can be connected to the flue of a furnace, and a gas outlet opening for discharging the flue gas from the flue gas chamber to a chimney. The air chamber has an air inlet and an air outlet, with a fan connected to the outlet for drawing air through the air chamber in a direction opposite to the flow of the flue gas in the flue gas chamber. A plurality of finned heat pipes extend through the corrosion-proof wall from the flue gas chamber into the air chamber. The chambers are mounted on a base tray with the air chamber supporting the heat pipes and separable from the flue gas chamber for cleaning.

Fig. 230 illustrates the described flue heat recovery device.

L. Kunsagi, et al., (Foster Wheeler Energy Corp.), disclose in Patent (64.1) a heat exchange apparatus in which each of a plurality of thermal siphon pipes has an upper portion extending in an upper heat exchange section and a lower portion extending in a lower heat exchange section. Each pipe is closed at its ends and contains a heat transfer fluid so that when a hot fluid is passed through the lower heat exchange section, the heat is transferred from the hot fluid to the heat exchange fluid. A cool fluid is passed through the upper heat exchange section to remove the heat from the heat exchange fluid. According to the inventors the apparatus can be manufactured at a relatively low cost.

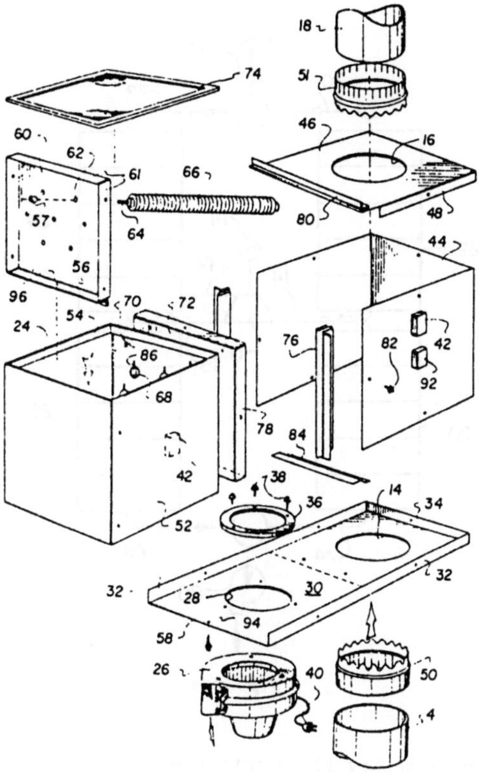

Fig. 230

Furthermore, the invention minimize the need of excessive fan power and the use of heat pipes, requiring internal wicking material. See Fig. 231 showing schematically a part of a power generation system 10, utilising the described heat exchanger.

Another Patent (64.2) assigned to Foster Wheeler Energy Corp., (inventors: R.A. McAllister et al.), refers to a heat exchanger having first and second fluid chambers 12, 30 respectively for passing primary and secondary fluids. The chambers are spaced apart and have heat pipes 56 extending from inside one chamber to inside the other chamber. The heat pipe portion between the first and second chambers lies within a third chamber 45 in which flows a purge fluid, and means 58 are provided to detect any leakage of

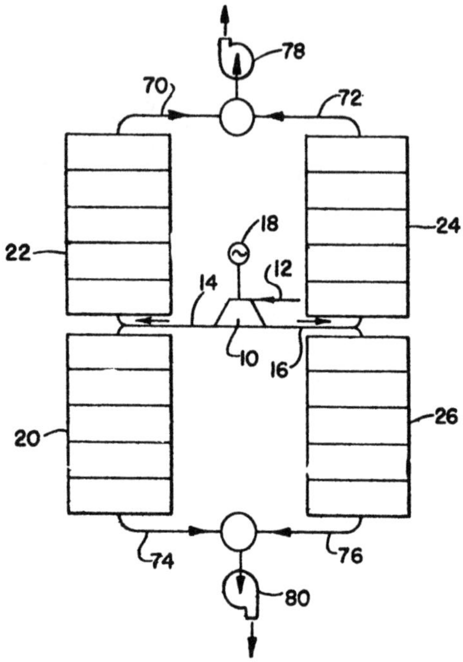

Fig. 231

primary and/or secondary fluid into the chamber 45 and into the purge fluid. Fluid leakage to or from the heat pipes is sensed by detection means 62. Fig. 232 is an elevational cross-sectional view of the heat exchanger of the invention showing the heat pipes within the chambers of the heat.

Patent (64.3) of Foster Wheeler Energy Corp., (inventor: B.M. Barnet), and essentially, Patent (64.4), (inventor: C.J. Brigida, et al.), relate to a steam condensing apparatus for providing a faster heat transfer rate than possible heretofore. The apparatus according to the inventions comprises a plurality of heat pipes 22 (see Fig. 233, each having a lower evaporator section in a steam plenum 14, 14' and an upper condensing section for contact with a cooling medium, whereby a heat transfer liquid (e.g. liquid ammonia) in each tube

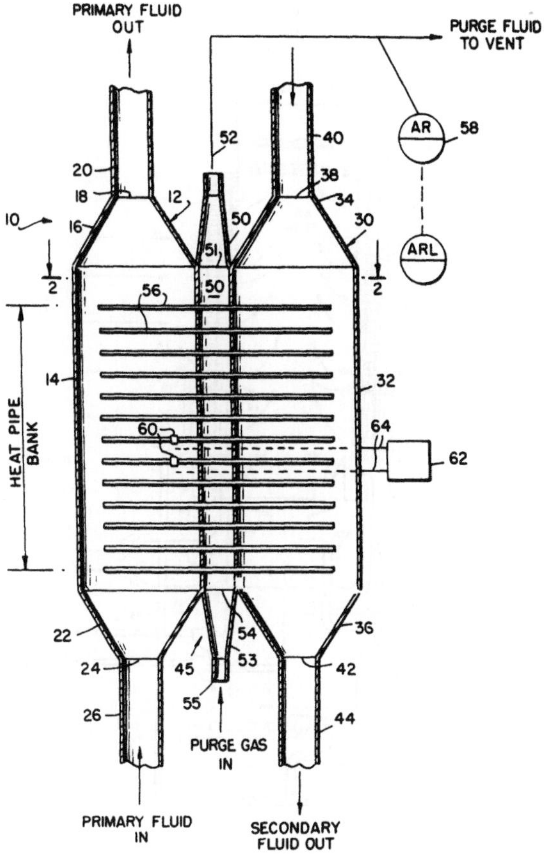

Fig. 232

is evaporated from the evaporator section upon receipt of steam in the plenum 14, 14', which steam is thereby condensed, and is condensed in the condensing section to flow-back to the evaporator section. Each condensing section has a first portion bearing fins 28 which will normally be cooled by a flow of air supplied by a fan 16, and cooling water from, for example, flood water troughs 18, may be applied to a second portion of the condenser section. The cooling air flow can be applied simultaneously to the second portion.

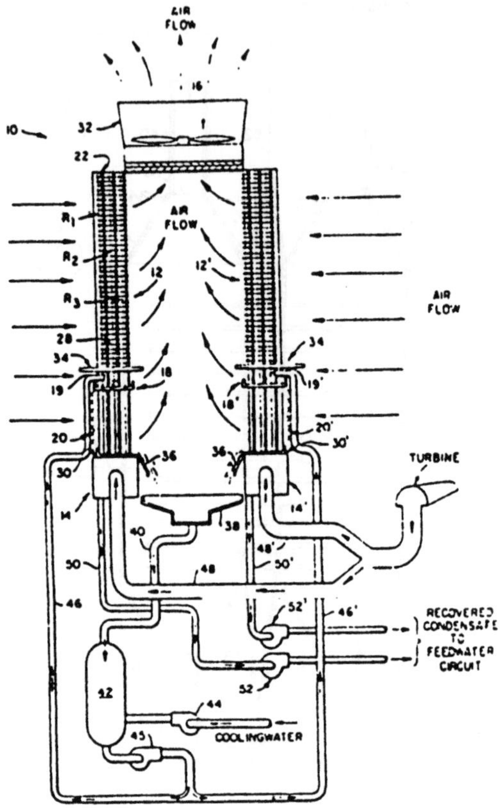

Fig. 233

In Patent (65) of R.M.V. Beith, (Foster Wheeler John Brown Boilers Ltd.), discloses a steam boiler having an air heater to preheat combustion air and extract heat from waste combustion gases, which air heater has passages for hot waste gases and for cool air to be heated and a plurality of heat pipes extending between the passages for extracting heat from the hot waste gases and supplying it to the cool air; the heat pipes having an extended surface area and any convenient cross-section (circular, rectangular). See Fig. 234.

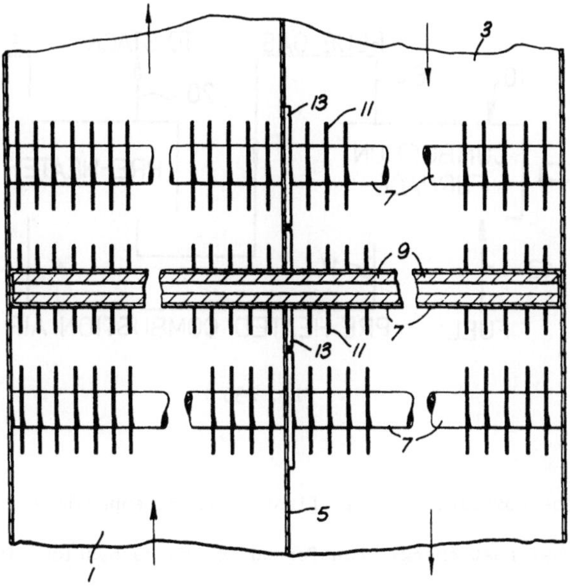

Fig. 234

The combustion air preheater for heating combustion air with heat energy provided from flue gases, described in Patent (66.1) of R.O. Barratt et al., (Foster Wheeler Power Products Ltd.), includes, according to Fig. 235, a first heat exchanger 26, having first and second heat transfer compartments 54, 56, with a plurality of heat pipes 60 extending therebetween and a second, connected heat exchanger 24, which includes a plurality of tubes 32, supported by tube sheets 36 within a shell 38. Ducting is provided to direct flue gas over the tubes 32 of the second heat exchanger 24 and then through one of the compartments 56 of the first heat exchanger 26 and additional ducting is provided to direct combustion air through the other compartment 54 of the first heat exchanger 26 and through the tubes 32 of the second heat exchanger 24. Heat energy is transferred through the walls of the tubes in the second heat exchanger

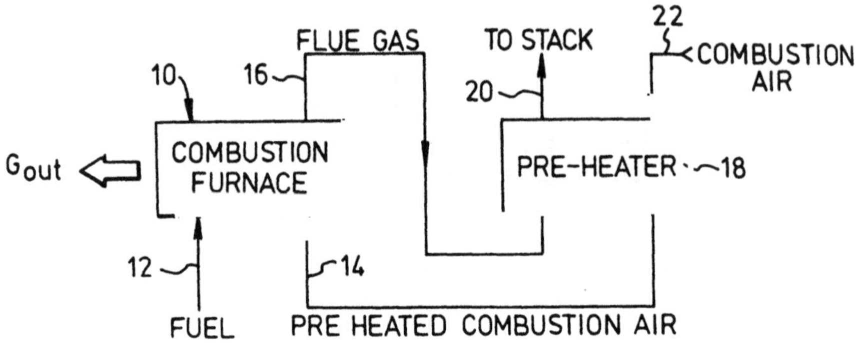

Fig. 235

24 to the combustion air in first, higher temperature range with additional heat energy transferred to the combustion air in the first heat exchanger 26 in a second, lower temperature range through the heat pipes 60. A method of fabricating the first "heat pipe" heat exchanger 26 is also disclosed. Fig. 235 is an overall system diagram of a combustion furnace comprising an air preheater according to the invention, Patent (66.1).

Fig. 236 is a side elevational view, in cross-sectional schematic form, of a combustion air preheater in accordance with the present invention.

T.K. Garberick claims in Patent (72) a heat recovery apparatus in which an exhaust gas section and a fresh air section are disposed in side-by-side relationship to form an integral unit, for recovering heat from a stack or flue and discharging fresh air into a space to be heated. Heat pipe units extend from the hot gas passage of one section into the fresh air passage of the other section to transfer the heat between the two passages, the

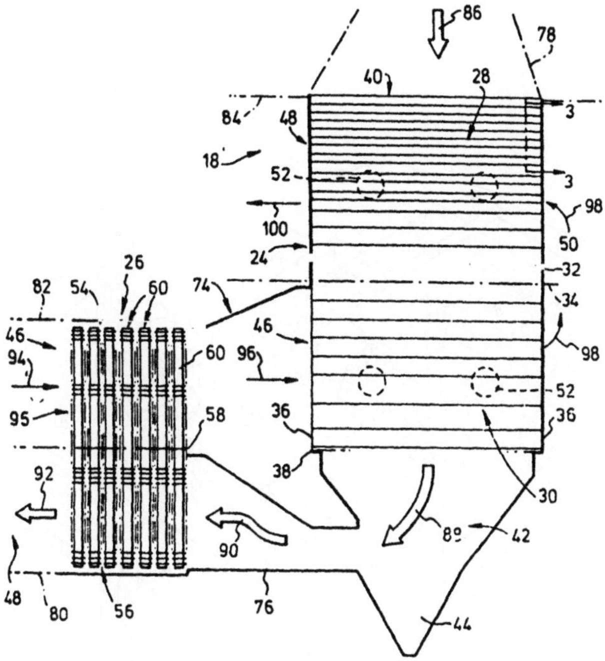

Fig. 236

air flow in the exhaust gas section being in a series flow relationship and the flow in the fresh air section being in a parallel flow relationship. Louvers are provided for controlling the flow of exhaust gases and fresh air through the respective passages, and blowers are used to force the exhaust gases and fresh air through the passage. The passage of each section are provided with a main and by-pass passages controlled by louvers which are used to regulate the heat recovery and output of the unit. Fig. 237 is a cut-away perspective view of the proposed apparatus.

Patent (100.6) of P. Rohner, (Kabel- und Metallwerke Gutehoffnungshütte A.G.), reveals heat pipe heat exchangers, wherein a spiral, wavy heat pipe 2 is arranged in the interior of the heat exchanger pipes 1; the spiral heat pipe being in an

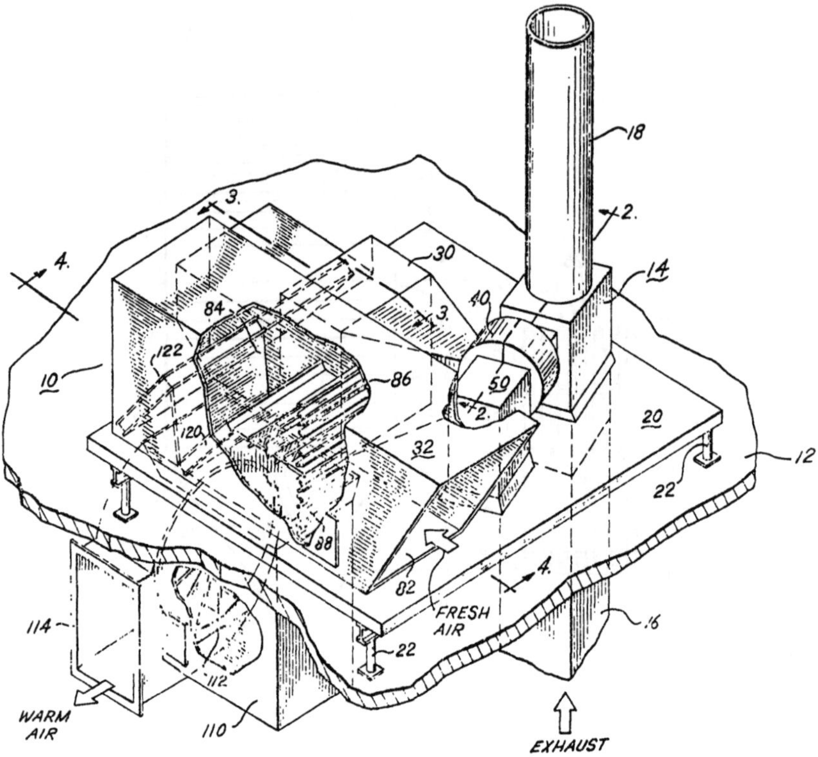

Fig. 237

intensive contact with the inner face of the pipe 1. Pipe 1 can have a hexagonal cross-section. The spiral pipes 2 should preferably be made of aluminium or copper. The windings of pipes 2 are fixed, at given distances from one another, by means of beam-like elements 3 of copper. Elements 3 are in heat-conductive contact both with the inner face of pipe 1, and with the windings of heat pipes 2, through, for example soldering. The proposed apparatus is mainly suitable for cooling water of power installations. See Fig. 238.

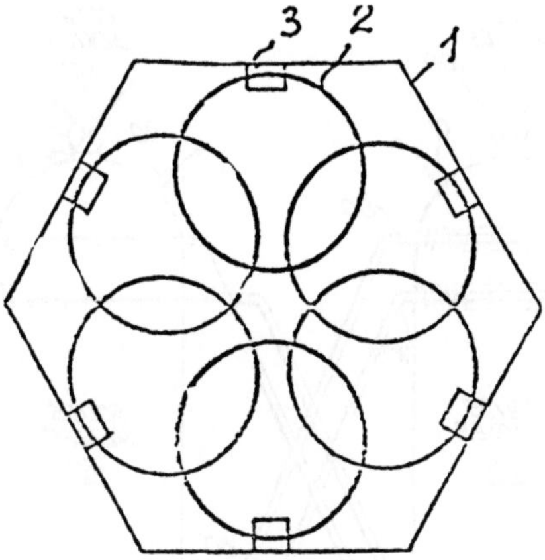

Fig. 238

O. Levenspiel, et al., provide in Patent (110) apparatuses and methods for transferring heat from a relatively hot stream of solid materials to a relatively cold stream of solid materials. Both streams cascade downward under the force of gravity over a series of heat pipes which are arranged between the streams to transfer heat therebetween. The heat pipes are arranged to provide countercurrent heat transfer between the two materials streams. The invention is shown in Fig. 239, applied to a process for recovering hydrocarbons from shale rock.

According to Patent (121.6) of E.D. Waters, (McDonnell Douglas Corp.), an effective dry cooling system for dissipating waste heat of a steam-electric generating power plant includes an assembly of cells or modules having passively acting heat pipes installed in a Y-configuration in each module. These heat pipes thermally couple the exhaust steam from a turbine flowing in a graded duct with the atmosphere. A large fan mounted at the top of each module is driven to induce air flow past exterior heat pipe portions of the module to

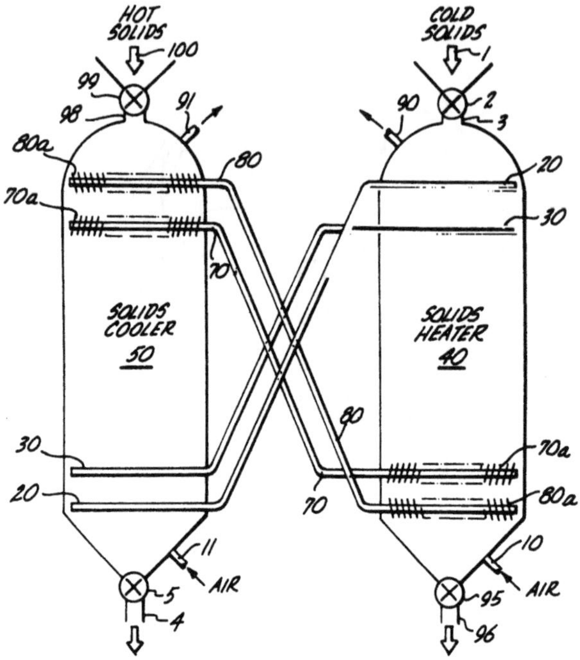

Fig. 239

dissipate heat picked up from the steam in the duct. The steam condensate flows down the graded duct and collects in a hot well for return as boiler feedwater. Steam from the boiler drives the turbine and is exhausted into the graded duct to repeat the cycle.

Fig. 240 is a simplified side elevational view, partially in diagrammatic form of the dry cooling system of Patent (121.6).

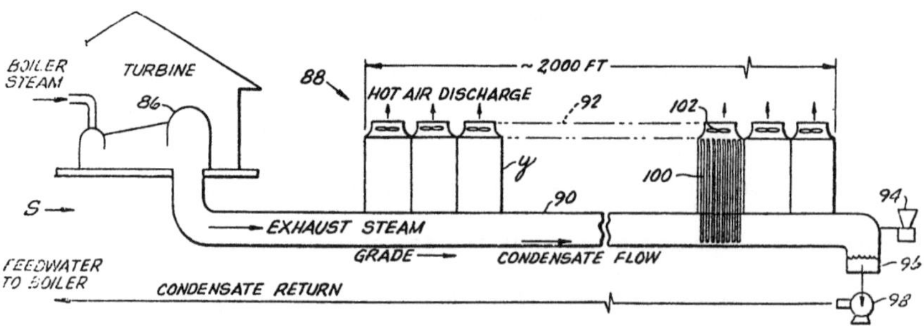

Fig. 240

In Patent (136.7), assigned to N.V. Philips Gloeilampenfabrieken, a heat exchanger is described, composed of two serially mounted sectors, the sector effecting the evacuation of exhaust gases having double separating walls, forming intermediary spaces, which contain an evaporable heat transferring fluid in order to render the separating walls isothermal in the direction of circulation during operation, by an evaporation and a condensation cycle. During operation the isothermal separating walls are heated to about 150°C, thus to a temperature, which is enough high to prevent the deposition of sulfur compositions and, consequently, the corrosion of the heat exchanger channel walls. As heat transferring fluid, water, acetone, benzene, ethanol, propanol, butanol can be used.

C.E. Gunter of PPG Industries Inc., provides in Patent (140) heat pipes, adapted to application in drawing dies for glass fibres, for evacuating heat from the drawing zone.
The proposed device comprises a collector, having an element permitting the circulation of a cooling fluid. Vanes are provided in parallel rows between the rows of apertures of the drawing die, to ensure heat exchange with the glass fibres, extruded through the said apertures. The vanes display at their longitudinal axes a wick material, which is in contact with a heat exchange fluid, which evaporates at a first temperature and condenses at a second temperature. In the vane there is also formed a central cavity, permitting the circulation of the fluid's vapours from the wick at one end of the vane towards the other end and the condensation of vapours on the wick at a point, which is distant from the point of evaporation.

G.A. Harrow of Shell Internationale Research Mij. B.V., reveals in Patent (160) heat pipes, which can be applied in furnace installations, either for assisting the heat transfer from one region thereof to another one or for heating another fluid than the process fluid. The proposed system contains a plurality of heat pipes, executed in the form of a sealed vessel, containing heat transfer fluid and a non-condensable gas. Each heat pipe comprises a heat receiving zone; a heat discharging zone, in communication with the former zone and a heat transfer liquid storage zone, in which condensed heat transfer liquid can be stored without being directly returned to the heat receiving zone. The amount of heat to be transported is a function of the amount of heat to be discharged to the heat discharging zone of the heat pipe, this amount varying depending on the amount of air to be heated per unit of time and its temperature.

Patent (165) of E.L. Rakoswky, et al., (The Singer Co.), concerns the adaptation of a heat pipe as a turbo-generator or other power output device for a reliable, quiet, lightweight high-endurance power source. The device requires input thermal energy from a burner, radioisotope (or solar heat) and also forced or natural heat rejection from condenser surfaces. Thermal energy conversion to a suitable power output is accomplished by encapsulating a turbine wheel within a heat pipe shell, located in an appropriately geometrical contoured section. Flow work extracted from the kinetic energy of the vapour flow provides rotary shaft power output. The shaft power can drive an electrical generator, pump, compressor, or similar

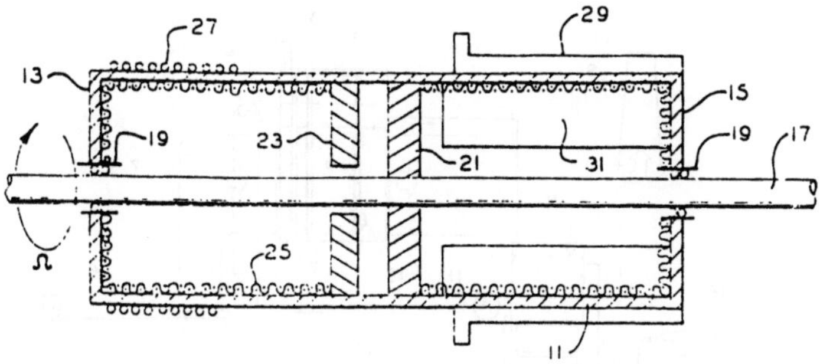

Fig. 241

device, also mounted within the heat pipe shell structure. A completely self-contained enclosed unit is provided which requires only external power connection at attachment terminals. Fig. 241 shows a cross-sectional view of an embodiment of the invention, wherein: 11 - a hollow cylindrical pipe structure, closed by end pieces 13 and 15, with a hole therein, through which a turbine shaft 17 may extend; 21 - turbine rotor; 23 - turbine stator; the inner surfaces of 11, 13, 15, being covered with a wick material 25, which can be saturated with ammonia, methanol, ethanol, acetone, water, Dowtherm, ethylene glycol, mercury, freon, cesium, naphthalene, potassium, sodium, indium, lithium, bismuth, lead, inorganic salts.
Patent (167) of Tstomu Naramura, (Snow Brand Milk Products Ltd.), and (Suzuki Metal Industry Co. Ltd.), describes a waste heat recovery device preventing corrosion by sulfur oxides. More particularly, in a boiler or industrial furnace installation comprising a device 1 which uses heat pipes 5 to recover heat from exhaust gas, a temperature detecting sensor 15 is mounted at the heat absorbing side of one of the heat pipes 5, i.e. at the side which is disposed in the path of exhaust gas. The output signal of the temperature detecting sensor is compared with

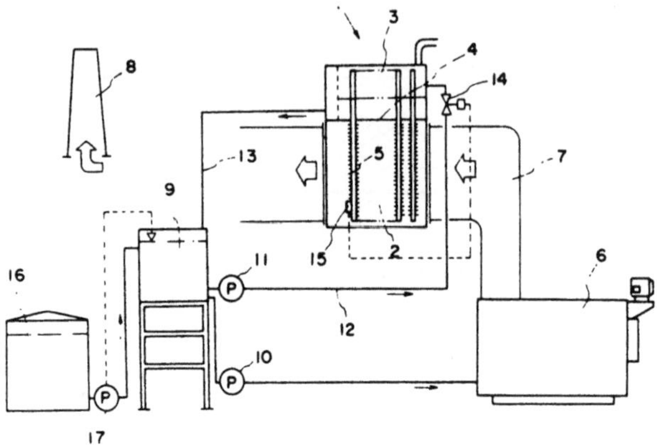

Fig. 242

a reference signal set to correspond with the dew point of a sulfur oxide in the exhaust gas, and the resulting difference signal is used to operate a flow rate control valve 14 which controls the flow of water to the other ends of the heat pipes 5. The pipes are led through a partition 4 of the device 1, from a heat absorbing (gas) section 2 to a water heating section 3. A discharging means is provided for controlling the level of liquid in the liquid heating section, according to the flow rate of the liquid supplied. Fig. 242 is a schematic diagram of the water heating equipment, employing the proposed waste heat recovery device.

According to Patent (173.2) of Stein Surface, the shaft of a rotary valve of heating installations can be cooled by means of a heat pipe system. More particularly, the shaft is cooled by evaporating a heat carrier fluid, by condensing it in a space, which is in connection with the said shaft and outside of the fire place. When the shaft is in contact with hot fumes, the fluid therein evaporates and the vapour forming condenses

Fig. 243

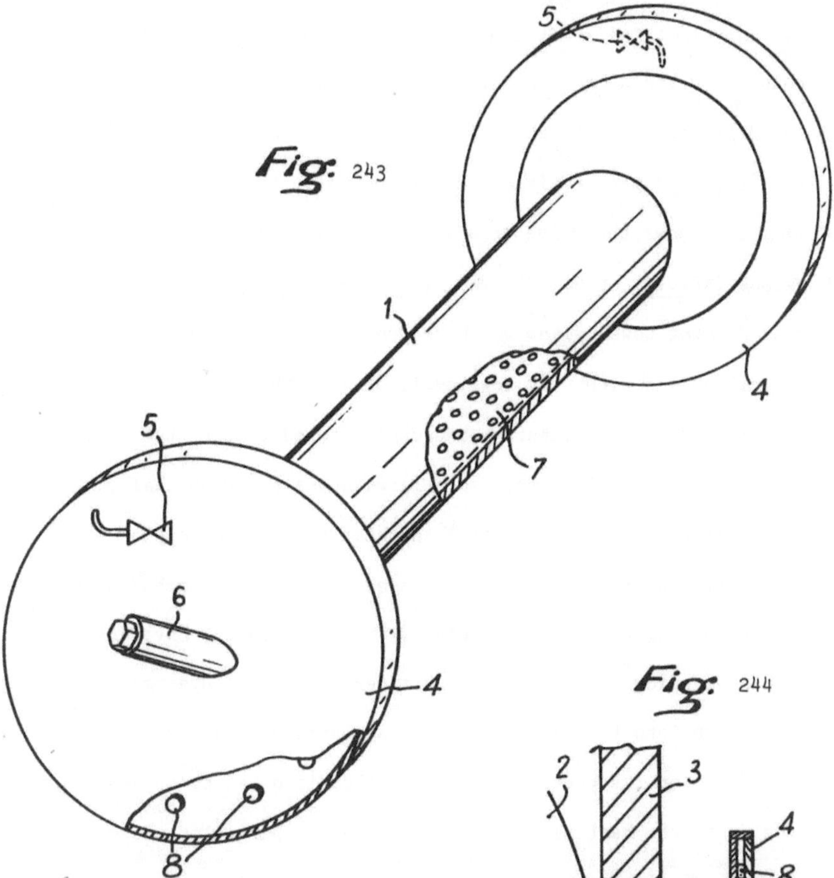

Fig. 244

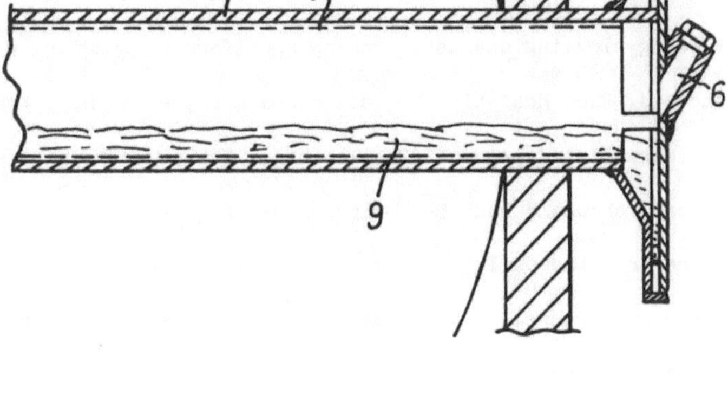

on the cold wall of the said space. The heat thus released therein is evacuated by natural convection into the atmosphere and by radiation. The shaft temperature depends on the fumes' temperature and the dimensions of shaft and space, which are so established, that the shaft temperature will not exceed 450°C. Fig. 243 is a perspectivic view of the valve shaft, while Fig. 244 shows the valve arranged in an air duct 3.

Patent (181.10) of G.Y. Estman, (Thermocore, Inc.), concerns a coal burning power plant which produces electricity at high efficiency and with very low levels of atmospheric contamination. A fluidized bed of limestone and coal is used to burn the coal with little air pollution, and a large quantity of specially constructed heat pipes transfer the heat to a closed cycle gas turbine with low temperature loss. The preferred embodiments of the heat pipes are constructed with steel casing, have a thin aluminium oxide layer on their surfaces to prevent hydrogen permeation into the casing and have a plasma-sprayed coating of ceramic on the portion within the fluidized bed to prevent erosion of the heat pipe casing by the solid particles within the bed. The array of simple, cylindrical heat pipes permits ready circulation of the particles of the fluidized bed, preventing clogging and establishing a uniform temperature within the bed. Each heat pipe is self-contained, sealed heat transfer element independent of all others. The power plant employs several thousand such heat pipe elements in the heat exchanger. Therefore, the failure of an individual heat pipe will not cause a system failure, as would be the case with a conventional heat exchanger. The effect of the loss of a single element will be to increase the thermal resistance of the heat exchanger very

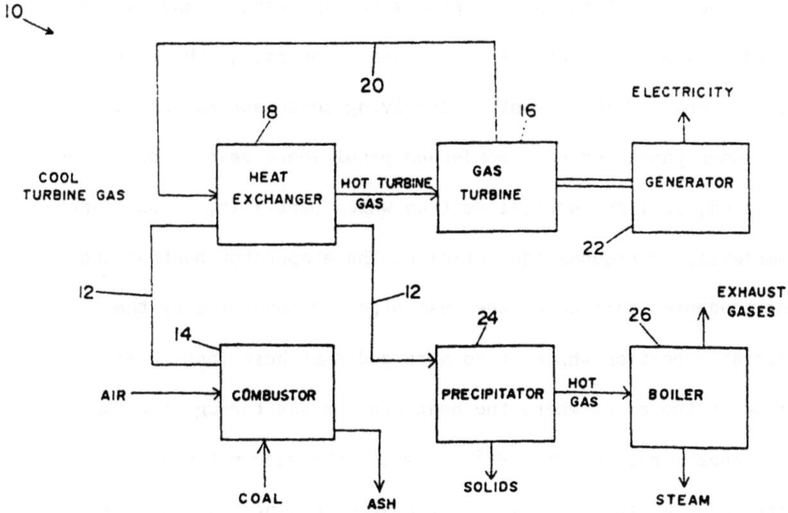

Fig. 245

lightly. For instance, if one thousand heat pipes are used, a single loss has an effect of only 0.1 percent, a negligible loss of efficiency. Failed heat pipes can be readily replaced during normal system shut-down periods when other regular maintenance is performed. Fig. 245 is a diagram of a power plant, utilising heat pipe heat exchangers.

According to Patent (182.1) of L.J. Lazaridis,(Thermo Electron Corp.), heat pipes can be adapted to application in heating workpieces or articles. For example a vacuum furnace heats articles by radiation from a condenser portion of a heat pipe. The heat pipe includes an adiabatic portion passing through the enclosing wall of the furnace and an evaporator portion outside the enclosure. The evaporator portion is preferably heated by a gas-fired burner which may surround it, and the heat pipe acts as a flux converter accepting heat outside the enclosure and delivering heat inside the enclosure to the articles to be heated. The condenser portion of the heat pipe may be panel-shaped and is disposed adjacent the articles to be heat-treated. In the more elaborate forms of the invention involving condenser panels, the condenser panel, or each condenser panel where several heat pipes are used, is a cylindrical section which partially, or substantially completely, surrounds the articles. The evaporator portion and the condenser portion of the heat pipe are connected by the adiabatic portion which is so arranged that heat loss is minimized at the point where the heat pipe passes through the furnace enclosure. See Fig. 246, in which the vacuum furnace incorporates an enclosure 1 having heat shielding 2 and a hinged cover 3. A diffusion pump 4 produces vacuum within the enclosure 1 and heat is applied by the heat pipe means 5. The heat pipe 5 is associated with a furnace 6 and includes an evaporator portion 7 extending into the furnace 6 an adiabatic portion 8 passing through the wall of the enclosure 1, and a condenser portion 9 within the enclosure. The adiabatic portion is

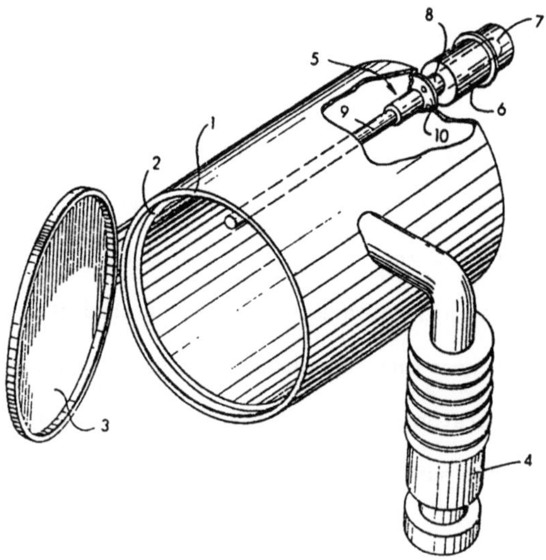

Fig. 246

equipped with sleeve and flange 10 where it passes through the wall of the enclosure 1.

According to Patent (192) of D.K. Edwards, (The United States of America as represented by the Secretary of the Army), heat-pipe-diodes can be used to charge a thermally-stratified heat reservoir to permit enhanced preheating of feed water. Maximum available energy is automatically extracted from waste and feed streams having intermittent flow and variable temperature. The thermal reservoir comprises a plurality of thermosyphons operatively disposed in a condenser unit through which the feed water passes, and a plurality of fluidically connected heat-receiving diodes operatively disposed in an evaporator unit through which the waste stream flows. The invention permits a waste stream exchanger to be positioned at a low level and at a short and somewhat remote location. It also permits a waste stream to have a variable temperature without degrading the desirable high temperature strata in a thermal reservoir. In addition, when high temperature waste becomes available, the device automatically charges the highest temperature strata first, thus conserving the greatest amount of available energy. Fig. 247 is a partial cross-sectional schematic view of a thermal reservoir condenser section of the heat-pipe-diode charged closed system.

Patent (194.2) of W.A. Ranken, (The United States of America as represented by the United States Energy Research and Development Administration), refers to a heat pipe methanator for converting coal gas to methane. Gravity return heat pipes are employed to remove the heat of reaction from the methanation promoting catalyst, transmitting a portion of this heat to an

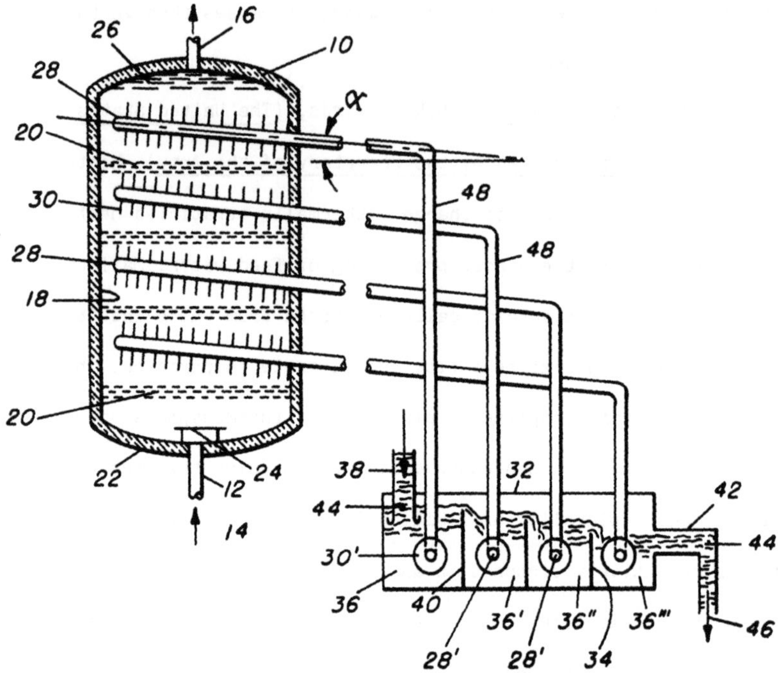

Fig. 247

incoming gas pre-heat section and delivering the remainder to a steam generating heat exchanger. It is evident that the invention takes advantage of the self-contained heat transfer capability of the heat pipe to provide a methanator with a simple configuration, combining the catalyst bed incoming gas preheat section and thermal recovery in one unit. The catalyst is sprayed on the outer wall of each removable heat pipe, enabling easy replaceability of the catalyst. As shown in Fig. 248 an array of gravity return heat pipes 11 are enclosed in a pressure shell 12. Heat pipes 11 are supported by a grid plate 13. The pressure shell 12 encloses a methanation section 14 extending from baffle 15 to grid plate 13. A preheat section 16 extends from the high pressure bulkhead 17 to baffle 15. Heat exchanger section 21 extends from bulkhead 17 to end of bulkhead 22. A start-up

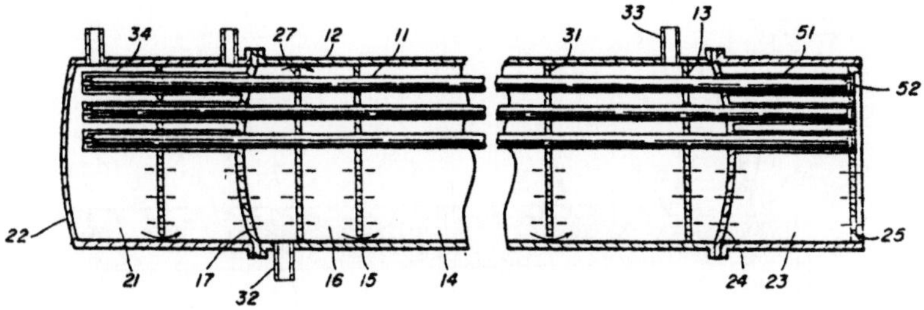

Fig. 248

heat exchanger 23 may be provided at the other end of the pressure vessel 12. Start-up heat exchanger 23 is defined by a bulkhead 24 and plate 25. Additional flow baffles 27, 31, are provided to lengthen the gas flow path between inlet 32 and outlet 33. In the methanation section incoming gas through inlet 32, including hydrogen and carbon monoxide passes through the preheat section 16, reaching the temperature of about 650°K, required for the methanation reaction. The gas is then passed into the methanating section 14. In the preheat section the heat pipes are roughened to facilitate heat transfer.

Patent (197.2) of G.A.A. Asselman,(U.S. Philips Corp.), provides a heating device of simple structure and of which the heating space is isothermal during operation. More particularly, the device is characterised in that the boundary of the heating space is formed by at least a first heat transmission wall which, with its side remote from the heating space, constitutes a part of the boundary of a closed container. In the container a heat transporting medium is present. Thermal energy is absorbed by the medium from the heat source through a second heat transmission wall at another area of said container. The medium changes from the liquid phase into the vapour

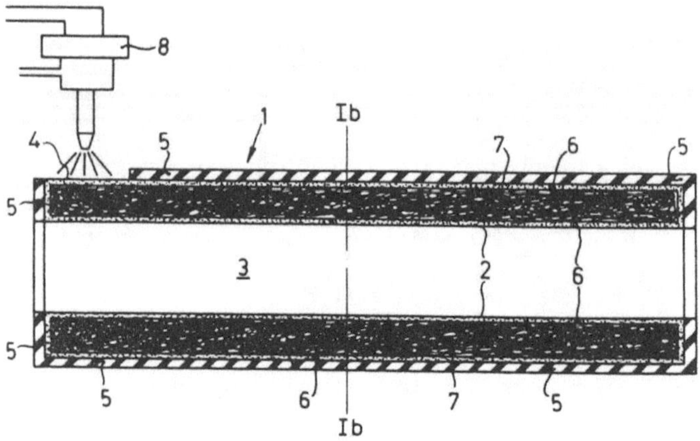

Fig. 249

phase, and supplies thermal energy to the heating space through the first heat transmission wall while changing from the vapour phase into the liquid phase. The container has a porous mass (steel wool), which connects the first heat transmission wall to the second heat transmission wall in such a manner that, due to capillary action, heat transporting medium condensed on the heat transmission wall can flow back to the further heat transmission wall through said mass. See Fig. 249.

The isothermal heating device, revealed in another Patent (197.3) of G.A. Asselman, (U.S. Philips Corp.), comprises an annular tubular body, providing a heating chamber for objects. Between the inner wall and the outer wall of this tubular body there is provided a plurality of separate ducts which are situated in a ring-shape about the heating chamber and which extends parallel to the tubular body axis. These ducts contain an evaporable heat transport medium and are in communication with each other preferably at one end of the tubular body. Such interconnection is in turn connected with a gas buffer reservoir containing an inert control gas.See Fig. 250.

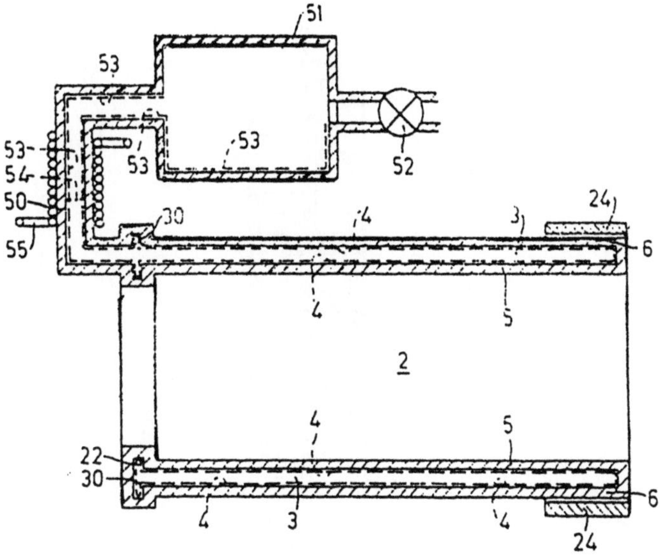

Fig. 250

5.6 Heat pipe heat exchangers adapted to domestic or workshop applications.

An energy saving liquid to liquid heat exchanger for a dishwasher, discharging hot waste water and designed by C.H. Bush, Patent (26), Fig. 251 comprises a hot water tank 6 for holding the waste water from the dishwasher and having inlet and outlet pipes 8, 10, a cold water tank 7 for holding the fresh water going to a waterheater and having inlet and outlet pipes 13, 15, the cold water tank 7 being disposed on top of the hot water tank 6, a bundle of heat pipes containing low boiling refrigerant disposed inside of the two tanks so as to extract heat from the hot water tank 6 and give it up to the cold water tank 7, whereby the temperature of the fresh water leaving the heat exchanger is higher than its entering temperature, each tank 6, 7, including a central vertical baffle 21, dividing the tank into elongate water inlet and outlet sections,

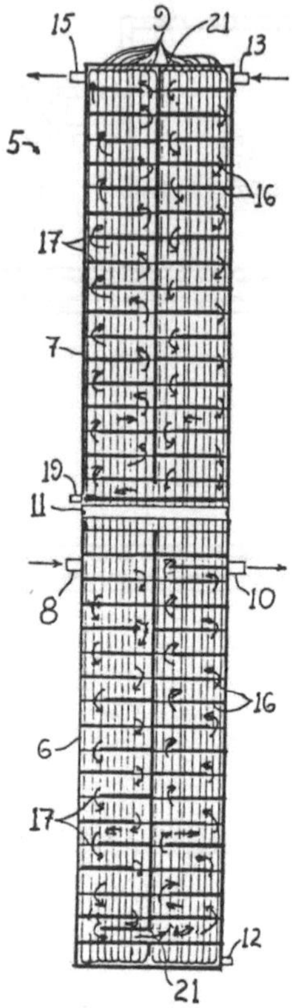

Fig. 251

and each section including horizontal baffles 16 which are staggered along the length of the respective section.

G. Häfner, et al., (Daimler Benz A.G.). disclose in Patent (40.14) a heat exchanger, having a structure wherein, in case of fluctuations of temperature from the heat supply side and/or heat consumption side, the reflux of heat from a secondary part to a primary part of the structure can be avoided, even if the temperature drop is inverted. The aims of invention can be realised by providing a heat accumulating mass, which is divided in several thermal zones, separated

and thermally insulated from one another and having different temperatures. The zones are arranged in the order of increasing temperature and are traversed by the heated conduit.
The heating conduit is isolated from the heat accumulation zones by a heat-insulated wall, while between the heating conduit and each of the heat accumulating zones a heat transfer relationship is realised, which acts only in one direction, i.e. from the heating conduit towards a heat accumulating zone.

Fig. 252, illustrates a longitudinal section of the proposed heat exchanger, 1 - thermal insulation; 2 - accumulating reservoir; 7 - heat insulated walls; 11, 16 - heat accumulation zones; 17 - conduit, inserted in the thermal insulation 1 and separated from the reservoir by an isolating wall 18.

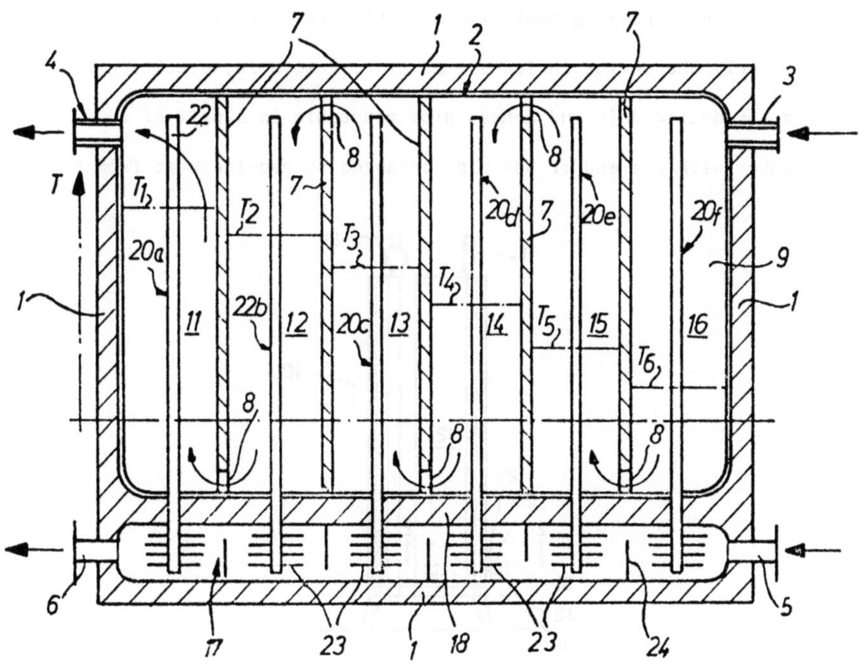

Fig. 252

The heat pipe type water heater, claimed by J.E. Kennedy, (Electro Therm, Inc.), in Patent (49.1) includes an elongate cartridge-type electric heating unit and at least one heat pipe, having a heat absorbing end and a heat transmitting end.

The electric heating unit is attached to the heat absorbing end of each heat pipe and the heating device is secured in an aperture in the wall of the container by means of a mounting flange. The electric heating unit is mounted in an aperture in the mounting flange and the heating unit and at least one heat pipe attached thereto extends into the interior of the chamber to apply heat to the liquid in the container.

Fig. 253 is a view of the proposed water heater, with a heating device mounted in a wall thereof.

Another Patent (49.2), assigned to Electro Therm, Inc., (inventor: J.R. Shutt), also refers to an electrically heated liquid tank employing heat pipe heat transfer means. More particularly the device includes a modular, removable, electrical, heat-producing unit and a heat pipe mountable in a wall of a chamber with one end of the pipe arranged to receive heat from the

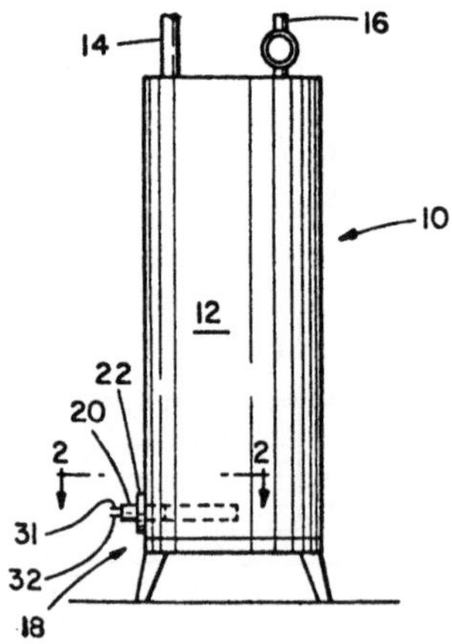

Fig. 253

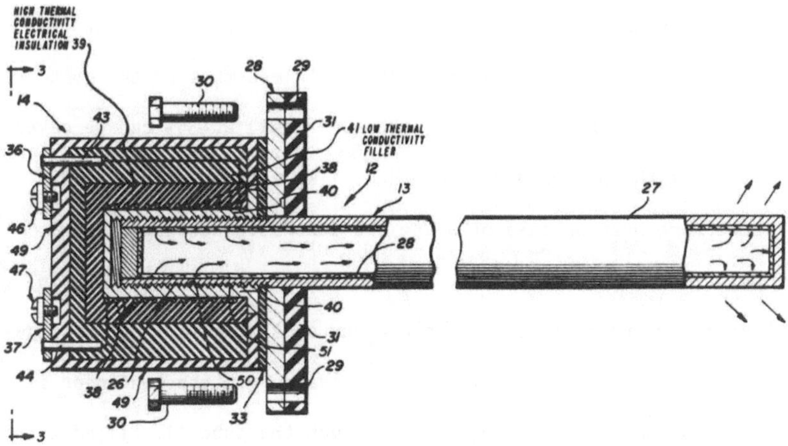

Fig. 254

electrical heat producing unit exterior of the housing and with another end of the pipe constructed and arranged to apply heat to the medium within the chamber. The heat pipe has high conductivity with a low temperature differential between the ends thereof and the heat producing unit includes an electric coil positioned about and removably secured to the one end of the heat pipe. The electric coil is embedded in a high thermal conductivity, low electrical conductivity filler material which is surrounded by a low thermal conductivity insulating jacket and which is received around a metal core member which is removably secured to the one end of the heat pipe. See Fig. 254.

According to Patent (59) of F. Stranicky, (Aktiebolaget Feros Patenter), a considerable part of heat which is entrained from a fireplace (hearth) by the exhaust gas conduit, can be retained and utilised by a heat recuperator, shown in Fig. 255, essentially comprising heat absorbing elements (evaporators) and heat distributing elements (condensers), with a heat carrier medium in circulation, for example water or another liquid, having at atmospheric pressure a boiling point higher or lower than that of for ex. trichloromonofluormethane. It is claimed that of the heat released by a kilogram wood (2000-4000 Kcal) 20-35% can be saved by using the proposed system; thus, 1 kg wood supplies by 1000 Kcal more heat in case of using the described heating system, which is shown in Fig. 255, with a water evaporating duct 2 in ascending arrangement and a condenser duct 5 in vertical arrangement, the fuel is wood 3. The condensate from the condenser 5 is returned to the evaporator 2 through the tube 11, filled with a porous material. The entire heating equipment is removable and mounted on telescopically adjustable legs 13, 14.

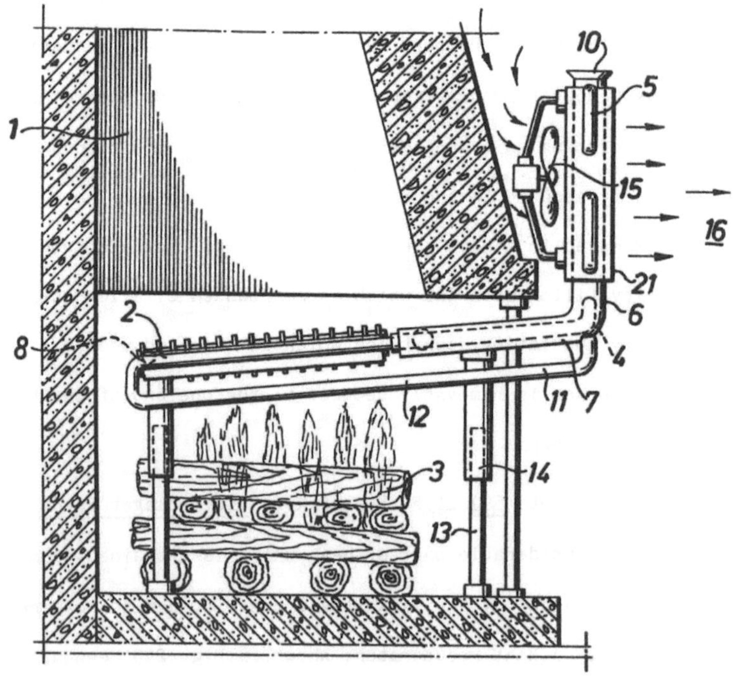

Fig. 255

Patent (120) of R. Sgroi, assignor to Kenneth C. McCord, refers to a heat transfer blanket, composed of a pair of juxtaposed rectangularly shaped fabric sheets having a plurality of flexible heat pipes disposed in an overall pattern thereinbetween. The heat pipes comprise a volatile liquid and a wick disposed within a flexible metallic sheath such that the heat transfer is dependent upon the phase state of the volatile material. More particularly, the blanket has a plurality of portions of flexible heat pipes sandwiched between the outermost layers thereof. The portions of the flexible heat pipes disposed about the blanket provide a uniform heating or cooling pattern therein. The ends of the flexible heat pipes that are free from the blanket are thermally coupled to a combination heating and cooling system. When utilising the heating system, the flexible heat pipes provide elevated temperatures at the blanket surfaces. When utilising the cooling system, the flexible heat pipes provide lower than ambient temperatures at the blanket surfaces. See Fig. 256.

In Patent (156.1) of H. Scurrah, reveals a heating apparatus, comprising a heat pipe having a sealed container enclosing a condensable vapour and means within the container for assisting transport of condensed vapour from a cooler portion to a warmer portion, a heat source, a material to be heated and a separating member separating the heat source and the material to be heated. The heat pipe is so received in an aperture formed in the separating member that one portion is located on one side of the separating member to receive heat from the source and another portion is located on the other side of the separating member, the heat pipe being in sealing engagement with the wall of the aperture and supported by the separating member, whereby, in use, heat absorbed

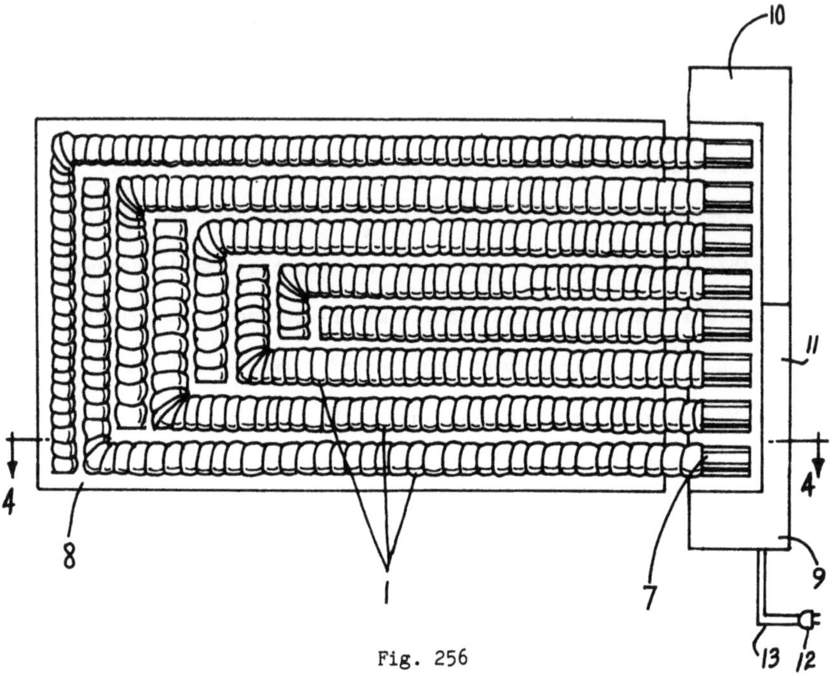

Fig. 256

from the source by the one portion of the device is transferred to the other portion and thence to the material.

The system may be used for water heating, in which case one end of the heat pipe may be placed in a fluid or encased in a fire brick and the other may be placed in an existing back boiler or simply be inserted into a large diameter pipe forming part of a central heating or hot water circuit.

In Fig. 257 there is shown one wall 10 of a flue, from a boiler of a central heating system for example through which flue gases, constituting a source of otherwise wasted heat, flow in the direction of the arrow 11. The flue wall 10 is transversed by a heat pipe 12 having a heat absorption portion 13, an intermediate portion 14 at which the pipe 12 is supported where it passes through the flue wall 10 and a heat dissipation portion 15. The heat pipe 12 comprises

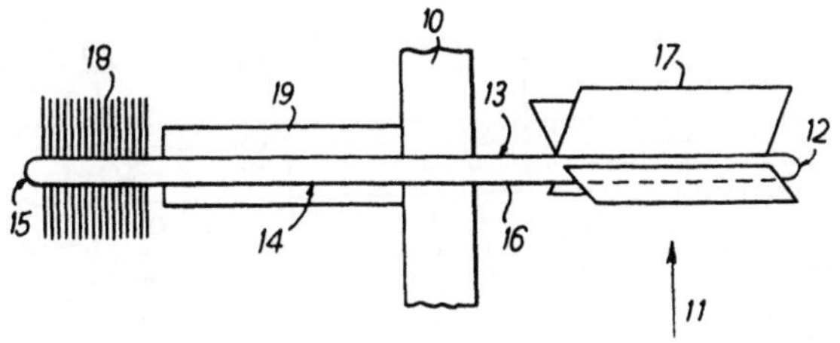

Fig. 257

a tube 16 closed at each end, lined with a layer of wick material (not shown), e.g. of metal and partially filled with a working liquid, e.g. water; 18 - convection fins; 19 - insulating jacket.

5.7 Heat pipe exchangers adapted to air conditioning systems.

Patent (1) of J. Raillon, (Agence Nationale de Valorisation de la Recherche (ANVAR)), concerns the natural air conditioning of buildings and, more particularly, the intensification of natural air conditioning systems, the operation of which is based on the consideration that heat transfer from a heat source to a heat receiver by means of a gaseous medium circulating between these two points, is more effective (per unit of time and unit of volume of the gaseous medium), if the heat carried by this medium is in the form of latent vaporisation heat and not a directly perceptible heat. An essential part of the system is a thermal storage reservoir, filled with a heat accumulating material, for example aqueous solutions of eutectic salts, the said reservoir being in direct or indirect thermal contact with at least a part of the wall, whereon a working fluid is condensed. The working fluid may be, for example trichlorofluorethane (with a boiling point at 47.6°C) or trichloroethane (with a boiling point at 74.1°C).

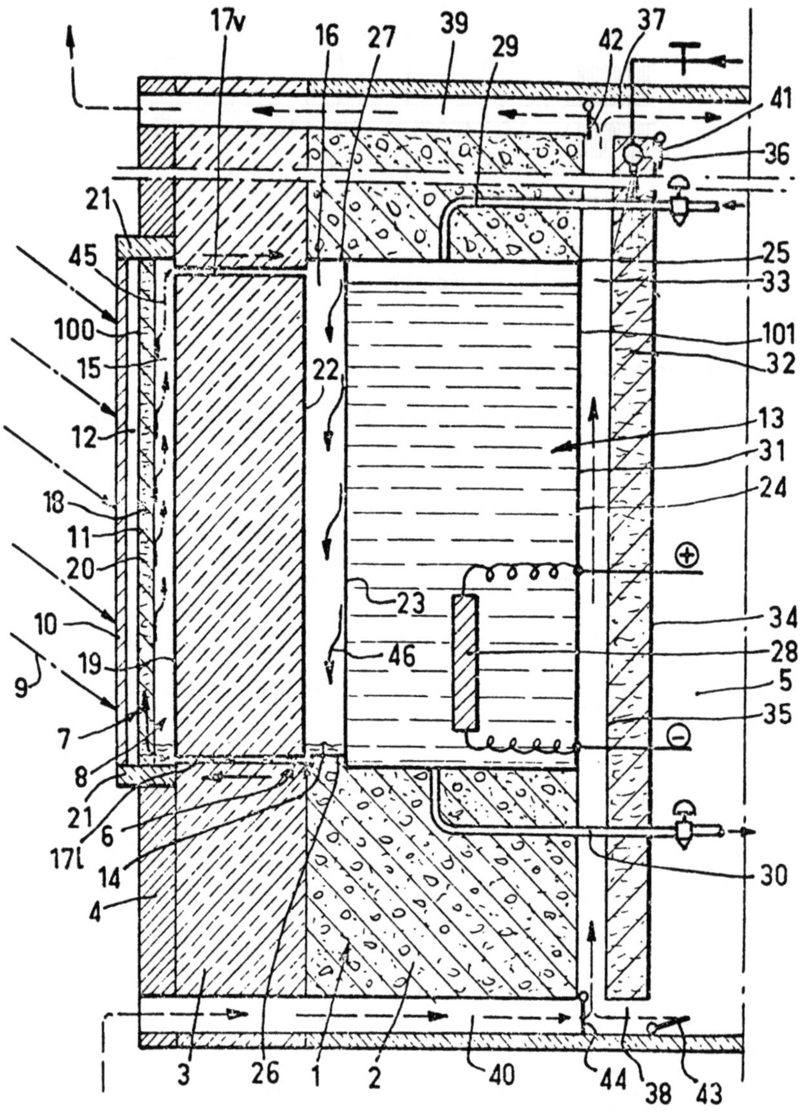

Fig. 258

Fig. 258 illustrates schematically (in vertical section) a building (of flats), provided with the proposed natural air conditioning system, the air conditioning system being indicated by 6.
The heat exchanger, incorporated in a room heating system and described in Patent (19) of D. Bouvier, consists of a closed hollow

body, provided with an external casing of heat conductive material, a tubular coating, placed within the said casing and composed of a porous material, with a central passage therein.

The closed body contains a fluid, which can evaporate under thermal effect and be condensed at the contact face of a cold zone. The fluid is stored in the porous material and on evaporating at the level of a heated zone, the said fluid will pass through the central passage in form of vapour and be transferred from the heated zone towards the cold zone, to be condensed therein and returned into the porous material. The movement of the condensed fluid takes place by capillary action in the porous sector of the exchanger, from the coldest zone towards the hottest zone. According to the invention, at least a part of the heat exchanger is placed in the wall of the room to be heated and is surrounded with an insulating coating for reducing heat losses.

Fig. 259 is a longitudinal section of the proposed heat exchanger, while Fig. 260 illustrates the installation thereof in a room to be heated.

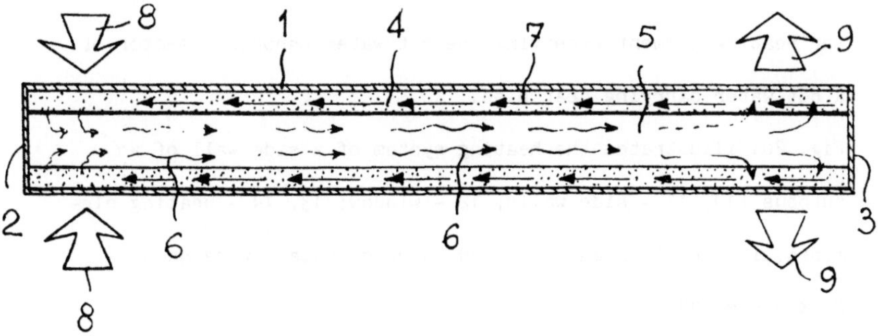

Fig. 259

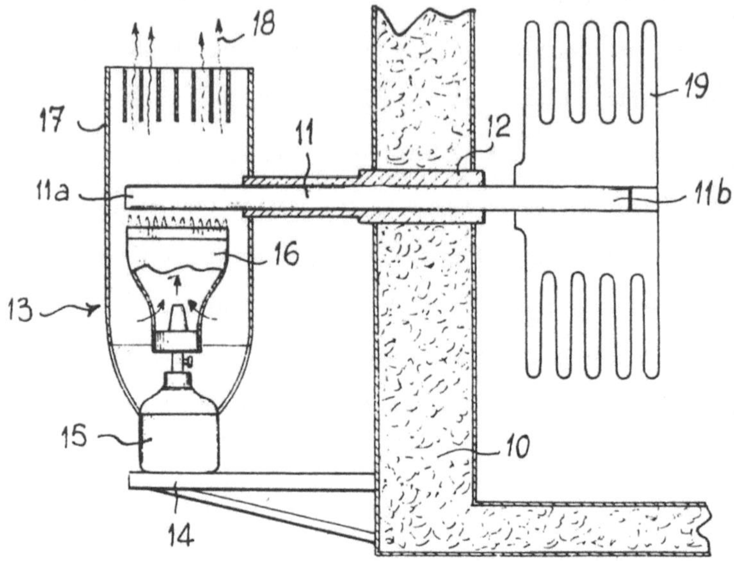

Fig. 260

Patent (40.4) of F. Bährle, (Daimler Benz AG), mainly relates to heating systems for autobuses, such systems being executed in the form of flat heating elements and arranged in the side walls of the vehicle.

The invention improves the effectivity of heating, by realising the heat supply to the flat heating elements by means of hot water, the heating element receiving the hot water through a sector of a hot water carrying heat pipe.

Fig. 261 illustrates the heating system of a side wall of an autobus (10, 11 - side walls; 12 - window; 13, 14 - heating elements, in serial arrangement and interconnected by means of flexible hoses 17, 18).

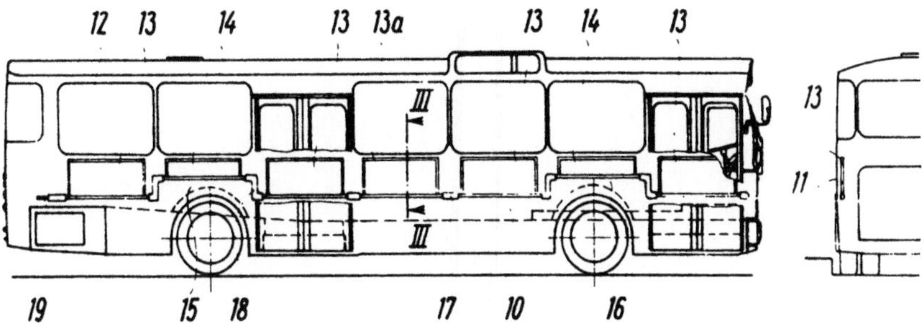

Fig. 261

Another Patent (40.7) of F. Bährle also claims a heat pipe based heating system for autobuses. See Fig. 262, arranged in the side walls 2 of the bus 1, and consisting of flat heating elements 4, receiving heat through a pipe 5. In order to construct all parts of the most suitable material, pipe 5 and heating element 4 are constructed as independent units, which are in heat conducting connection through clamping springs 9. See Fig. 263.

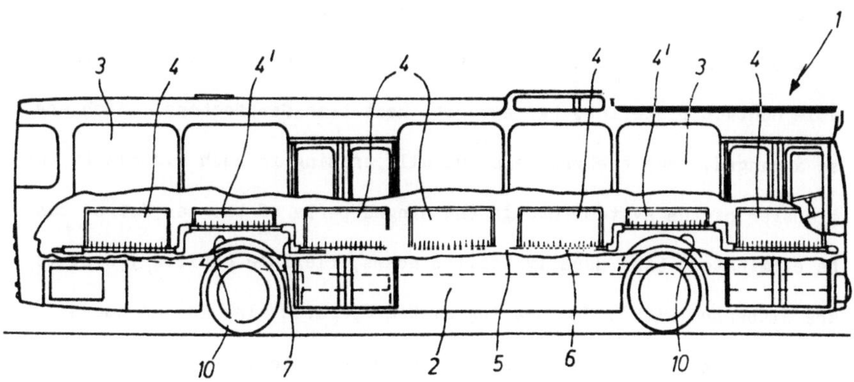

Fig. 262

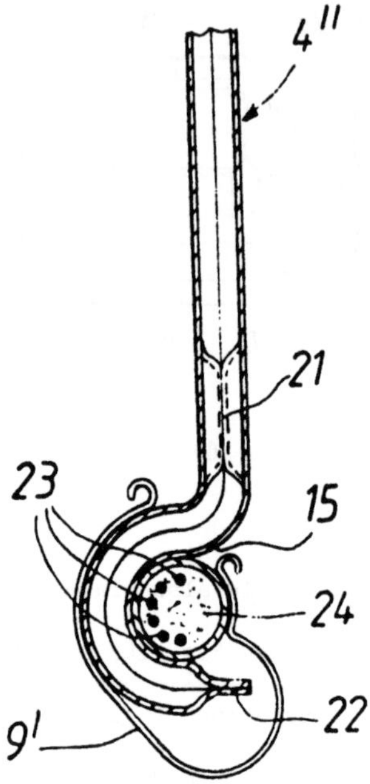

Fig. 263

R. Andres, (Daimler Benz AG.), provides in Patent (40.5) a heating system which can be disposed in the door of cars and which is composed on the basis of heat pipes. An advantage of (40.5) according to the inventor is, that the heat transfer is subdivided at a detachable contact plate into two separate heat transfer sectors, which are so arranged, that the problems, usually arising in such systems by the transfer of the system through the hinged parts of the car doors, can be eliminated, (the usually applied hoses, flexible capillary systems are dispensed with). The heating system operates also in car doors, which open by a vertically outward swinging motion.

Fig. 264 illustrates a contact place within a heat transfer system, while Fig. 265 represents a heat transfer system to the rear door of a car.

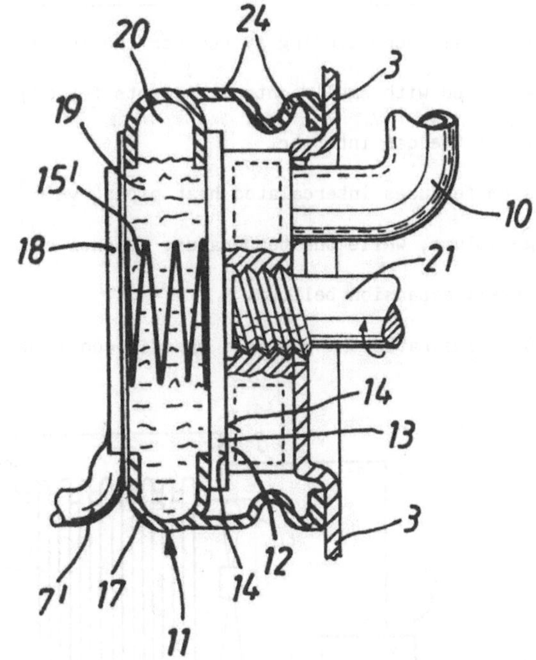

Fig. 264

Fig. 265

Another Patent (40.6) of R. Andres, (Daimler Benz AG), provides an improved air conditioning system for the interior of passenger cars, supplied with small control elements for adjusting the temperature of the car interior.

The system features intercalated heat pipes, which must not be of large volume, while the temperature control can be effected with a small expansion bellows.

Fig. 266 illustrates the proposed car air conditioning system.

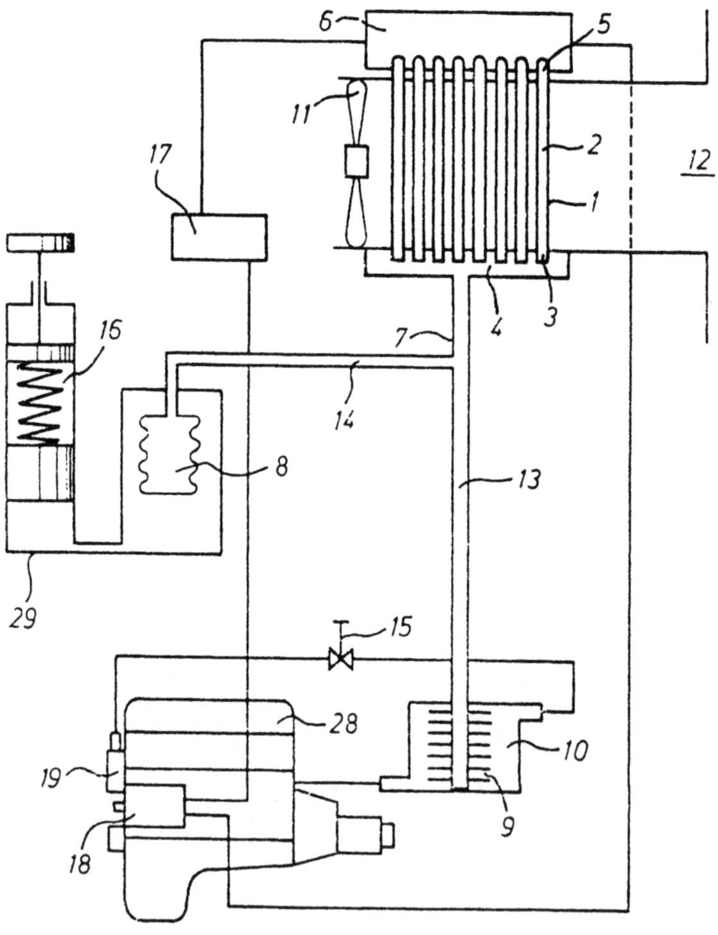

Fig. 266

Patent (43) of M. Delery et al., concerns a device for heating the interior of a motor car, in connection with an air cooled motor and a radiator, disposed in the proximity of the car interior or therein. The device features a heat pipe, which is in direct connection with, on the one hand, the said radiator and, on the other hand, with a heat source, issued from the motor.
In case the motor is an internal combustion type motor, the heat source is provided by the exhaust pipe.
In case of an electric motor, the heat source consists of the motor proper, around which a heat pipe is wound.
Fig. 267 illustrates the invention, wherein: 1 - internal combustion motor; 2 - inlet pipe; 3 - motor block with air cooled cylinders 4, around which ventilating elements 5 are disposed; 6 - exhaust pipe of hot exhaust gases, connected via pipe 7 to sound absorber 8.
The end 9 of a heat pipe 10 (of copper: Ø = 15/16 mm), envelope 7, while the other end 11 of the heat pipe 10 is connected to a heating

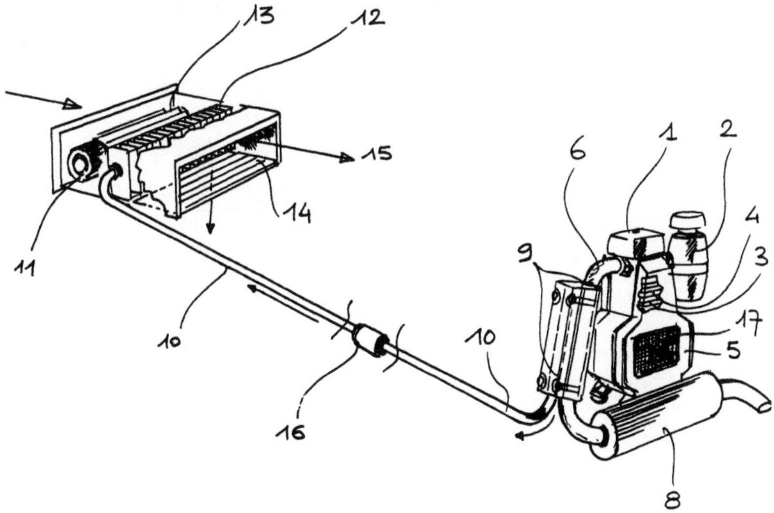

Fig. 267

radiator 12, connected to ventilator 13, placed before a shutter 14, opening into the interior 15 of the car.

16 indicates a valve, by which heat transfer towards the radiator can be switched off.

Patent (66.2) of A. Bell, (Foster Wheeler Power Products Ltd.), relates to air heaters and provides a means by which the operation of heat exchangers therein can be sustained at an optimum efficiency. The invention proposes the incorporation of a preheater 16, (see Fig. 268 for incoming unheated air which can be selectively thermally coupled to heated air in an outlet duct 12) from the heat exchanger 4. During normal operation the heated air bypasses the thermal coupling, but when the temperature of the incoming unheated air is particularly low, the coupling is made and the incoming air thereby preheated to ensure efficient operation of the heat exchanger. As shown in the figure the preheater 16 comprises a plurality of heat pipes 20 which extend into the outlet duct 12. A damper 22 is selectively movable between a first position in which the heated air is directed over the heat pipes and a second

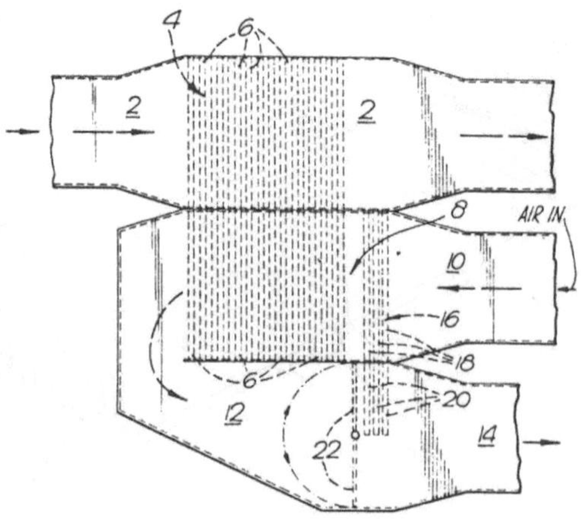

Fig. 268

position in which the heated air bypasses the heat pipes. Heat is exchanged between hot waste gases flowing through duct 2 and the air by way of heat pipes 6.

V.H. Gray, provides in Patent (81) rotating heat pipes and, more particularly, air conditioning apparatus, incorporating a rotary heat pipe for transferring heat from a heat source to a heat sink. The heat pipe has a tapered condensing surface, which is curved preferably to provide uniform pumping acceleration, the heat pipe being rotated at such a velocity that the component of centrifugal acceleration in an axial direction parellel to the tapered surface is greater than 1g and so that the condensing surface is kept relatively free of liquid at any attitude. The heat pipe may be incorporated in an air conditioning apparatus so that it projects through a small wall opening. In a preferred air conditioning apparatus, a hollow hermetic air impeller is provided which contains a liquefied gaseous refrigerant, such as freon, and means are provided for compressing the refrigerant in the evaporator region of the heat pipe.

Fig. 269 is a longitudinal sectional schematic view of a rotating heat pipe according to Patent (81).

The object of Patent (89.1) of Michio Yanadori, (Hitachi Ltd.), is a ventilating system including a heat exchanger, utilising boiling and condensation of a liquid, mounted in a cabinet formed with a first air passage and a second air passage by a partition plate extending horizontally across the heat exchanger substantially in the center thereof. At least one air sucking and exhausting fan is mounted in one of the first and second air passages, and the heat exchanger is provided with at least one auxiliary heat exchanger

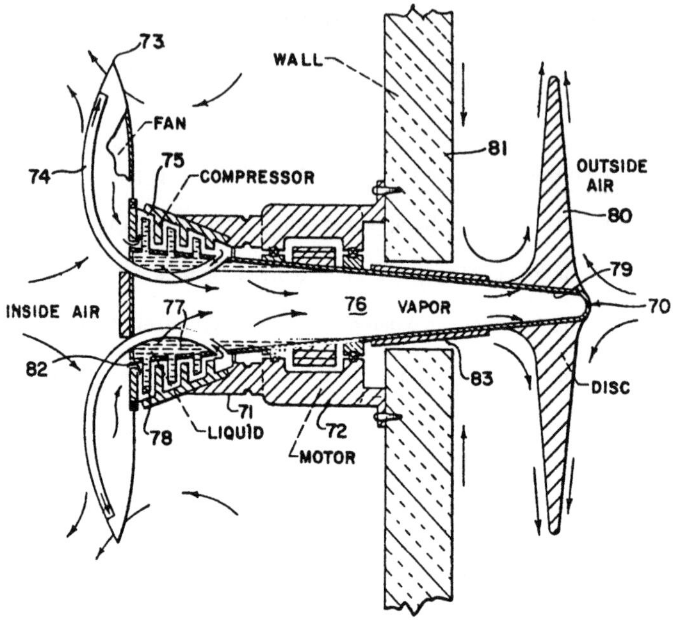

Fig. 269

operative to perform space cooling or space heating. The first and second air passages are each provided with ventilating ports for effecting heat recovery, vertilating ports for effecting cooling by ventilation and a ventilating port for allowing cooled air of heated air to flow therethrough, so that the ventilating system can be operated in a heat recovery mode, a cooling by ventilation mode and a space cooling or space heating mode by actuating the auxiliary heat exchanger and suitably opening and closing various ventilating port.

Fig. 270 is a sectional view of the upper and lower portions of the ventilating system described in Patent (89.1).

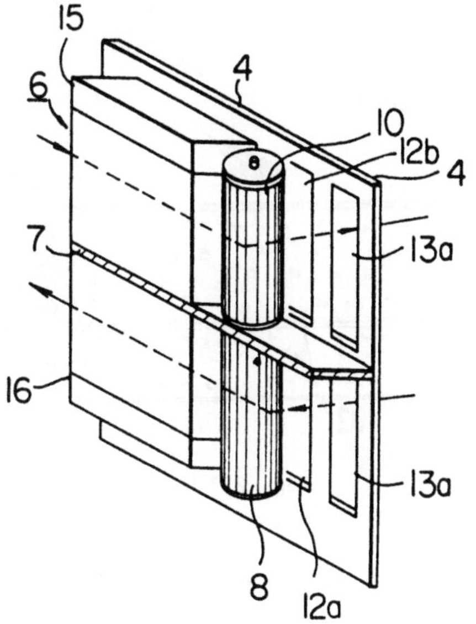

Fig. 270

N. Laing reveals in Patent (108.3), and essentially in Patent (108.4) heat rectifier plates (hollow plates), comprising a fluid and gaseous heat carrier medium. If the plates are used as horizontally arranged roof plates, the fluid phase of the heat carrier medium is in the lower zone of the plate in heat conductive contact with a latent storage mass. Depending on whether the plates are used as roof plates for heating or cooling a building, the condensate of the heat carrier means must be provided for the transport of the condensate. For this purpose Laing proposes various methods.

According to Patent (108.4) the plates referred to can be used as insulating plates without any thermal effect. In this case special heat carrier substances are recommended like 1,2,4,5, - tetraethylbenzene (smelting point: + 13°C) or 1,2,3, - tribromopropane (smelting point: + 16°C) or bromophorm (smelting point: + 9.2^{o}C) or 1, - bromonaphthalin (smelting point: + 6,2^{o}C).

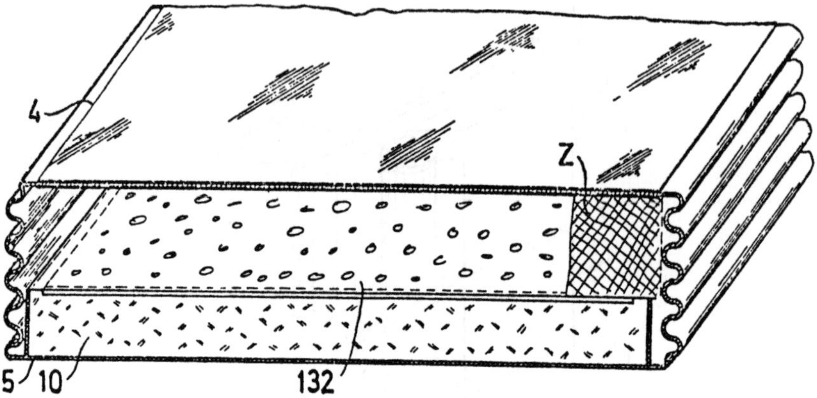

Fig. 271

Fig. 271 shows a perspectivic view (partly in section) of a roof plate according to the invention, while in Fig. 272 a modified embodiment thereof is represented.

In Patent (108.5) and essentially in Patent (108.6) of N. Laing reveals the operation of heat pipes or heat rectifier roof plates, used for heating or cooling of buildings and supplied with heat carrier substances of different boiling point.

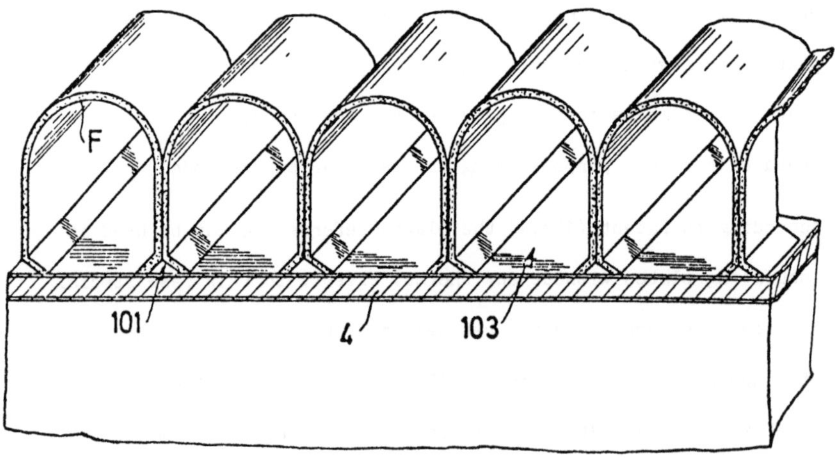

Fig. 272

According to these patents this is advantageous for the control of the pressure conditions in the interior of the plates or pipes. Also the use of a binary mixture of two heat carrier substances of different boiling point is recommended. By increasing the share of the higher boiling point or lower boiling point component in the mixture, it is possible to vary the vapour pressure within the plate (pipe) continuously between the vapour pressures of both pure substances. If, for example, the pressure within the plate increases, due to the overloading of the storage tank, and its temperature rises, the vapour pressure can be reduced by the addition of the higher boiling point component, or vice versa.

Patent (108.7) of N. Laing describes a thermal system, consisting of a heat pipe in combination with a heat storage element. The invention features a special heat pipe container, into which, during the charging period, heat transfer takes place from a heat source towards the storage element, while during the discharge period, heat transfer is effected in opposite direction. The system can be applied both to heating or cooling of rooms, spaces, etc.
Fig. 273 illustrates a cooling roof plate, the upper wall 1 of which forms a solid roof cover, while the lower wall 5 forms the roof of a room, from which heat should be removed.
A heat transferring system, mainly for the air conditioning of mines, tunnels, refrigerating chambers, is the object of Patent (113) of G. Sinreich, (Mabag Luft- und Klimatechnik GmbH.),
In contrast to usual systems, consisting of three circulations: one for cooling water, one for cooling agent and one for cold water, the proposed system consists of heat pipes, which are in parallel arrangement with one another between the cooling water feed conduit,

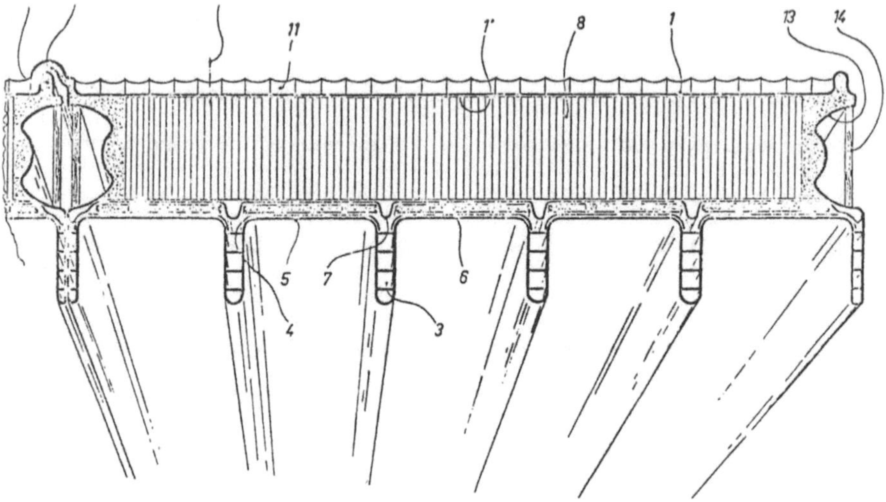

Fig. 273

and the cooling water return conduit. It is possible to arrange the heat pipes according to various patterns, several pipes in group-like arrangement, which are housed in a casing, provided with an air input, an air filter between air input and heat pipe (groups), a fan and an air output. Since the system requires only one circulation, it is mainly suitable for mines, where it can be continuously moved ahead, according to the process of the underground working.

Fig. 274 illustrates schematically the cooling (air conditioning)

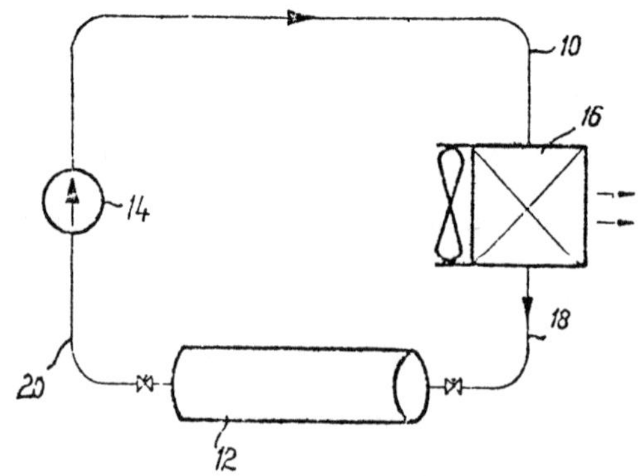

Fig. 274

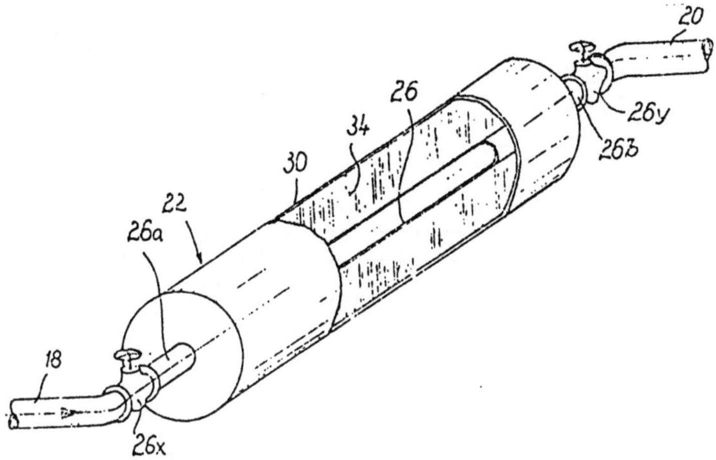

Fig. 275

system of a mine gallery, while in Fig. 275 a cylindrical heat pipe is represented, in combination with a central cooling water conduit.

D.T. Morgan, (Thermo Electron Corp.), discloses in Patent (182.5) a heat-actuated air conditioner/heat pump, which includes a sealed rotatable tube. The sealed tube contains a working fluid and includes an evaporating leg and a condenser leg, the condenser leg extending a greater distance from the axis of rotation than the evaporator leg. As the tube is rotated, a vapour pressure differential is created between the evaporator and condenser legs with the higher pressure in the condenser leg. Because of this pressure differential, the working fluid will evaporate in the evaporator leg at a lower temperature than that at which it will condense in the condenser leg. The evaporator leg thus can be used for cooling a stream of house air (house air conditioning) while the condenser leg rejects heat to a stream of ambient air. When all of the working fluid has evaporated, the system may be recharged for another cooling cycle by supplying heat to the

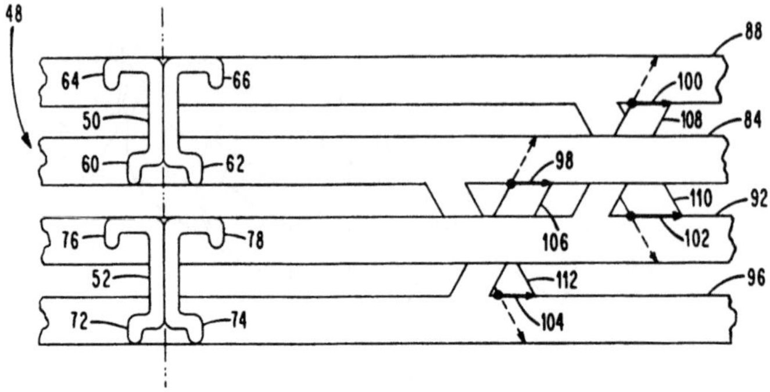

Fig. 276

condenser leg to drive the working fluid back into the evaporator leg. The sealed tube may also be operated as a heat pump for heating house air.

Fig. 276 illustrates an air conditioner heat pump assembly 10, including a pair of heat pipes 12 and 14. The heat pipe 12 has an evaporator leg portion 16 and a condenser leg portion 18. Similarly, the heat pipe 14 has an evaporator leg portion 20 and a condenser leg portion 22. The evaporator legs 16 and 20 are located within a channel 24 and the condenser leg portions 18 and 22 are located within a channel 26. The heat pipes 12 and 14 are made of material with good heat transfer properties and adequate strength, such as stainless steel. See Fig. 276: schematic side elevation view of the proposed system.

5.8 Heat exchangers having a heat pipe or heat pipes, mounted in refrigerators, cooling systems and heat pumps.

5.8.1 Refrigerator defrosters equipped with heat pipe

Patent (30) of N.S. Cristopher, relates to a passive cooling system. More particularly, the invention relates to a solid state passive cooling system for cooling an enclosure sheltering heat producing equipment such as, for example, a microwave repeater station. According to Patent (30) a cold bank tank in an enclosure for storing thermal energy in a very small area contains a liquid heat transfer fluid and has a top which is the top of the enclosure and a spaced opposite bottom. An inside heat exchanger extends from the bottom of the tank in the enclosure for cooling warm air in the enclosure rising by natural convection, returning the cooled air to the enclosure and transferring the warm air to the tank wherein it warms the fluid in the tank. An outside heat exchanger outside the enclosure is spaced from the top of the enclosure and coupled to the tank.

A plurality of energy storage rods in the fluid in the tank store thermal energy in a very small area. The heat warming the fluid is transferred to the energy storage rods and when the outside temperature decreases below a predetermined magnitude, heat is released from the rods to the liquid fluid and vaporizes the fluid.

The vaporized fluid flows upward through piping to the outside horizontal heat exchanger, through the outside heat exchanger, whence its heat is dissipated in the cooler outside air, and the vapour is recondensed and returns by gravity force to the tank via piping thereby storing cold energy in the rods for release when the air in the enclosure is next warmed.

Fig. 277 is a schematic diagram (partly in section) of the proposed passive cooling system.

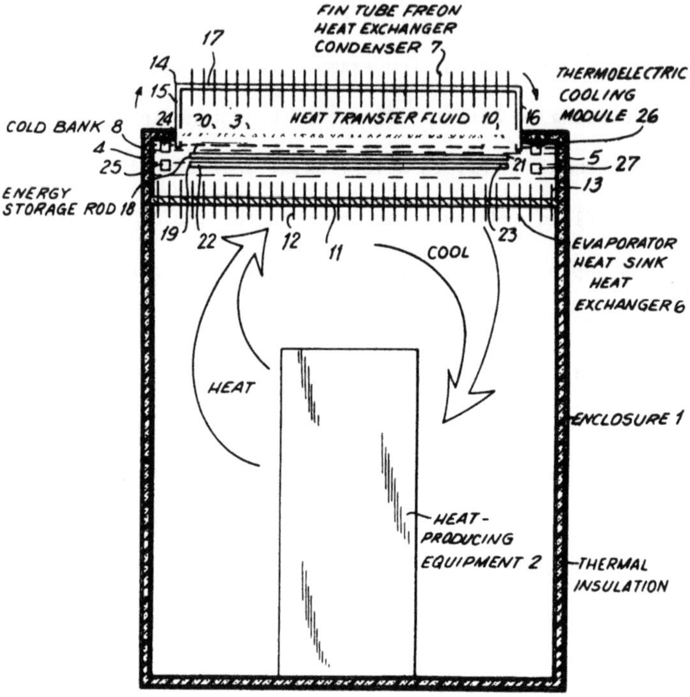

Fig. 277

A. Bricard, (Commissariat à l'Energie Atomique), provides in Patent (34.1) a cooling installation, which is based on the principle, that black (or nearly black) bodies absorb radiation, independent of the wave length. According to the invention, the body, exposed to cooling is thermally connected to a material, which presents a transition between the solid and liquid phase, at the temperature at which the installation operates. The thermal connection is realised by means of thermal diodes, ensuring a connection exclusively in the material ⟶ body direction.
The result is that the material's temperature is lowered, effecting its solidification without an inverse transformation from the solid to the liquid phase, since the transfer of calories which might be issued from the radiating body is blocked by the thermal diodes.

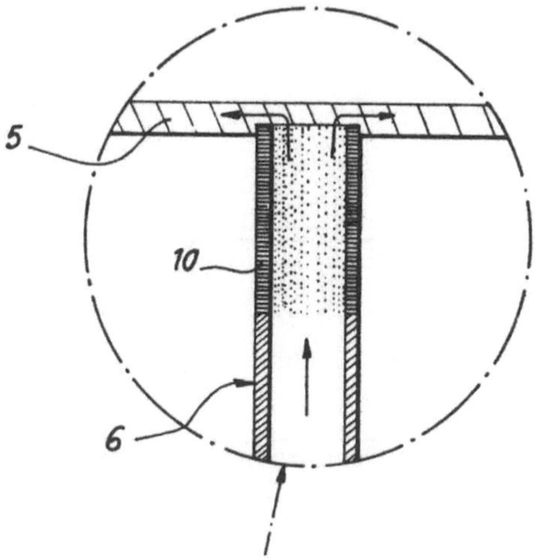

Fig. 278

Preferably, the thermal diodes have a discontinuous capillary structure.

The installation displays a radiating surface, a material, a storage space of a solid/liquid phase material; means thermally connecting the said space and the said surface (diodes), and insulating elements, separating the installation from the environment, without weakening the radiation from the surface.

See Figures 278 and 279.

According to Patent (57) of E.C. Faccini, a dual mode heat pipe for roadways, bridges, etc. includes an auxiliary evaporator formed concentrically with the upper end of a vertically disposed primary evaporator portion. The auxiliary evaporator portion comprises an annular sleeve disposed about the upper end of the primary evaporator portion and arranged between the primary evaporator portion and the condenser portion of the heat pipe such that all of the condensed working fluid returning to the primary evaporator portion must enter and overflow the auxiliary evaporator portion prior to return to the primary evaporator

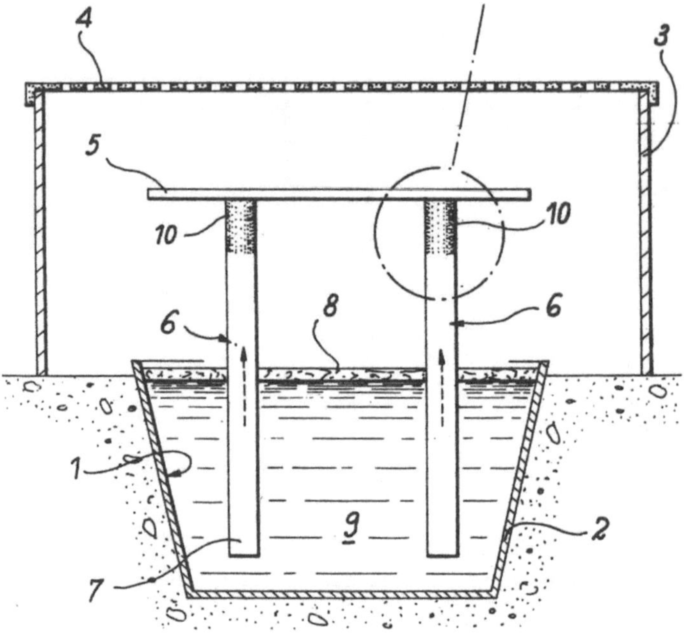

Fig. 279

portion. The auxiliary evaporator portion is provided with heat input means whereby the auxiliary evaporator may function even in the absence of heat pipe function by the primary evaporator.

Figures 280, 281 illustrate the operational environment of the proposed heat pipe; a section along line 2-2 of Fig. 280 and a detail of the auxiliary evaporator.

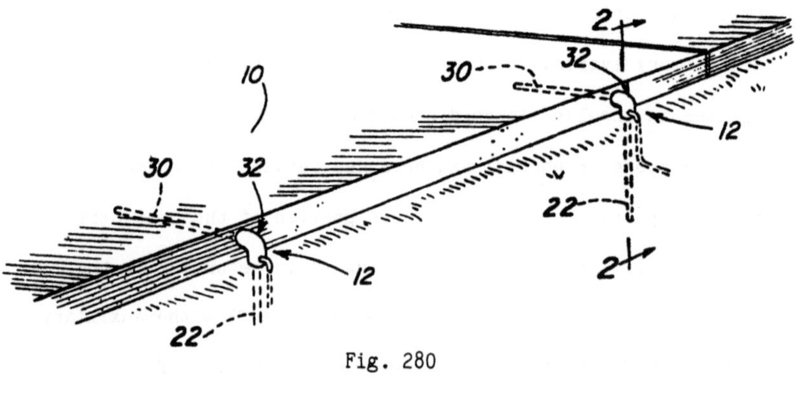

Fig. 280

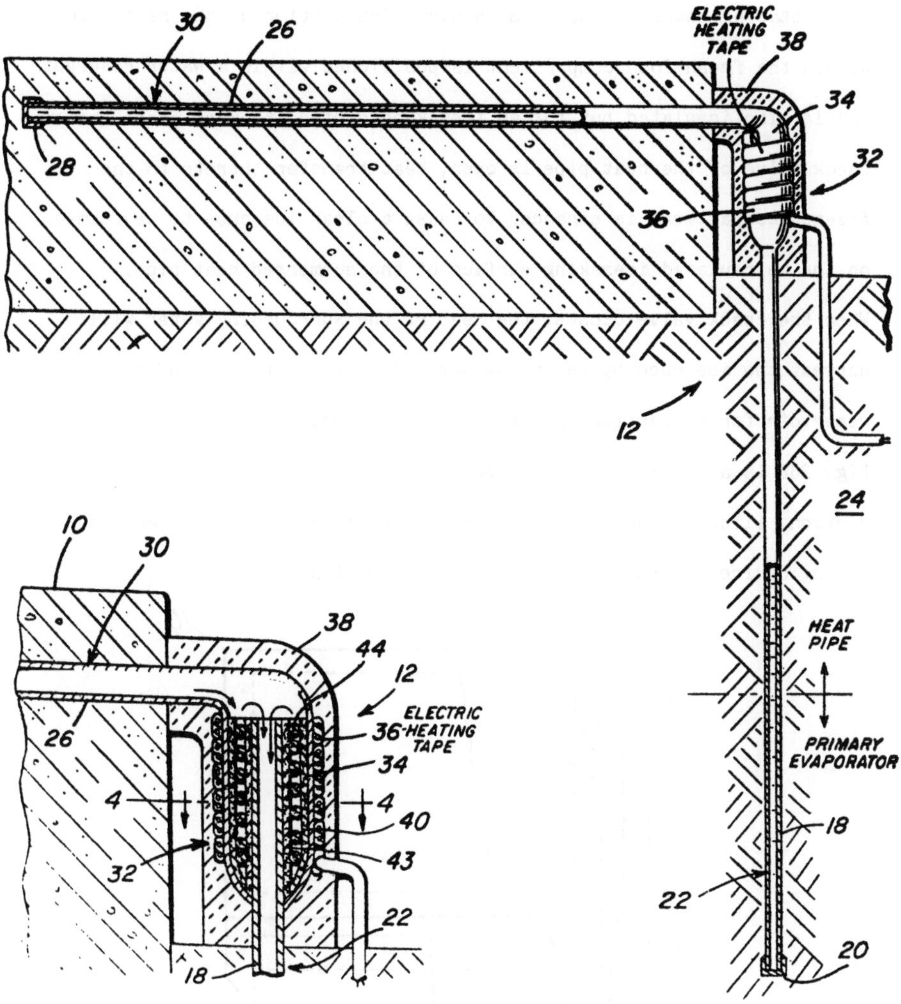

Fig. 281

An automatic icemaker, disclosed by F.A. Schumacher, Patent (76.2), (General Electric Co.), includes a heat pipe in the form of a sealed tube containing a refrigerant which extends from the freezer compartment to the fresh food compartment of a household refrigerator. Positioned within the heat pipe and located intermediate the freezer and fresh food compartment is a valve, preferably in the form of a steel ball which is periodically unseated from an annular valve seat

by a rotating magnet actuator mechanism. The portion of the heat pipe within the fresh food compartment is positioned within a water reservoir and is also insulated but for an icemaking surface thereof. When the check valve of the heat pipe is open, heat transfer thereto to the freezer compartment is enhanced to cause a clear ice cube to be formed on the uninsulated icemaking surface of the submersed heat pipe.
When the check valve closes, the heat of the surrounding water will harvest the ice cube by releasing same to float to the top of the reservoir for collection and storage by conventional means.
Fig. 282 is a schematic front view, partially broken away and in partial section, illustrating a household refrigerator equipped with an ice making device according to the present invention.

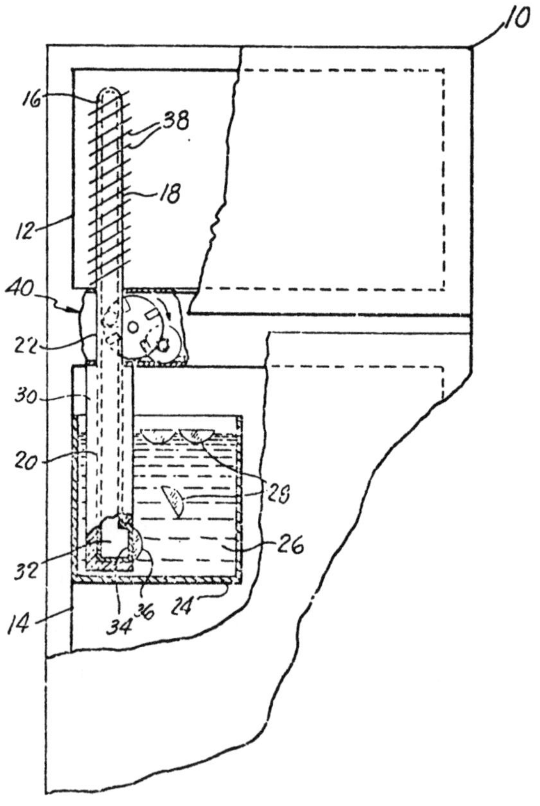

Fig. 282

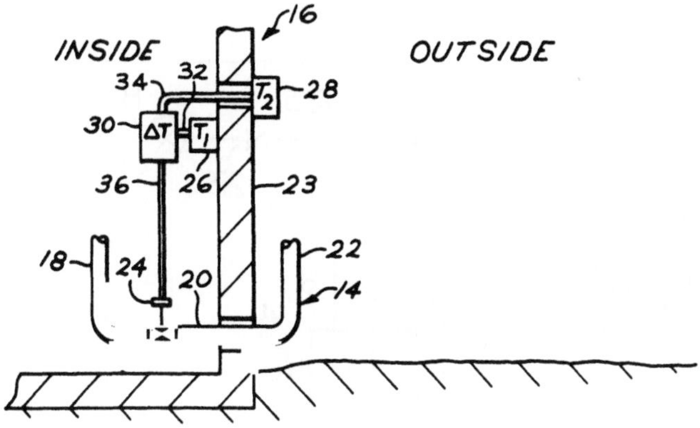

Fig. 283

Another Patent (76.3) also assigned to General Electric Co., (inventor: J.B. Horvay) concerns a heat pipe for controlling the temperature of a refrigeration system of a refrigerator and is of a general "U" configuration. A means is provided within the heat pipe to terminate the passage of gas and/or liquid through the heat pipe when the ambient temperature of the second end of the heat pipe is greater than the temperature of the portion of the refrigerator contacting the first end of the heat pipe.

Fig. 283 is a diagrammatic view of the proposed control and heat pipe system.

Asai Setsuro, (Hitachi Ltd.), reveals in Patent (89.3) an absorption cooling and heating system, as represented in Fig. 284, wherein 30 - a heat storage element; 302 - heat pipes; 1 - generator; 2, 3 - external heat exchangers; 7 - relief valves for cooling 8, heating 9, and defrosting; 4, 5 - inner heat exchangers; 10 - a pressure reducer valve; 11 - a solution pump; 11, 12 - valves for switching over the channels. 30 is arrayed downwards of the valve 9, (defrosting), as duct for the cooling agent. The heat pipes 302 have

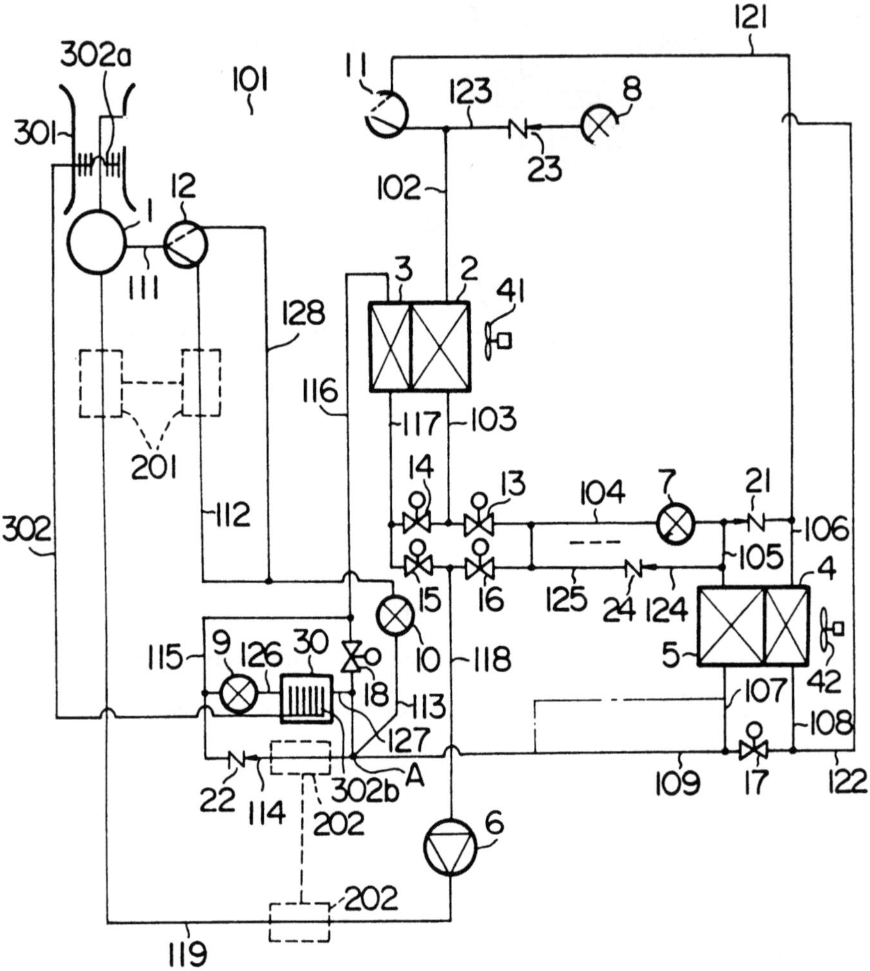

Fig. 284

their heat output end 302b in the storage element 30 and their heat source end 302a at 301 - a burner of generator 1.

Fig. 284 illustrating a flow line diagram of the proposed cooling-heating system.

Another Patent (89.5) assigned to Hitachi Ltd., (inventor: Yoshinori Hakuraku, et al.), discloses a magnetic refrigerator, comprising an operative material, destined to magnetic cooling

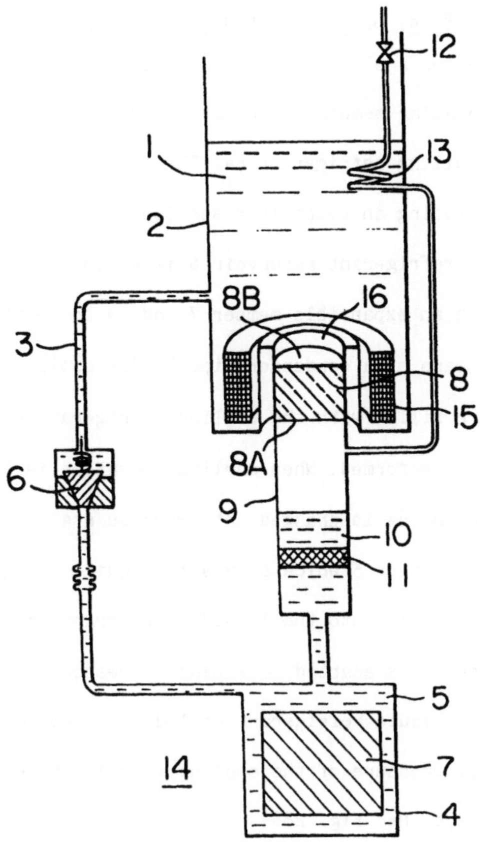

Fig. 285

and a system of magnetic field, capable of modifying the distribution or intensity of the applied magnetic field. A feature of the invention is, that the operative material is arranged in a fixed position and the distribution or intensity of the applied magnetic field is periodically modified, in order to create a refrigeration cycle. The advantage claimed by the inventor is that the operative material is in fixed relationship with regard to two heat exchange sections (i.e. sections of high temperature and low temperature). As a rule, the refrigerating agent is liquid helium ($\sim$ 4,2 K, $\sim$ 1013 mbars). See Fig. 285.

Yet another Patent (89.6) of Hitachi Ltd., (inventor: Okada Sada-Yuki, et al.), concerns a closed type boiling cooling apparatus for cooling a heat generating member 10, by utilizing the latent heat of vaporization of a boiling refrigerant 8a. The apparatus comprises a cooling vessel, including an evaporator section 1 and a condenser section 3. A liquid refrigerant reservoir 6 is enclosed at its upper opening portion with an expansible member 7 and is connected with the cooling vessel by means of a condenser pipe 5. The cooling and the condenser pipe 5 are filled with the boiling refrigerant 8a when no cooling operation is performed. When cooling operation is initiated the boiling refrigerant 8a is changed into gaseous state 8c and then led into the condenser pipe 5 which serves to liquify the gasified refrigerant 8a and feed it into the liquid refrigerant reservoir 6. The expansible member 7 is adapted to expand in response to the change in the amount of the liquid refrigerant 8b led into the reservoir 6 so that the internal pressure of the cooling vessel can be maintained substantially constant. See Fig. 286.

Patent (105) of C. Grant, (Kooltech Ltd.), provides an apparatus for cooling liquids, as represented in Fig. 287.

The apparatus 10 for cooling liquids flowing in a pipeline 5, for example potable liquid such as beer, stored in a bulk container 6 and dispensed at a font 7, comprises a product coil 3, connected in the pipeline 5 and immersed in a liquid body 4 of a volatile liquid such as the refrigerant R12, body 4 being in fluidic communication with a vapour body 9 of the same volatile liquid in which there is immersed a refrigerant coil 2, connected to a refrigeration system 8. Bodies 4 and 9 may be housed in a single container 1 or individually housed in respective containers which are interconnected by a heat pipe. See Fig. 287.

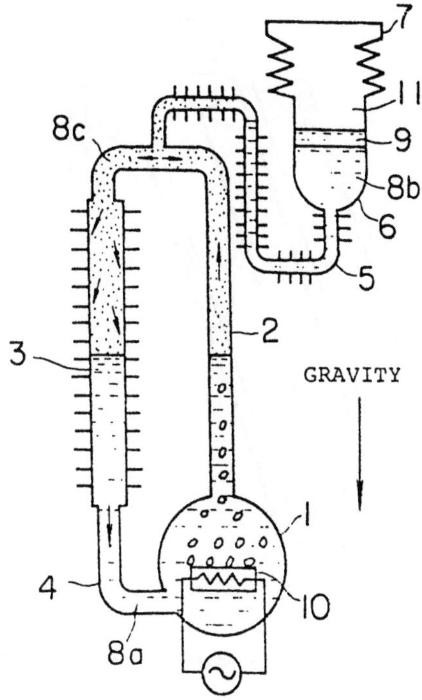

Fig. 286

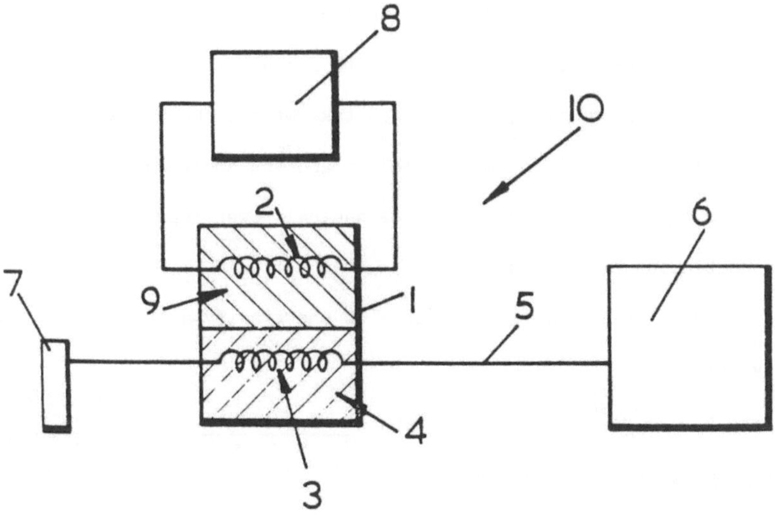

Fig. 287

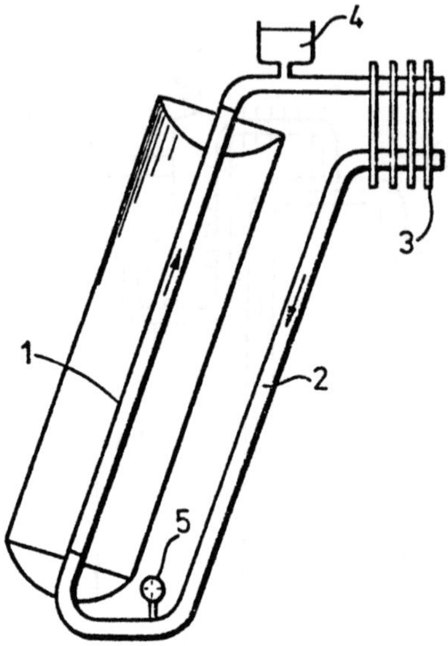

Fig. 288

The process and equipment developed by H. Pfeiffer, Patent (114), (MAN Maschinenfabrik Augsburg-Nürnberg AG), for cooling an absorbing agent in a closed cooling circulation system is based on the principle, that heat removal from the absorbing agent takes place under the effect of gravity force. The absorber is arranged vertically in the closed circulation system, with a heat exchanger in raised position. Consequently, circulation takes place due to the different specific weight of the cooling medium and the cooled medium, without having to apply any electric or mechanical power means to maintain a continuous circulation.

Fig. 288 illustrates the principle of the gravity-activated cooling system.

Patent (136.6) of G.A.A. Asselman, (N.V. Philips' Gloeilampenfabrieken), relates to a refrigerator (illustrated in Fig. 289) having a freezing compartment 2 and a refrigerating compartment 3, and comprising a primary refrigerating system which contains a refrigerant and a primary evaporator 7 arranged in the freezing compartment 2, and a secondary refrigerating system which also contains a refrigerant and which comprises a secondary evaporator 8, arranged in the refrigerating compartment and a secondary condenser 10 which is in heat-exchanging contact with the primary evaporator 7, to which secondary refrigerating system a reservoir 12 is connected which contains an adsorbent 13 for the refrigerant, the adsorbent being a porous solid substance which can be heated by means of an electric heating element 14 included in an electric control circuit, which control circuit comprises a temperature-sensitive element arranged in the refrigerating compartment, which temperature-sensitive element controls the heating element so as to maintain a specific temperature level in the refrigerating compartment. Suitably, the adsorbent is silica gel and the used refrigerant for the secondary refrigerating system is CF_2Cl_2. See Fig. 289.

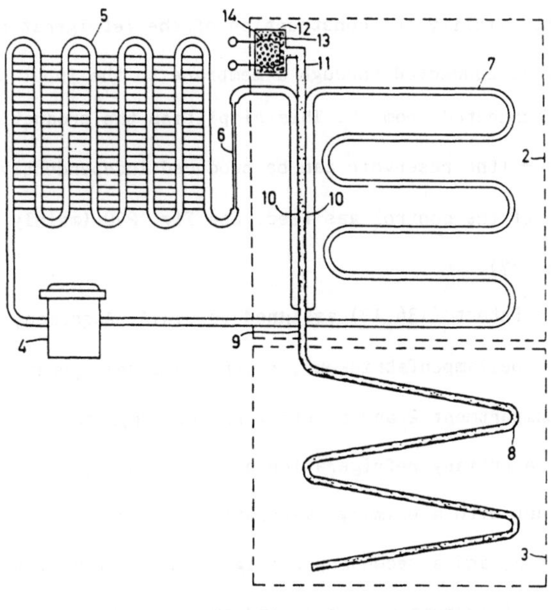

Fig. 289

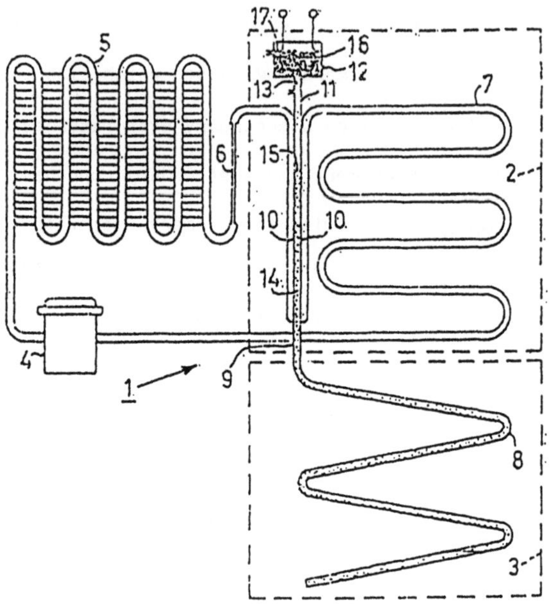

Fig. 290

A modified embodiment of (136.6) is the object of Patent (136.10) of G.A.A. Asselman, et al., (Philips' Gloeilampenfabrieken), wherein, in order to simplify the construction of the refrigerator, the reservoir 12 is connected through a membrane to the input channel and can be disconnected from it. This simplifies the production of the refrigerator (the reservoir can be produced independently) and the adjustment of the control gas feed. See Fig. 290 (mainly parts 11, 12, 13, 16, 17).

Yet another Patent (136.12) assigned to G.A.A. Asselman, et al., (Philips' Gloeilampenfabrieken), refers to a refrigerator 1 with a freezing compartment 2 and a refrigerating compartment 3 and provided with a primary refrigerating system 4, 5, 6, 7, containing a refrigerant with a primary evaporator 7 disposed in the freezing compartment 2, and a secondary refrigerating system 8, 9 which also contains a refrigerant with a secondary evaporator 8 disposed in the refrigerating compartment 3, and a secondary condenser 9 which is in heat-exchanging contact with the primary evaporator 7. The condensor 9 has a condensation wall 10 on whose surface the refrigerant condenses during operation, means (12, 13, 16, 17) being provided for varying the available condensation wall area 10, so as to control the temperature of the secondary evaporator 8. Preferably, the secondary condensor 9 is provided with a reservoir 12 containing a control gas 13, which during operation constitutes an interface 15 with the refrigerant vapour 14 at the location of the condensation wall 10, the interface 15 being movable along the condensation wall 10 with the aid of a reversible control gas getter 16, which can be heated and which is located in the reservoir 12 and enables the amount of free control gas 13 to be varied. The getter 16 can be heated by means of an electric heating element 17. See Fig. 291.

In Patent (161) of M.W. Shepherd, reveals a heat transfer device, suitable for constant temperature refrigeration. The device comprises an insulated container defining an internal space for containing low temperature matter, an external heat transfer structure, and a heat pipe containing a heat transfer fluid, the heat pipe being of closed form, with one end extending into the internal space and the other end in heat transfer relationship with the heat transfer structure, whereby ambient heat is carried from the heat transfer structure to the internal space by way of the heat pipe, the heat pipe comprising a casing, housing a tubular wick structure of porous material, so as to define an annular space bonding a central space. The heat transfer structure may be situated in an insulated chamber so that, with a cryogenic fluid in the container, the insulated chamber has its

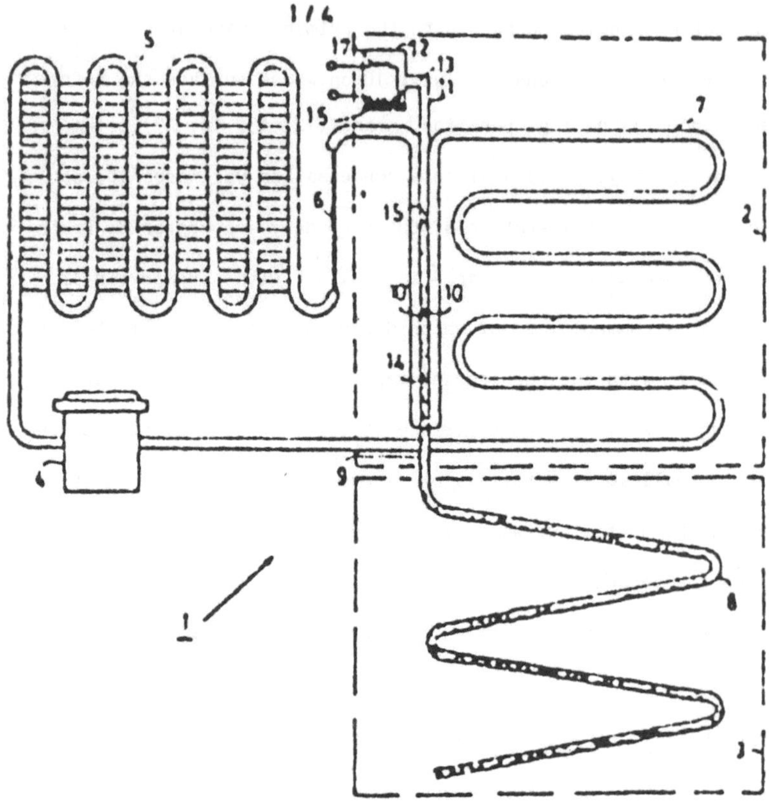

Fig. 291

temperature automatically and continuously reduced towards a constant value.

One form of heat transfer structure may comprise a cold plate, with or without extended heat transfer surfaces, to transfer heat from a heat insulated compartment. A further such structure may be a grid or jacket to remove heat from a solid article which has to be cooled to reduce its dimension for shrink or expansion fitting.

Typical heat pipe fluids are: water, glycol mixtures, nitrogen gas, gas mixtures, salt water, CO_2 gas, helium gas, freon (R.T.M.) and

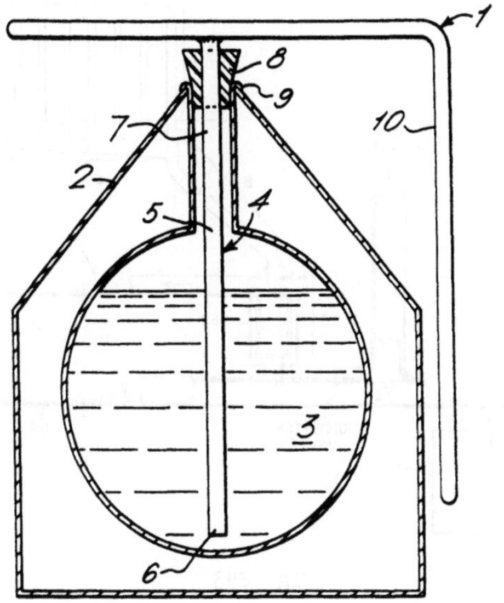

Fig. 292

typical cryogenic liquids: liquid nitrogen, oxygen, argon, helium, hydrogen.

Fig. 292 is a side view in medial section of the proposed cryogenic device.

In Patent (182.4) of Seigo Matsuda, (Thermo Electron Corp.), a heat pipe deicing system is diclosed for preventing accumulation of ice on portions of structures such as the deck, hand-rails, and ladders of a fishing vessel operating in very cold oceans. The heat pipe includes an evaporator in contact with a source of heat such as exhaust gases of an engine, one or more condensers positioned above the evaporator to cover surfaces to be deiced, and an insulated transition section for transporting a heat transfer fluid between the evaporator and condenser. The transition section and condenser have legs arranged to provide multiple passageways which assure continuous gravity-assisted return flow of condensed heat transfer to the evaporator

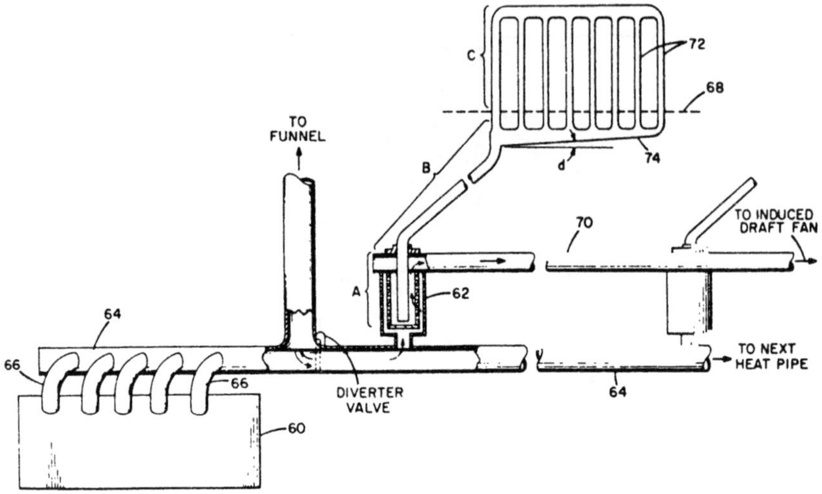

Fig. 293

and a high level of deicing capacity even during pitching and rolling of the structure on which the deicing system is installed.

See Fig. 293.

Patent (186.1) of Miyazaki Yoshiro, et al., (Tokyo Shibaura Denki K.K.), relates to a cooling apparatus for a closed housing cut off from the outside atmosphere, and, more specifically, to a cooling apparatus provided with a heat pipe heat exchanger and adapted to be mounted on a closed housing which contains electronic equipment to be protected from the outside atmosphere.

Fig. 294 shows a cooling apparatus 42 for a housing 40 which is provided with an opening at the upper part thereof and comprises a casing 44 located on the upper part of the housing 40 and including a bottom plate to close the opening in a chamber, a partition wall 46 provided in the casing 44 to divide the chamber into a first chamber section 48 and a second chamber section 50, heat exchanger 68 including at least one heat pipe 70 provided in the casing 44 through the partition wall 46 with coolant therein for exchanging

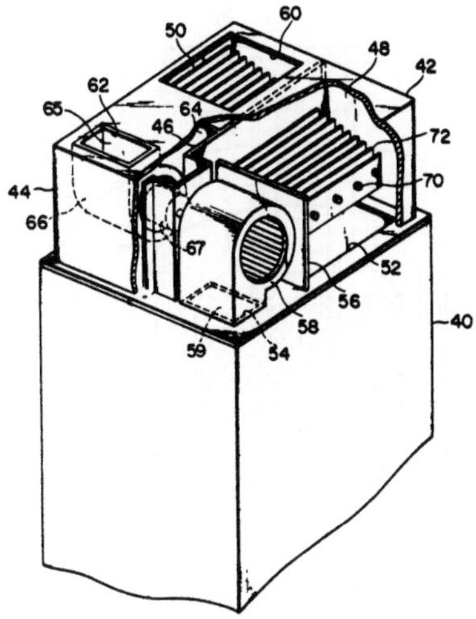

Fig. 294

heat in the first chamber section 48 with heat in the second chamber section 50. The heat pipe 70 has one end portion in the first chamber section 48 and the other end portion in the second chamber section 50, and inclines with the one end portion located lower than the other end portion.

The device is mainly suitable for protecting computers operating in dusty factory workshops.

A magnetic refrigerator, claimed by Nakagome Hideki, (Tokyo Shibaura Denki K.K.), in Patent (186.3), comprises a working material 23, radiating heat when a magnetic field is applied thereto and absorbing heat when the magnetic field is removed therefrom, and magnetic coil 23 which is movable for selectively applying or removing the magnetic field from the working material. Directional heat pipes are disposed between the working material and an object 28 of cooling, whereby heat from the object of cooling is transmitted to the working material when the working material absorbs the heat, and whereby heat from the working material is prevented from being

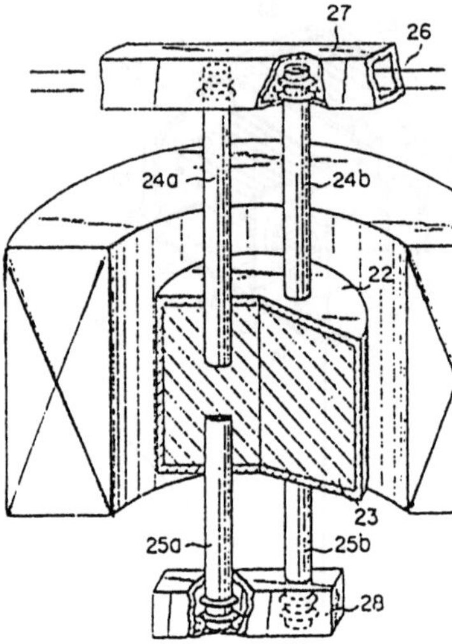

Fig. 295

transmitted to the object of cooling when the working material radiates the heat. See Fig. 295.

In Patent (193.1) of W.W. Schertz, (The USA represented by The United States Department of Energy), a device is provided for storing coolness in a material and for later removal of this coolness for desired applications. The device includes a tank, containing fluid and at least one reflux condenser heat pipe which penetrates the tank. The lower end of the heat pipe is immersed in the fluid and preferably extends along the bottom of the tank. The upper end of the pipe extends above the tank and is preferably exposed to the ambient environment. When the external temperature of the ambient environment is lower than the internal temperature of the fluid stored in the tank, the heat is transmitted by action of the pipe from the fluid to the ambient. In the reverse situation, when the temperature of the ambient is higher than that of the fluid, no heat is reverse-conducted and therefore the condenser pipe acts as a thermal diode. This action of continual

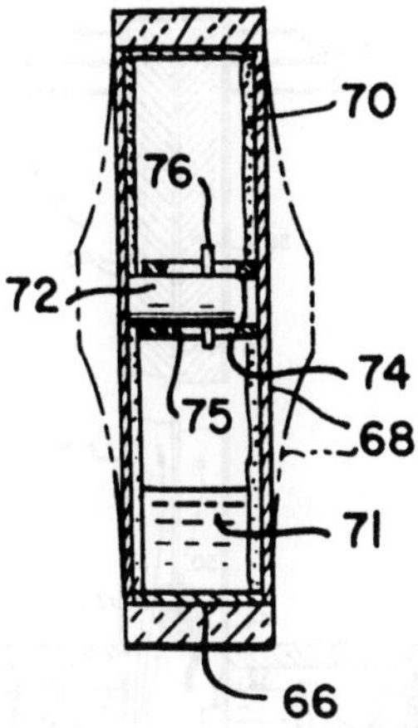

Fig. 296

removal of heat from the fluid will result in the relative coolness being stored in the fluid. Maximum cooling capability may be stored if the fluid, such as water, is actually frozen by this action. At some desired time this cooling of the fluid can be utilized as a source of coolness such as for summer air conditioning.

Fig. 296 is a cross section of the proposed device.

A cold storage device is also the object of Patent (193.6) of A.J. Gorski, assigned to The Department of Energy of the USA, which is able to freeze and, periodically, release ice for storage of coolness in a tank. This may be accomplished in a reliable manner without requiring the ambient temperature to rise above 0°C, and is carried out by a passive means without any need for valves, pumps, or other mechanical devices. The device includes a main heat pipe within which is: an internal heat pipe to promote reliable release of ice and an on-off timer structural system which periodically attenuates the normal working cycle of the main heat pipe to allow the internal heat pipe to effect release of ice formed on the main heat pipe exterior.

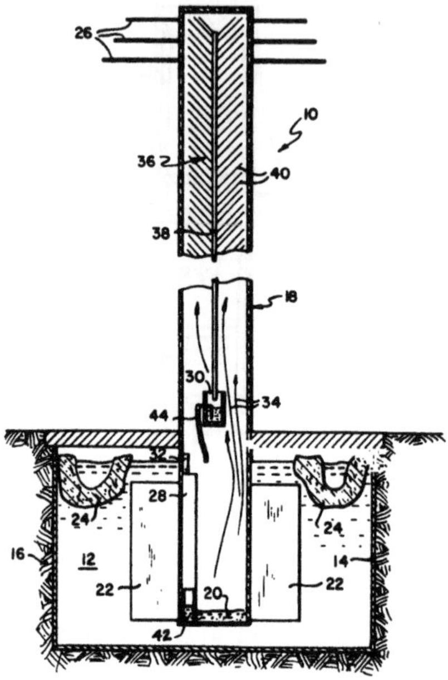

Fig. 297

Fig. 297 illustrates an ice formation system with a heat pipe design able to effect completely passive ice-freezing and release.

5.8.2 Heat pumps equipped with heat pipes.

J. Nadot presents in Patent (151), (SARL Carriservice), a heat exchange system, involving two independent fluid currents, using an equipment, combining at least one heat conduit with at least one heat pump, wherein the factor of recuperation from the heat conduit is varied in order to adjust the temperature of fresh air blown in, while maintaining the temperature of the air current cooled in the conduit and directed to the evaporator of the heat pump at a temperature, which is sufficiently high for preventing the freezing thereof. Fig. 298 represents the principle of the

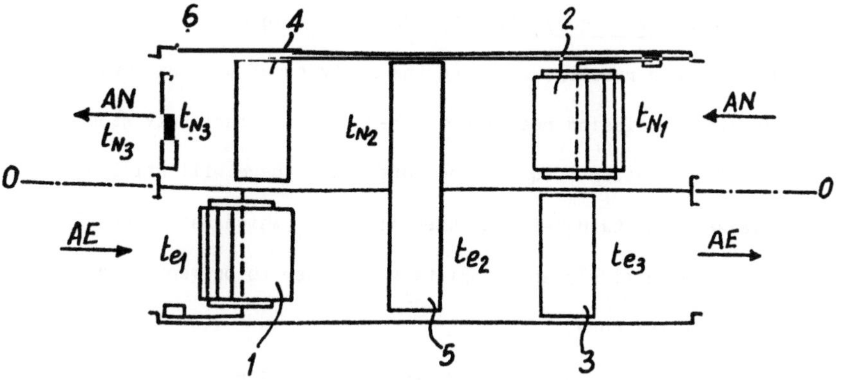

Fig. 298

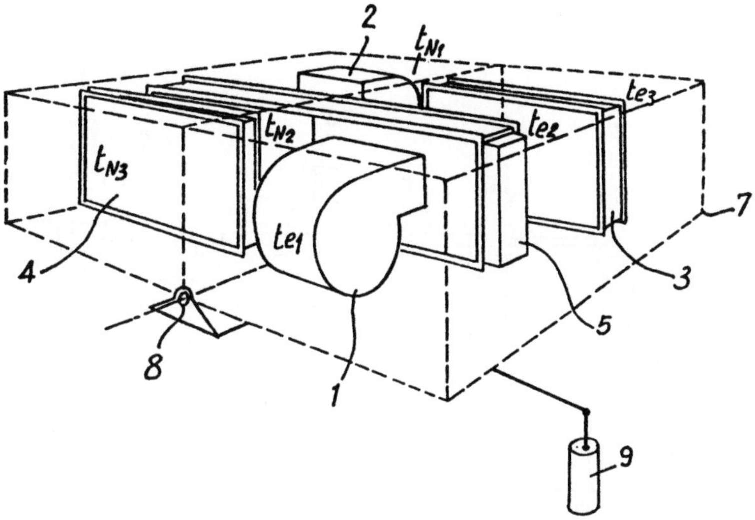

Fig. 299

heat pump - heat exchange device, while Fig. 299 is a perspective view of an embodiment of the proposed device.

5.9 Heat exchanger, with heat pipes, applied in nuclear fission or fusion plants

A device for the evacuation of residual power from fast nuclear reactors is described in Patent (34.3) of A. Bret, (Commissariat à l'Energie Atomique). It is evident, that in case of an accident in a nuclear reactor, stopping the activity of primary pumps and wherein the fission reaction in the core is immediately stopped by the activation of quick shut-off rods, there will still be released a certain residual thermal power, which has to be eliminated effectively and completely in order to prevent a local fusion of the core.

The device developed in (34.3) for this purpose comprises an evaporator 42 (see Fig. 300), consisting of a bundle of tubes in

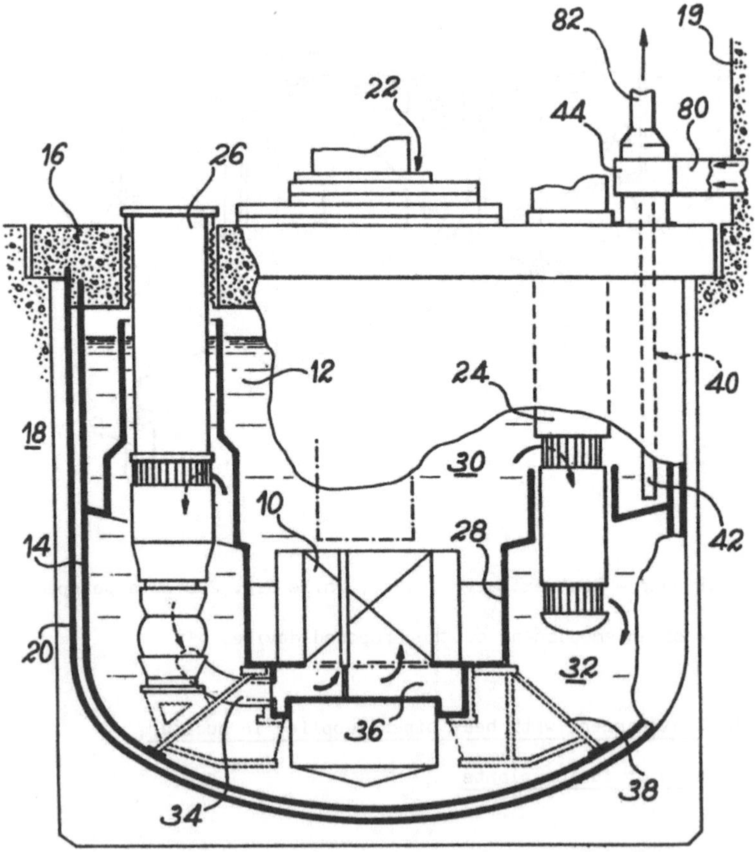

Fig. 300

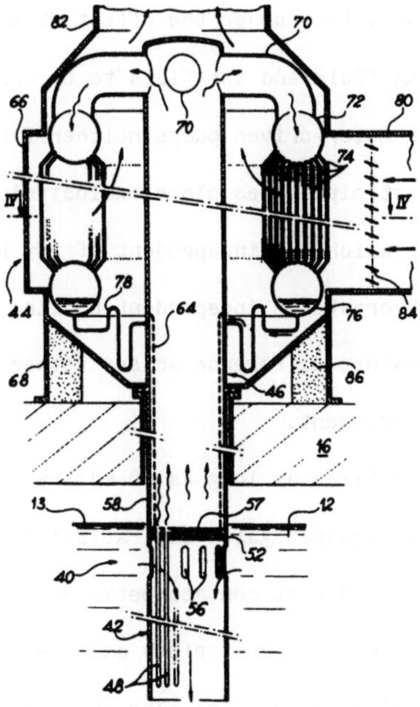

Fig. 301

"glove finger" arrangement, immersed in liquid metal, contained in the reactor vessel; an adiabatic collector 46, composed of a conduit 58, traversing a slab 16, which closes the reactor vessel and a condenser 44, wherein the heat transfer medium, (for example mercury), is condensed by heat exchange with atmospheric air, introduced through a tunnel 82. See Fig. 301.

P. Paikert, (GEA GmbH.), provides in Patent (74.3) a heat exchange system for the storage place of used radioactive fuel rods, which cannot be made inactive, under no circumstances, by any kind of power failure or by any authorized or unauthorized human action. The proposed system, corresponding to the rigorous security requirements, is based on heat pipes, for removing heat to the environment. In each heat pipe an independent cooling process takes place, based on the free convection of fluid and/or gaseous heat carrier agents, completely under the effect of the density difference between the cooling fluid and the fluid to be cooled. Thus the system requires no motor-driven pumps neither fans. The cooling process in each heat pipe takes place exclusively under the effect of natural forces, which are independent of any kind of human activity and which create an independent cooling process, if a perceptible temperature difference arises between a fluid to be cooled and the environment.

The heat pipes (sometimes as long as 10 m) are inserted in tubes, which are insulated against gases and radiation and which, consequently, impede any direct contact between the heat pipes and the storage space. Both the heat pipes and the enveloping tubes should be made of high-grade steel. The tubes are introduced into a cooling medium (air or fluid) filling the storage space, taking over the heat thereof and transferring it to the heat pipe therein, whereafter evaporation takes place in the capillaries of the heat pipes, the steam transporting the heat into the airy part of the heat pipe and depositing it through condensation, while the condensate returns through gravity and capillary force into the evaporation part of the heat pipe.

If a sufficiently large number of heat pipes is arranged in the storage space walls, any inadvertent, violent destruction of some pipes will not affect the effectiveness and safety of the proposed system.

Fig. 302 is a schematic vertical cross section of a storage space for used radioactive fuel elements.

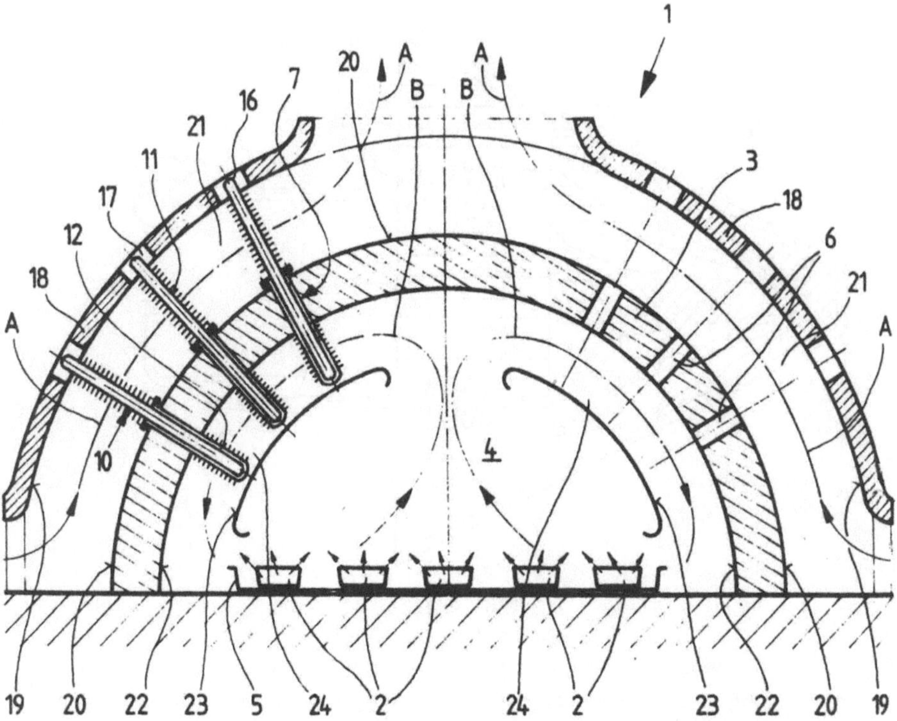

Fig. 302

In Patent (86) of V.E. Hampel, a fast nuclear reactor is described, which comprises a conical reactor core surrounding an embedded array of heat pipes of, per se, novel structure, carrying either moderator or nuclear fuel material as part of their working fluid. This reactor system is self-regulating, because an exccessive increase in reactivity drives the fuel or moderator working fluid out of the conical core region, thereby reducing reactivity. The heat pipes are protected against burnout by a novel heat pipe envelope shape and internal wicking structure designed to increase the working fluid circulation speed with increasing heat transfer loads.

See Fig. 303.

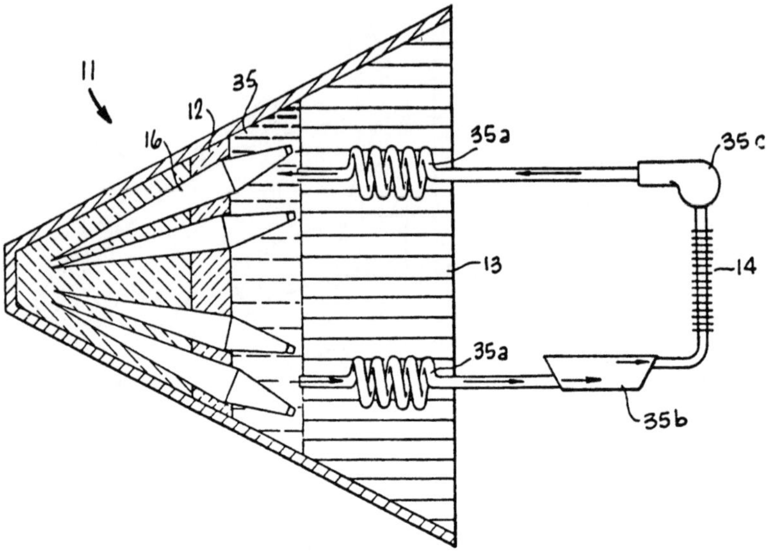

Fig. 303

Patent (186.2) of Yamamoto Setsuo, (Hitachi Ltd.), concerns a reactor vessel, comprising heat-propagating means (see Fig. 304), such as a heat pipe 10 which is provided in contact with or apart from a peripheral wall 11 of a reactor vessel body or inside of the peripheral wall 11, whose upper portion lies above a free liquid level 13 of a coolant and whose lower portion lies below the free liquid level 13, thereby enabling temperature prevailing in that portion of the reactor vessel peripheral wall 11 which lies near the free liquid level of a coolant heated by a reactor core 12 to have a gentle gradient.

Fig. 304 is a schematic sectional view of a part of a fast breeder reactor vessel according to an embodiment of the present invention, comprising a plurality of rod-like heat pipes arranged in contact with the inner surface of the peripheral wall of the reactor vessel.

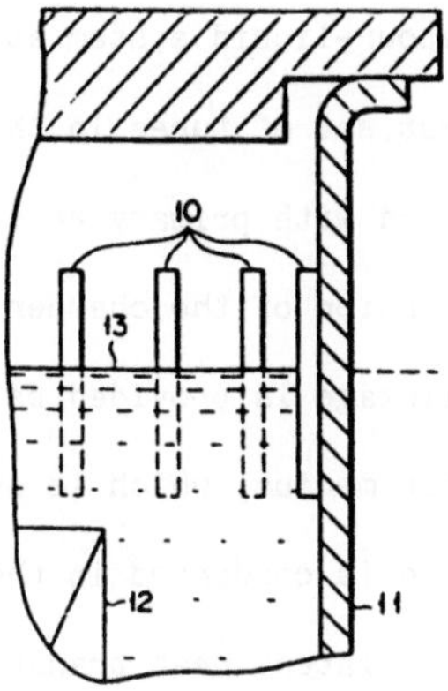

Fig. 304

Patent (193.8) of J.C. McGuire, (US Department of Energy), relates to a sealed refluxing heat transfer device, adapted to replace secondary or intermediate heat exchanger in a nuclear reactor used for generation of steam for power purposes. The proposed intermediate heat exchanger provides the required physical isolation between the primary reactor coolant loop and a secondary liquid loop in which steam is generated. It utilises the available heat transfer rates common to a heat pipe, but requires no wicking materials. It further serves to physically isolate non-condensible gases, which can be readily recovered.

These results are achieved by use of a sealed vapour chamber where the primary loop and the steam line are in proximity to one another, but not in contact. Heat transfer occurs within the vapour chamber by use of the two phase liquid-vapour-liquid process common to heat pipes.

More particularly the proposed heat exchanger is contained within a sealed vapour chamber that includes a bottom interior portion and an adjacent upper interior portion in vertical communication with one another. The chamber is exhausted of all non-condensible gases at ambient temperature. A heat transfer medium within the chamber maintains a two phase liquid-vapour-liquid system at the design heat transfer temperature. A first set of tubes in the bottom portion of the vapour chamber is supplied with primary reactor coolant. A second set of tubes in the upper portion of the chamber is supplied with water or steam. A thermal linkage is provided between the two sets of tubes by the heat transfer medium, which is evaporated in the vicinity of the first set and is condensed in the vicinity of the second set. This results in a latent heat transport system, condensate return being accomplished by gravity.

It is claimed that the proposed heat exchanger effectively isolates steam generating equipment from possible contamination with radio-active materials. It also reduces the danger of catastrophy which would accompany any leak between the primary loop and the steam generating piping were they to be directly coupled to one another. This is of special significance in the design of liquid metal cooled reactors.

Furthermore, this heat exchanger requires no secondary liquid pumps and contains no moving mechanical elements. As a rule the working pressure within the sealed vapour chamber will be less than atmospheric pressure. Rupture of the chamber will therefore not result in an explosive condition, since the reduced pressure within it will contain its elements and materials within its normal confines.

Fig. 305 is a schematic diagram of the heat exchanger according to (193.8),

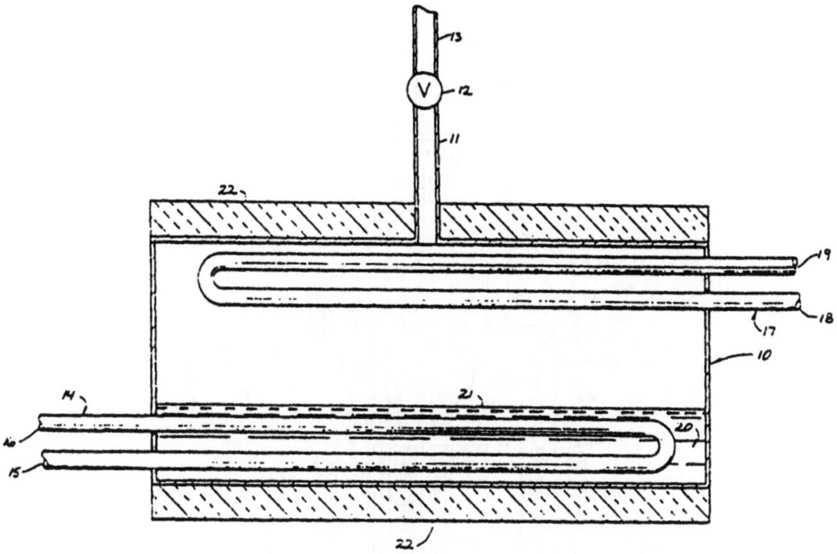

Fig. 305

wherein: 10 - sealed vapour chamber; 20 - heat transfer medium; 14 - tubes from which heat is transferred to tubes 17; condensable gases being removed from 10 through a vertical extension 11.

The cooling system for a power reactor should ensure to the utmost extent possible adequate heat removal from the reactor core following an accident or the occurrence of abnormal conditions in the reactor.

Thus, it is a primary object of Patent (194.1) of T.L. Dunckel, (The USA as represented by the United States Energy Research and Development Administration), to provide an emergency heat removal system for use on nuclear reactors.

The proposed heat removal system, serving as a supplement to an emergency core cooling system during a loss of coolant accident comprises a plurality of heat pipes having one end in heat transfer relationship with either the reactor pressure vessel, the core support grid structure or other in-core components and the opposite end located

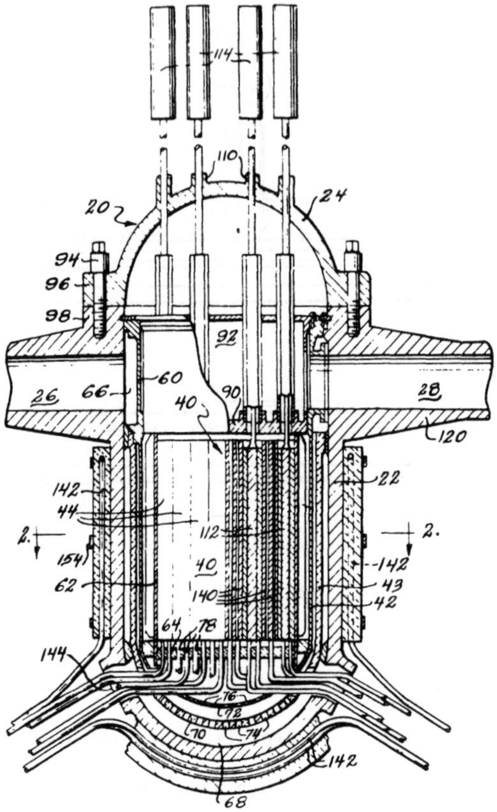

Fig. 306

in heat transfer relationship with a heat exchanger having heat transfer liquid therein. The heat exchanger is located external to the pressure vessel whereby excessive core heat is transferred from the above reactor components and dissipated within the heat exchanger fluid.

Fig. 306 is a vertical cross sectional view of a nuclear reactor, incorporating the features of (194.1). 142 are the heat pipes.

Patent (<u>207</u>) of R.W. Werner et al., (<u>US Department of Energy</u>), reveals a heat pipe configuration for use in a magnetic field environment of a fusion reactor.

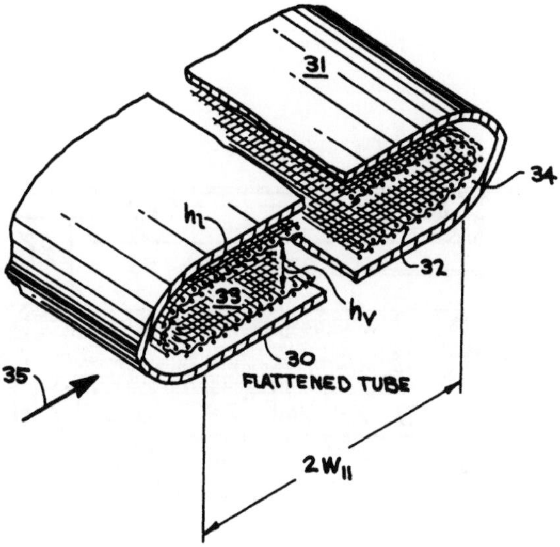

Fig. 307

Heat pipes for operation in a magnetic field, when liquid metal working fluids are used, are optimised by flattening of the heat pipes having an unobstructed annulus which significantly reduces the adverse side region effect of the prior known cylindrically configured heat pipes. The flattened heat pipes operating in a magnetic field can remove 2 - 3 times the heat as a cylindrical heat pipe of the same cross sectional area, it is claimed. See Fig. 307.

5.10 Heat exchangers used in space craft and missiles.

Patent (28) of G. Cambon, (Centre Nationale d'Etudes Spatiales), refers to a heat shaft, comprising a center punch 62, (see Fig. 308), more or less closing an aperture 60, entered by a cooling fluid, circulating through a coil system. The punch is in connection with a blower 56, filled with a fluid from the reservoir 54.

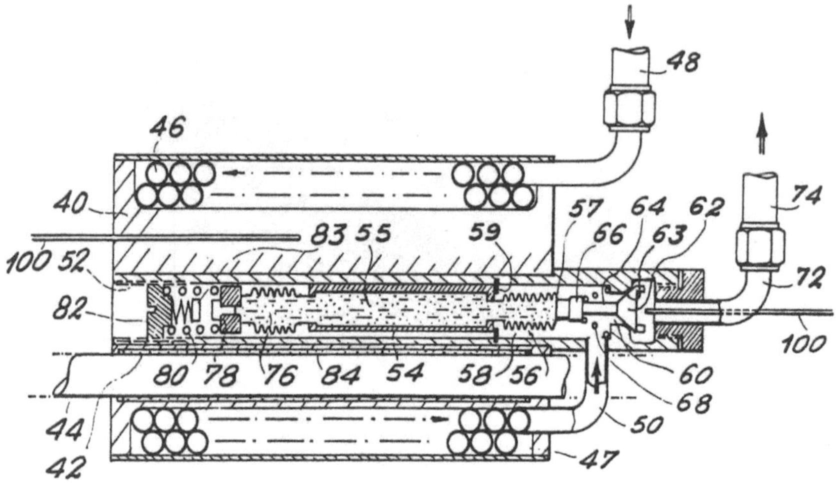

Fig. 308

An increase of temperature in the shaft increases the delivery of cooling fluid and vice versa. Heat transfer takes place at low temperature and the thermal contact is realised by a liquid metal. The device can be utilised in processing material (information) in space craft or rockets, or on the ground, in the range of natural or artificial gravity forces.

Patent (116.2) of S.J. Hightower, (Martin Marietta Corp.), concerns a pneumatically controlled wide heat load space radiator assembly for a space craft, an earth satellite, etc., comprising a heat radiator, mounted on the space craft. At least one tube array is fitted to the heat radiator in thermal conductive relationship and comprises concentric, radially spaced inner and outer heat conductive tubes and means defining a sealed annular gap between the tubes over a given longitudinal extent thereof, further means for selectively supplying gas pressure to the sealed annular gap between the concentric tubes for venting this gap to space in order to vary the thermal impedance

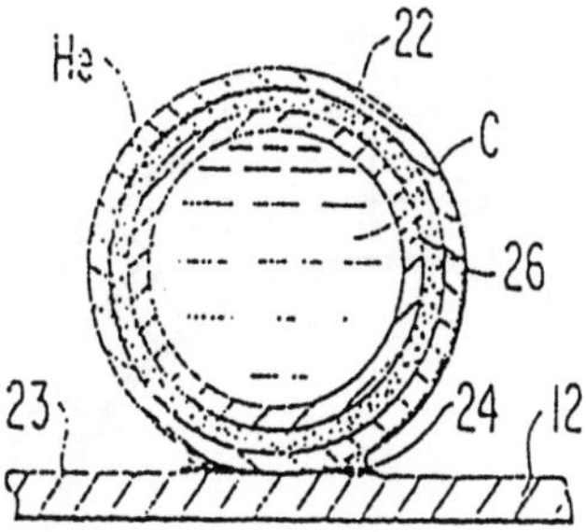

Fig. 309

between the tubes and the heat rejection rate of the spacecraft radiator. Fig. 309 illustrates a spacecraft, employing the proposed heat radiator.

J.R. Welling, (Motorola Inc.), reveals in Patent (124) a passive, self-contained cooling system for missile electronics which minimizes weight, eliminates external coolant supply requirements, and maximizes the allowable period of operation.

More particularly the invention provides a plurality of heat pipe/thermal diode devices which transfer heat efficiently from the electronics of a missile to the skin and which effectively prohibit heat transfer in the opposite direction, and a heat storage medium of the heat of fusion type in thermal communication with the electronics and the heat pipe/thermal diode devices.

Also disclosed is a method of cooling electronic components contained in a missile, comprising the steps of establishing an efficient, low thermal mass heat path from the electronics to the skin of the missile, severing the heat path when the skin temperature exceeds a predetermined level, providing a heat storage medium in thermal contact with the electronics to remove heat energy therefrom, re-establishing the thermal path when the skin temperature falls below a predetermined

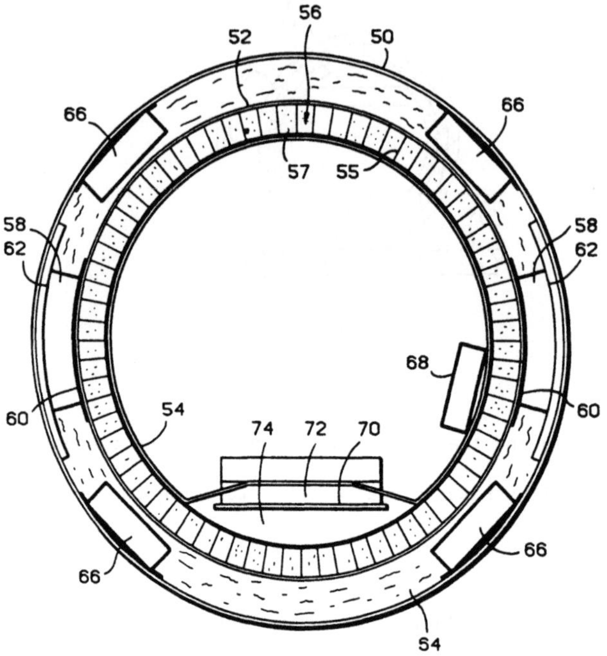

Fig. 310

level, and removing stored heat energy from the heat storage medium to recycle the system.

Fig. 310 shows a block diagram of the proposed cooling system.

A passive cooling device especially suitable for application where unattended high performance is required for long periods, such as with space satillite sensors, optics and electronics systems, is also the object of Patent (148) of S.I. Eilenberg, (Rockwell International Corp.). The device contains multiple stages of heat radiators, insulated from each other and from the supporting structure, with heat pipes thermally attached to some or all of the stages of the radiators to transport heat from the heat producing sources. The multi-stage radiator system utilizes intermediate radiator stages to intercept heat loads conducted

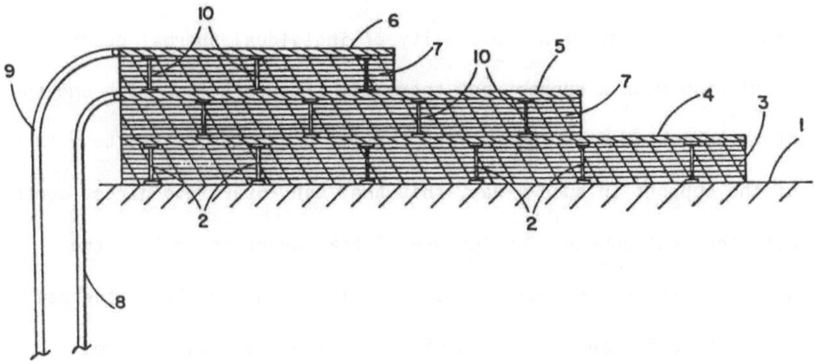

Fig. 311

through the insulation and supports in order to reduce such conductive loads on the outermost stage and to permit that stage to reject substantial heat loads at extremely low temperatures. These intermediate stages can also provide efficient thermal rejection at the different temperature levels of various elements to be cooled. Heat pipes of various characteristics such as the variable conductance and diode types can be used individually or in thermal combination to suit the needs and constraints of the space systems applications.

Fig. 311 is a simplified cross-sectional view of a preferred embodiment of the invention.

A patent (169) assigned to the Société Nationale Industrielle Aerospatiale provides a heat radiator, based on the principle of the evaporation-condensation cycles of a fluid, which is an essential element in spacecrafts for the temperature control of the electric, electromagnetic, electronic instruments applied in the spacecraft, further for the shielding thereof, in order to transfer heat from the sunshine-exposed parts of the shielding to its shadowed parts.

The proposed radiator, destined to evacuate heat from at least one heat source, comprises a plurality of individual thermal ducts in connection with a support and thermally interconnected with one another by a heat distributor. The thermal ducts are fitted to a face of the support in such a way, that they can enter into direct contact with the heat source. The feature of the invention is that the heat distributor is constructed of at least one auxiliary thermal duct. It is claimed that such radiator displays higher thermal activity, due to the direct contact between the thermal ducts and the heat source, by eliminating all intermediary thermal elements. Fig. 312 illustrates a plan view of the described radiator, (4, 5 - heat sources, for example electronic equipments); 2 - a heat conductive pad; 6, 7 - first and second system of main thermal ducts, which are fitted to the supports 1, 2, 3 by bolts, gluing, etc.

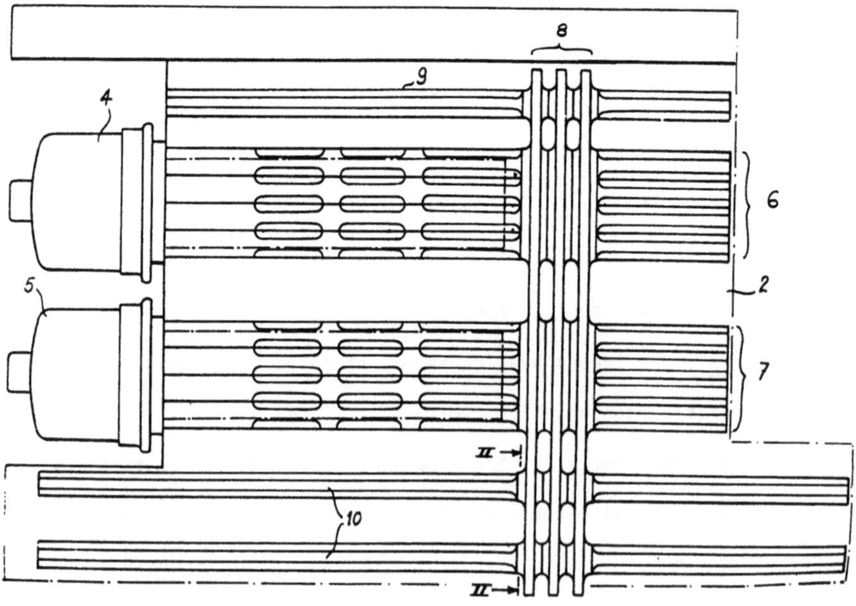

Fig. 312

Patent (191.1) of H.I. Kroebig, (The USA as represented by the Secretary of the Air Force), is related to a device which transfers heat from a component, such as Vuillenier refrigerator crankcase, to the skin of a missile which limits the reverse flow of heat when excessive skin temperatures are encountered during flight.
More particularly, a heat pipe, having one wall formed by the component to be cooled and the other wall formed by a cover plate, which is a portion of the missile skin, wherein the component wall normally forms the condensing section, has a conventional wire mesh wick, connected between the condensing and the evaporating section. A support plate is attached to the wick, adjacent the cover plate. A bellows is connected between the cover plate and support plate to move the cover plate and wick against a spring and away from the cover plate, if the skin temperature becomes excessive.
Fig. 313 is a view (partially in section) of the proposed heat pipe system.

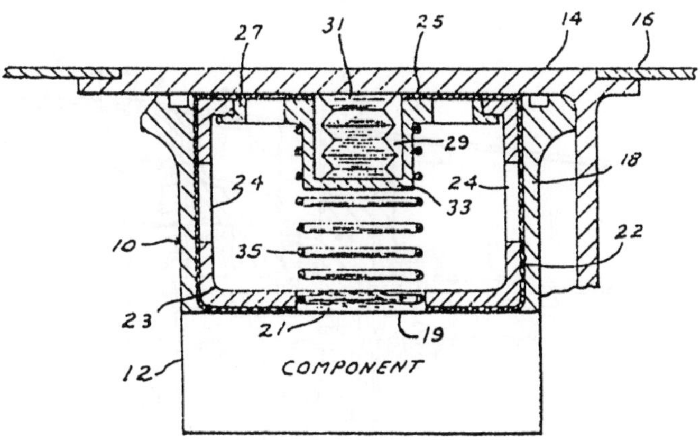

Fig. 313

According to another Patent (191.3) of H.I. Kroebig, (The USA represented by the Secretary of the Air Force), a fusible heat sink for a cryogenic refrigerator is used to provide cooling for a detector in the guidance system of a missile. The cryogenic refrigerator has a cold cylinder in contact with the detector and a hot cylinder. The hot cylinder and cold cylinder are connected to a crankcase housing.

A heat pipe is connected between the crankcase and the missile skin for providing primary cooling for the crankcase housing. The fusible heat sink is connected to the crankcase with the crankcase forming part of the wall of the heat sink housing. A fusible material is located within the housing. The inside surface of the heat sink housing is coated with nickel and silver to increase the heat transfer between the crankcase and the heat sink.

Fig. 314 is a partially schematic sectional view of a missile guidance assembly using the device of the invention.

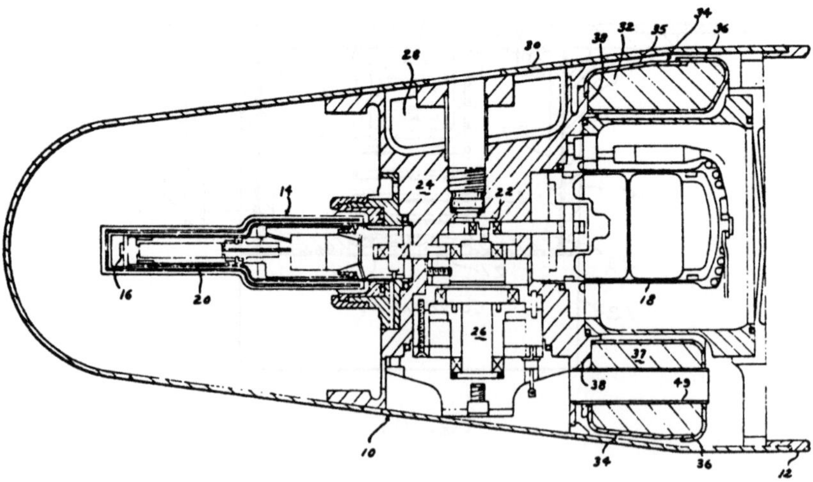

Fig. 314

5.11 Heat transfer, heat exchangers, heat pipes used for various (special) purposes

Heat pipe, heat exchangers in electric systems.

Heat pipes, heat exchangers in electronic systems, (mainly for cooling).

Heating systems for hot gas engines.

Heat transfer, heat pipes in internal combustion engines (vehicles).

Heat exchangers, heat pipes in turbines, compressors.

Heat recuperation systems.

Molding, extrusion.

Soldering elements.

Composite glass structure.

Catalytic cartridge SO_3 decomposer (hydrogen production).

Cooling of wheels.

Heat pipes used in combination with geothermal systems.

Support structures, using heat pipes, in arctic or subarctic areas.

Using heat pipe type heat exchangers in lakes.

Fabrication process of heat pipe panels.

Thermal conditioning of garments.

Heat pipes, heat exchangers in electric systems.

The object of Patent (21) of D.V. Shepherd, (British Aircraft Corp.), is to provide cooling means for a discharge lamp.
According to the invention a discharge lamp, of the kind in which discharge electrodes extending through the wall of the lamp envelope

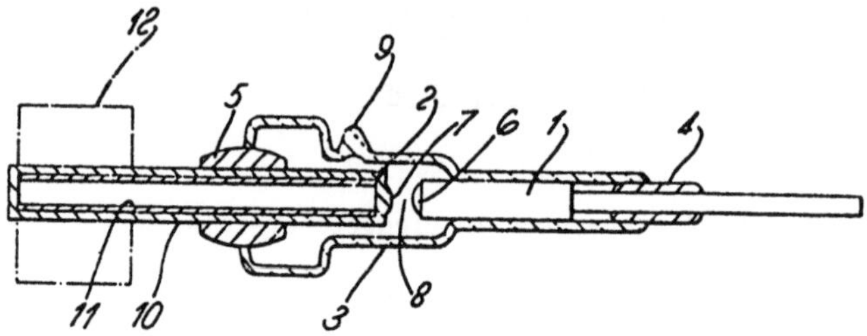

Fig. 315

are arranged to form a discharge gap within the envelope, includes a heat pipe having its condenser end outside the envelope and its vaporiser end either within the sealed space bounded by the envelope or within a re-entrant part of the envelope for transmitting heat from the discharge gap to a position outside the envelope.

Fig. 315 shows a xenon arc lamp, comprising an anode 1 and a cathode 2 contained within a glas envelope 3. The electrodes 1 and 2 extend into the envelope through glass seals 4 and 5 and positioned with their end faces 6 and 7 a small distance apart, thus forming an arc gap 8. Air is evacuated from the lamp envelope, and replaced by xenon gas at low pressure, via a hole in the envelope 3 which is subsequently sealed leaving an evacuation pipe 9. In a modified form of the lamp the envelope 3 includes a short length of capillary tube interposed between the end faces of the electrodes 1 and 2, and the xenon arc is struck between the electrodes in the bore of the tube. The cathode 2 comprises a tube 10 closed at both ends and containing a porous wick 11 in contact with the inner surface of the tube 10. The tube is evacuated and filled with sufficient working fluid to wet the porous wick 11 and fill the remaining space with vapour. The cathode 2 thus constitutes the heat pipe containing a liquid and its vapour at its saturation temperature and pressure.

The aim of Patent (76.5) of L.W. Pierce, (General Electric Co.), is to provide means for sensing the difference in temperature that exist between the heat exchanger and the transformer tank to determine the presence of admitted air within the heat exchanger assembly as well as the presence of excess moisture.

For this purpose a vaporization cooled transformer of the type comprising a transformer tank in combination with a heat exchanger, employs temperature sensing means in both the tank and the heat exchanger for sensing a predetermined temperature differential to indicate the presence of air in the transformer. When the temperature differential exceeds a predetermined range, an alarm and a transformer disconnect relay become actuated. The temperature differential sensor can also be calibrated to provide excess moisture indication above

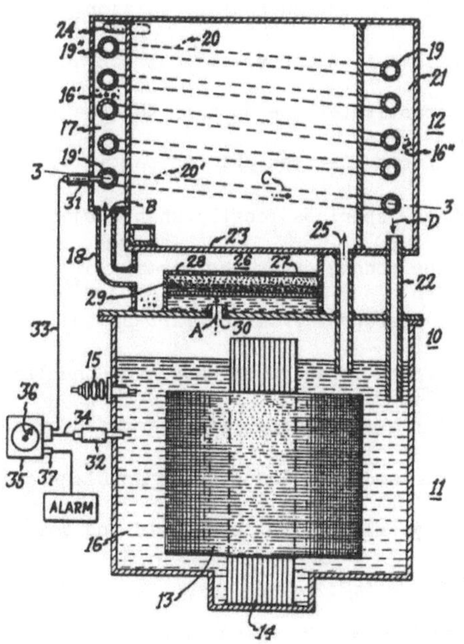

Fig. 316

a threshold moisture content caused by the saturation of the molecular sieve water scavenger within the transformer assembly.

See Fig. 316.

Patent (146) of J.D. Milburn et al., (Reyrolle Parsons Ltd.), aims at reducing the overall volume of space taken up by switch gear constructions, consistent with insulation limitations and minimum flashover spacing of adjacent phase-conductors. The electrical switch gear proposed in Patent (146) incorporates an electric-current-carrying conductor provided with a heat-transfer device which comprises either a heat pipe or a Perkins tube and is constructed and arranged to withdraw heat generated locally in the conductor by the passage of electric current therein and to transfer the heat to another and cooler locality. The heat transferred from a locally-heated part of the conductor may be released to a cooler region of the conductor, and/or it may be dissipated by means such as a radiator.

In one form of the invention the electrical conductor comprises a tube which itself constitutes the duct of the heat-transfer device, and whose interior forms a sealed chamber partially filled with a volatile liquid which vaporises locally when subject to localised heat generated in the conductor by the passage of current, whereby localised heat is transferred from one part of the conductor to another as a result of the local vaporisation of the liquid. In another form of the invention the conductor is surrounded by a container defining between itself and the conductor an annular-section chamber or chambers partially filled with the liquid and comprising the duct of the heat-transfer device.

Fig. 317, wherein: 10 - a current carrying conductor; 11 - tubular sealed vessel, with end walls 12; 13 - capillary surface layer (wire

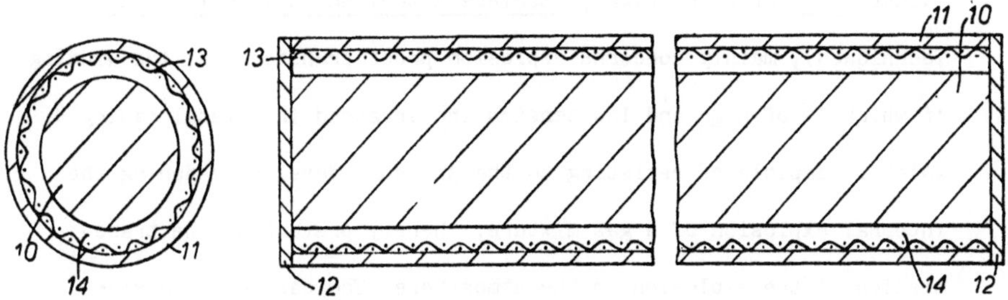

Fig. 317

mesh, capillary apertures); 14 - space. The working fluid filling the space 14 between 10 and 11 can be water, methanol or ammonia.

Patent (<u>164.3</u>) of M. Gunter, (<u>Siemens AG</u>), provides a heat exchanger for cooling electric machines, according to Fig. 318, comprising a shaft 1, carrying a short-circuited rotor 4, composed of rotor plate packets 3, forming between the shaft 1 and themselves a cooling channel 2. Between the packets 13 there are cooling channels 5, in connection with the cooling channel 2 and an air gap 6, furthermore with radial cooling channels 7 in operative connection therewith. A paraxial casing 9 encloses the electrically active parts of the machine.

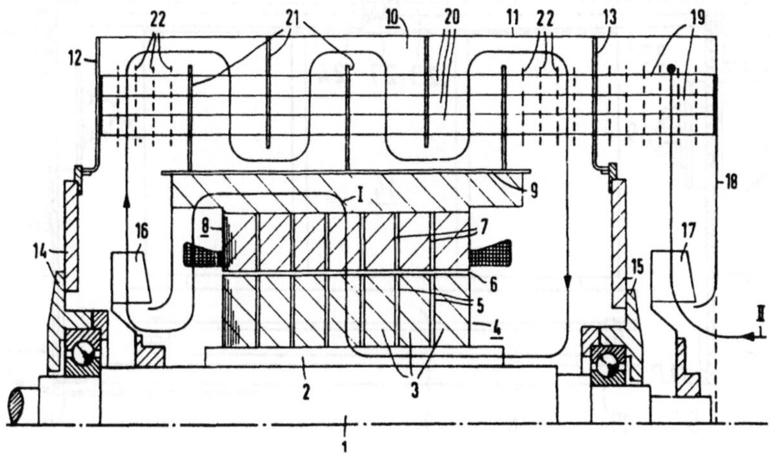

Fig. 318

Patent (168) of J. Montagne, (Société Alsacienne d'Installations Techniques), mainly concerns explosion-proof transformers, the windings of which 14 of high and low tension are arranged in a metal casing 4, which is capable of resisting to the pressure developing during the internal explosion of a gas mixture, thereby preventing the transmission of the explosion to the atmosphere. The casing has a side wall 5 and two covers 8, 9, supporting casings of high and low tension 20, 22. Heat pipes (23, 24, 25, 26) are arranged alongside one another, about parallel to the axis of the side wall 15 in such a way, that their receiving side is in the free space separating the windings from this wall, between a central bearing 27, receiving the ends of two heat pipes and flanges 41, fitted to the covers 8, 9, while the dissipating sides of the heat pipes are arranged beyond the flanges, outside of the casing, the pipes being provided with receiver members 33 and dissipating members 46 at the corresponding sides.
See Fig. 319.

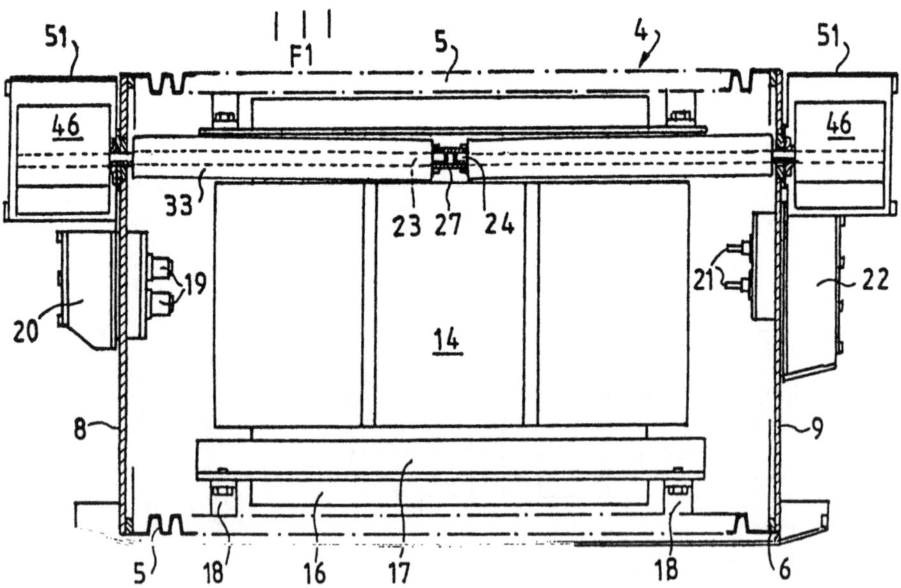

Fig. 319

Heat pipes, heat exchangers used in electronic systems (mainly for cooling).

In Patent (22) of M.E. Clarke, (The British Oxygen Co.), there is provided a superconducting device including a winding for current flowing through the device, a chamber adapted to receive helium surrounding the winding and a heat pipe in thermal contact with a part of the device adjacent to the winding and with a cooling system, wherein the working fluid in the heat pipe has a triple point above 4.5° K and below the working temperature of the cooling system. In order to achieve superconducting conditions in the winding it is necessary to add liquid or cold gaseous helium to the chamber in which the winding is located. When such cold helium is present the working fluid in the heat pipe freezes and very little heat is transmitted along the heat pipe. When the device is not being operated under superconducting conditions, however, the helium is removed and the winding is maintained at the temperature of the cooling system by transmission of heat along the heat pipe. Accordingly, when it is required to bring the device back into use, it is merely necessary to cool the winding down from the refrigerator temperature.
See Fig. 320, showing a superconducting magnet, with winding 2, core 1, radiation shields 4, 5, cryogenic vessel 6; inner tube 7, outer tube 8, (for helium); refrigerator 9, heat pipe 10.

Patent (60) of F.B. McCall, (Ferranti Ltd.), aims at providing a heat transfer device, having improved coupling means for ensuring reliable connection with a detachable heat sink.
The heat transfer device includes a tube arranged to be in thermal contact with a heat source, a fluid contained in and partly filling the tube, an expansible bellows attached to one end of the tube

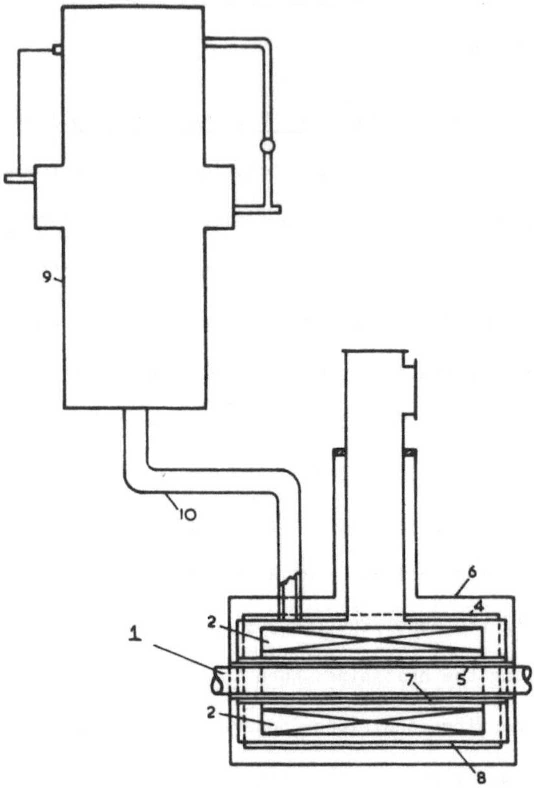

Fig. 320

in such a way that the interior of the bellows is in communication with the space inside the tube, and of such flexibility that the application of heat to the tube causes the bellows to expand longitudinally due to the increase of the vapour pressure within the bellows, and a heat sink, located adjacent to the bellows in such a way that the expansion of the bellows ensures thermal contact between the bellows and the heat sink whereby heat applied to the tube by the heat source may be transferred to and dissipated by the heat sink.

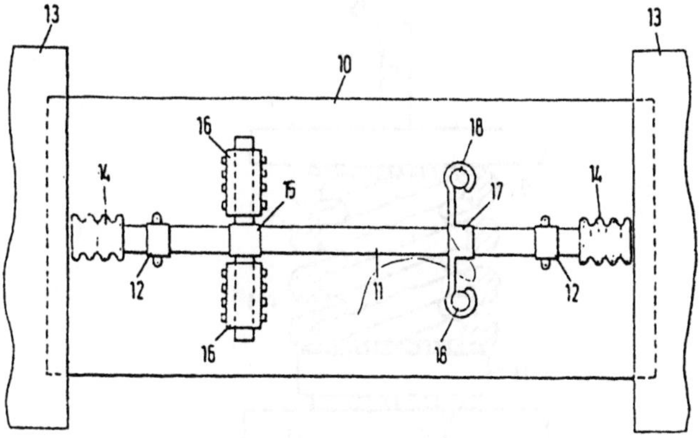

Fig. 321

See Fig. 321, wherein 11 - heat pipe, 12 - clamps, 13 - heat sink, 14 - bellows, 16 - semi-conductor, 17 - clamp, 18 - transistor. The system is applied to printed circuit boards.

Patent (95) of F.E. Andros, (International Business Machines Corp.), refers to a closed two phase liquid-vapour heat transfer device, which is independent of orientation and requires no wick 5. A helix bellows container 10 contains a liquid coolant. Heat input at the evaporator section of the container 10 results in evaporation of the coolant. The vapour flows through the bellow and condenses on the cold end where heat is released. Coolant in the evaporator section is replenished by flow through capillary passages 14 formed by the convolution of the bellow. See Fig. 322.

It is sometimes desired to have sharp demarcation between full and no transfer of heat to a heat sink at a specific operating temperature. For example, certain electronic components, such as diodes and gun devices, operate best at a constant temperature. At lower temperatures their output is less than optimal, and at higher temperatures they may be damaged. To prevent such overheating they are generally coupled to a heat sink.

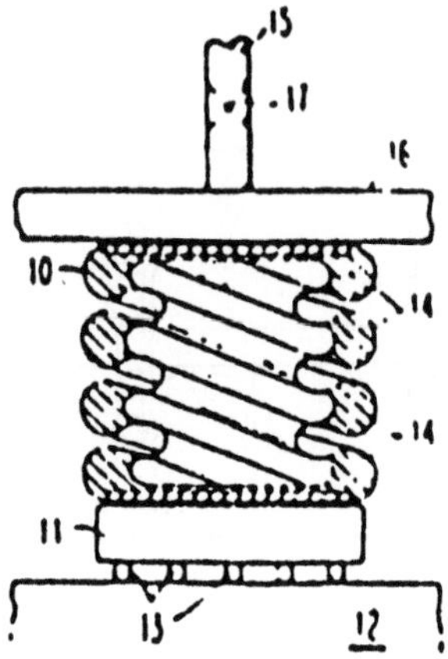

Fig. 322

Patent (90.9) of A. Basiulis, (Hughes Aircraft Co.), provides a heat pipe, which is claimed to improve the temperature control of electronic components. The proposed heat pipe (10 in Fig. 323), includes an evaporator 18, to which a heat generator is coupled. A heat dissipating condenser 13 is coupled to the evaporator via insulating tube 46. A capillary is provided on the interior faces of evaporator, condenser and insulating tube. Reservoirs 16 for receiving inert gas are connected to the condenser. The condenser and the gas reservoirs are thermally connected to a heat sink 28.

The cooling installation, claimed in Patent (118) of D. Minesi, (MATRA), is mainly applicable to electronic circuits and works through a change in phase. The installation has a chamber, connected to an expansion vessel through a valve calibrated at a determined opening pressure depending on the nature of the heat-carrier fluid and on the maximal temperature desired within the chamber and/or on the element to be cooled, the expansion vessel carrying a means for draining off the contents of the expansion vessel. See Fig. 324, wherein chamber 1 surrounds the element 2 to be cooled. The cooling liquid 3 contained in

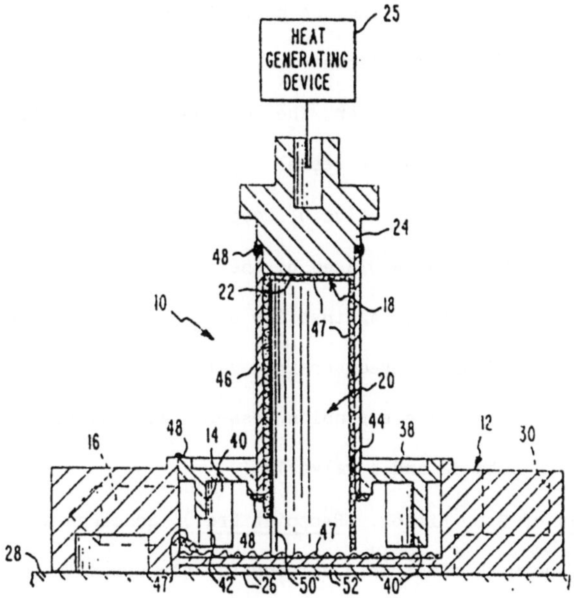

Fig. 323

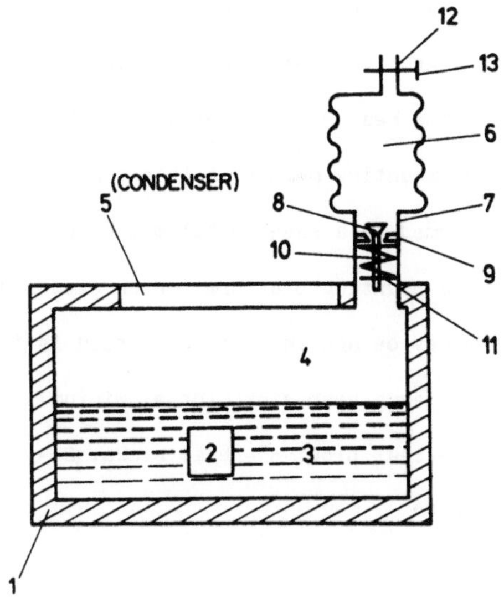

Fig. 324

chamber 1 vaporizes by receiving the heat of the element 2 to be cooled and the vapour 4 thus formed cools and condenses by transmitting the heat to the condenser. The condensed liquid falls down into the liquid 3. An expansion vessel 6 is connected to the chamber by a pipe 7, which carries a valve 8, cooperating with a valve-seat 9. The rod 10 of the valve 8 is biased by the action of a drawback spring 11 keeping the valve 8 flush against its seat 9 as long as the pressure inside the chamber 1 has a predetermined value lower than that inside the expansion vessel 6. The expansion vessel 6 also carries a vent valve shown by a tube 12 and a valve 13.

Patent (143.1) of G.L. Fitton, (Redpoint Ltd.), concerns a heat sink assembly for use with electronic components, the assembly comprising: Three integers, a thermally conductive plate mounting means for one or more electronic components, the means in use being in thermal contact with each component, a heat pipe to which the mounting is connected and a heat sink connected to the heat pipe to receive and, subsequently, dissipate heat carried thereto from the electronic components by the mounting means and heat pipe, the mounting member being integral with and forming part of the heat sink. According to the invention the thermally conductive mounting member is itself part of a heat sink and is preferably formed as a sheet metal member which in one portion is flat to serve as a mounting platform for an electronic component such as a semi-conductor device and in another portion is formed with twisted fins to serve as a heat sink, (of aluminium).

Fig. 325 wherein: 1 - heat pipe; 3 - heat sinks; 2 - fins; 5 - depression receiving the heat pipe 1.

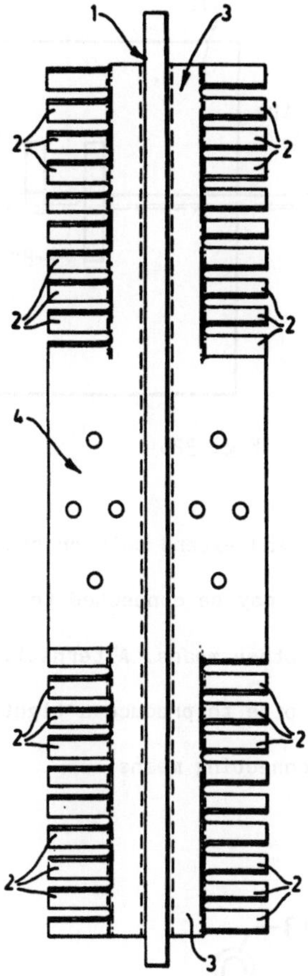

Fig. 325

Another Patent (143.2), assigned to Redpoint Ltd., also refers to heat sink assemblies for use with electronic components. The proposed assembly, shown in Fig. 326 comprises: a tubular heat pipe 6, a heat sink 7, two pairs of mounting strips 8 and 9 for dual line semi-conductor devices. The semi-conductor devices 10 are secured to a printed circuit board 11 and serve to space the strips 8 and 9 from the board 11. The mounting strips 8 are formed with a central channel 12 for receiving the heat pipe 6 and extend to either side of the heat pipe 6. The mounting strips 9 have a bent-over portion 13 for

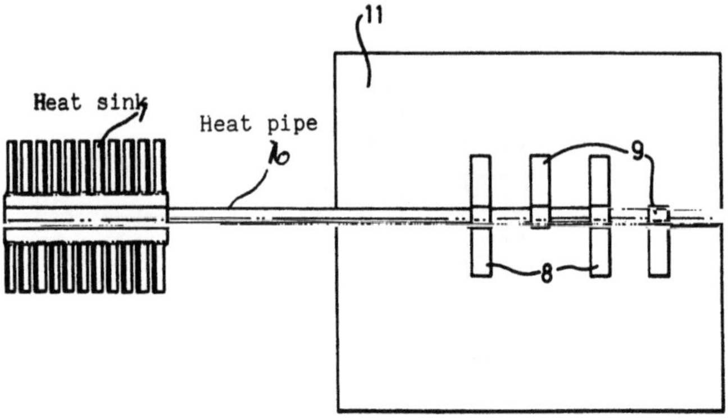

Fig. 326

receiving the heat pipe 6, and extend only on one side of the pipe 6. The mounting strips 8 and 9 may be connected to the heat pipe 6 by aluminium soldering or by other means. Alternatively, the strips may be clamped round the heat pipe to produce a tight fit which obviates the need for additional connecting means.

See Figures 326, 327.

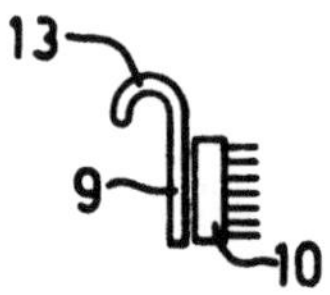

Fig. 327

The cooling body, revealed in Patent (163.1) of M. Munekawa, (Showa Aluminium Corp.), for cooling semi-conductors, 4, is provided with cooling ribs 5, 21. In order to ensure high cooling effect, in combination with small size and low weight, a part 1 of the cooling body serves to receive the semi-conductor 4 and displays at least one (7, 22, 31) heat pipe or heat sink. See Fig. 328.

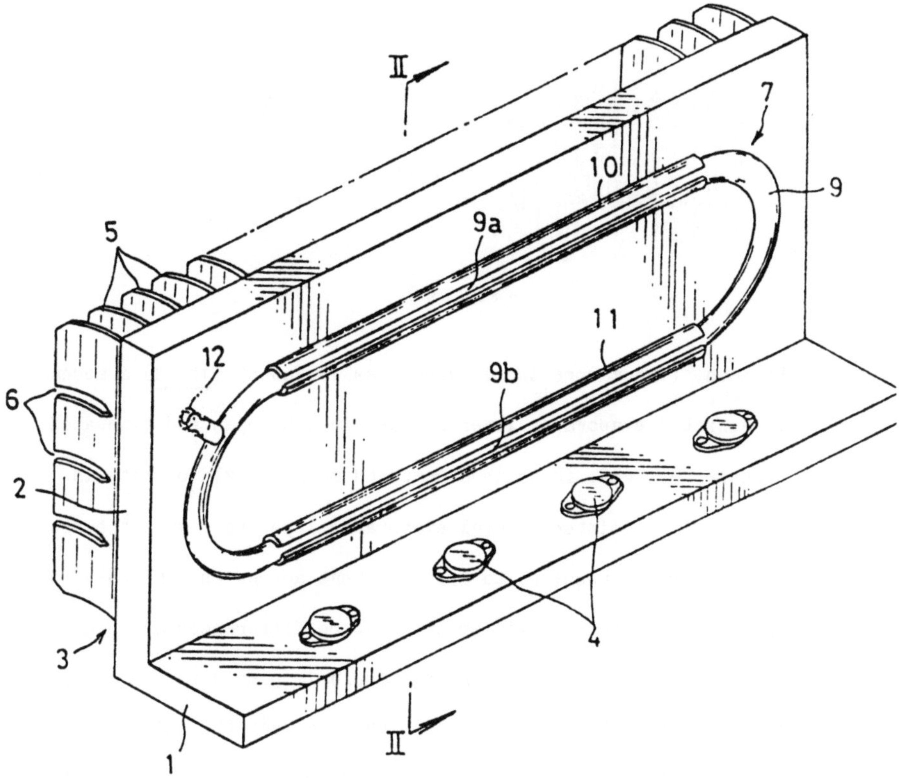

Fig. 328

Another Patent (163.2) of Showa Alluminium Corp. (inventor: M. Munekawa), refers to a heat releasing device comprising a heat transmitting block 10 having a heat source attaching portion and a bore 11 extending therethrough, a looped heat pipe 1 having at least one pair of straight tubular portions 1a, 1b, the first of the straight tubular portions 1b being intimately fitted in the bore 1, and a multiplicity of radiating fins 4, 5 attached to the heat pipe 1. See Fig. 329.

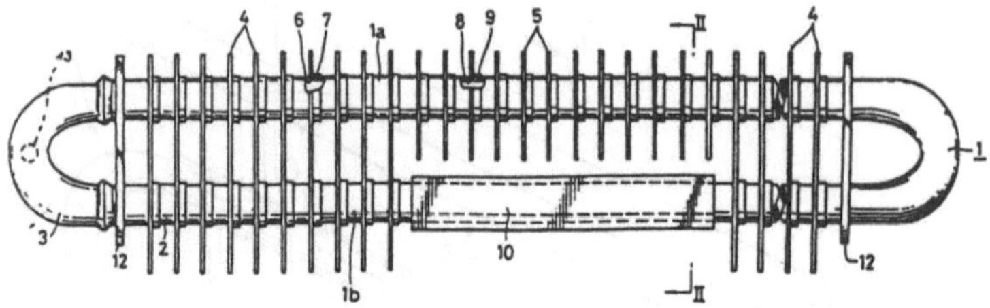

Fig. 329

G.Y. Eastman, (Thermacore Inc.), discloses in Patent (181.9) a mechanically assisted evaporator layer for use in both open and evacuated heat transfer systems, in which a pump and spray nozzle operate in conjunction with a sintered metal evaporator layer to reduce the temperature difference required to transfer heat across the thickness of the surface and to permit smaller temperature differences and higher power densities in transferring heat. Liquid is pumped to and sprayed from a nozzle onto the sintered metal layer to keep the entire surface wetted at all times so as to permit uniform thin film evaporation from the surface. The continual presence of liquid at the outer evaporative boundary reduces the likelihood of surface dryout while the thermal conductivity of the sintered metal promotes more effective vaporization. The device, having no limitation due to vapour movement interfering with liquid transfer back to the evaporative section, is particularly well suited for the high power density applications of some of the more sophisticated modern technologies such as cooling of X-ray tubes, electron tube electrodes, plasma arc electrodes, and high power laser mirrors. For instance, the device permits the transfer of heat from a small surface heated by an electronic device and efficiently transfers the heat to larger surfaces, thus in effect acting as a power density transformer, moving heat from a high power density surface to

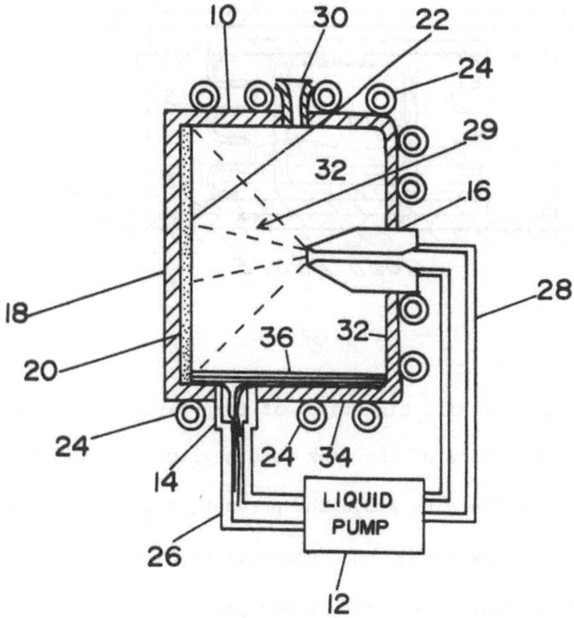

Fig. 330

a larger surface area which operates at a lower power density and is cooled by more conventional means. See Fig. 330.

In Patent (189) of C.E. Simmons, (United Aircraft Products, Inc.), there is provided a heat transfer system in which a fluid of high heat absorption capacity is flowing in successive heat absorption and heat dissipation phases. A cold plate-liquid boiler assembly provides flow paths for the successive phases, the differential heat dissipation modes being realised by the liquid boiler and by external cooling means, alternatively included in a heat rejection circuit. Vent control means, associated with the liquid boiler permit to maintain a pressure therein, promoting boiling at selected temperatures.

In Patent (191.2) of F.E. Altoz, (The US Secretary of the Air Force), a thermal switch is provided which changes thermal impedance between

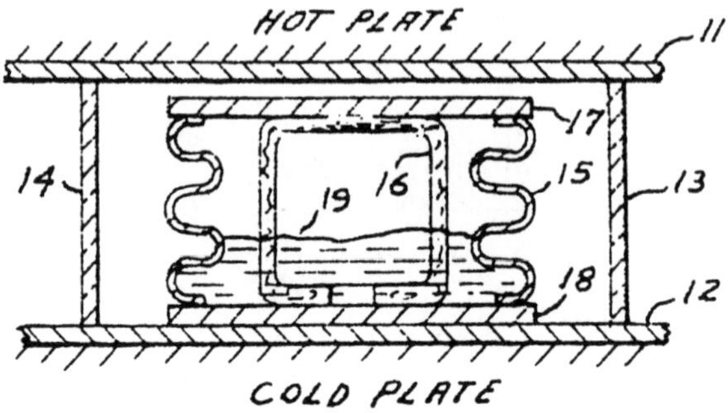

Fig. 331

a source and a sink from a high value at temperatures below a pre-set value of the temperature of the sink to a negligible (low) impedance at temperatures of the sink above the pre-set temperature. The switch is small, and requires no external connections or power sources and ensures a fast warm-up of electronic equipment.
See Fig. 331.

Patent (209.3) and essentially (209.4) of R.W. Garrett, (Western Electric Co.), relate to a method and apparatus for removing heat from a workpiece during processing in a vacuum chamber, and, more particularly, to micro-electronic device manufacturing fabrication involving ion-beam processing of wafers.
More particularly (see Fig. 332), a workpiece 16, for example a magnetic bubble memory wafer, is held in place, and heat is removed from it by an expandable heat pipe 22 during a process performed in a vacuum as, for example a sputter etch process. The expandable heat pipe 22 includes two plates 24, 26, joined by a bellows 28 to form a sealed chamber containing a vaporizable liquid (water). During loading of the wafers the heat pipe 22 is compressed by atmospheric pressure and thereby facilitates easy assembly of the wafer 16 and heat pipe 22 into a holding fixture 10, which, in turn, is bolted

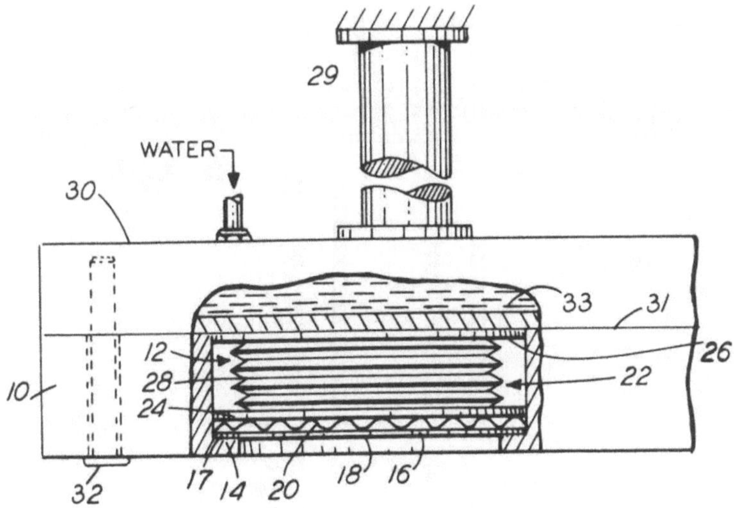

Fig. 332

to a cooling plate 30 situated inside a vacuum chamber. When the chamber is evacuated, the heat pipe 22 expands to hold the wafer 16 in place and forms a good thermal path from the wafer 16 to the cooling plate 30. In an alternate embodiment the heat pipe 22 does not completely fill the gap from the wafer 16 to the cooling plate 30 in a vacuum at room temperature, but rather expands to fill the gap when the heat pipe 22 is heated by the wafer 16. The expanding heat pipe 22 holds the wafer 16 at a constant, predetermined temperature above the temperature of the cold plate.

Cooling elements for the electronic systems of spacecrafts.

Cooling elements destined for the electronic systems of spacecrafts have to comply with the special conditions in and around the spacecrafts (high speeds, aerodynamic pressures, confined space, etc.). The Patent (180) granted to Telefonaktiebolaget L.M. Ericsson, takes into consideration these special conditions and comprises (see Fig. 333)

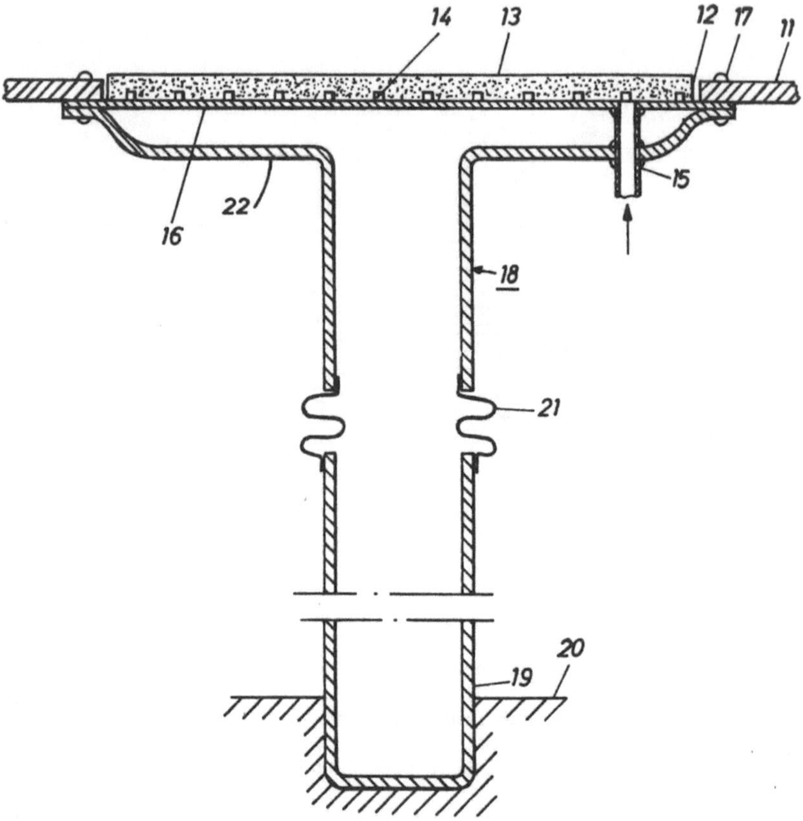

Fig. 333

an aperture 12 in the external wall 11 of the fuselage, the aperture 12 being filled with a sheet 13 of porous material, for example sintered copper. In the sheet 13 several channels 14 are made, interconnected with one another and receiving a fluid through a duct 15. The sheet 13 is fitted to a plate 16, of high thermal conductivity, (copper). Moreover, the plate 16 is fixed to the upper (condensation) sector 22 of a heat pipe 18, the lower sector 19 of which is in contact with a device 20, releasing heat. The heat pipe is provided with a bellows 21 for preventing the transmission of vibrations.

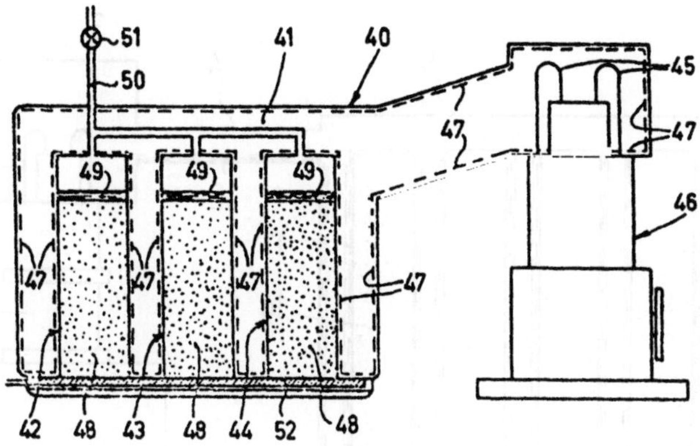

Fig. 334

Heating systems for hot gas engines.

Patent (197.4) of M.L. Hermans, (US Philips Corp.), relates to heating systems for hot gas engines and hot gas turbines, the system including a heat pipe containing evaporable heat transporting material, a closed reservoir containing heat-accumulating material within the heat pipe, and a quantity of the evaporable material in the reservoir for maintaining the same pressure in the reservoir and in the heat pipe. See Fig. 334.

Patent (197.5) of G.A.A. Asselman,(US Philips Corp.), also provides a heating system for supplying thermal energy to a hot gas engine, this system using a heat-transporting medium which flows within a closed space between a heat-receiving portion to a heat-discharging portion. Thermal energy is supplied from a heat source through a wall in the heat-receiving portion with the transporting medium changing from the liquid phase to the vapour phase, and the absorbed thermal energy is transferred to the heater in the heat-discharging portion

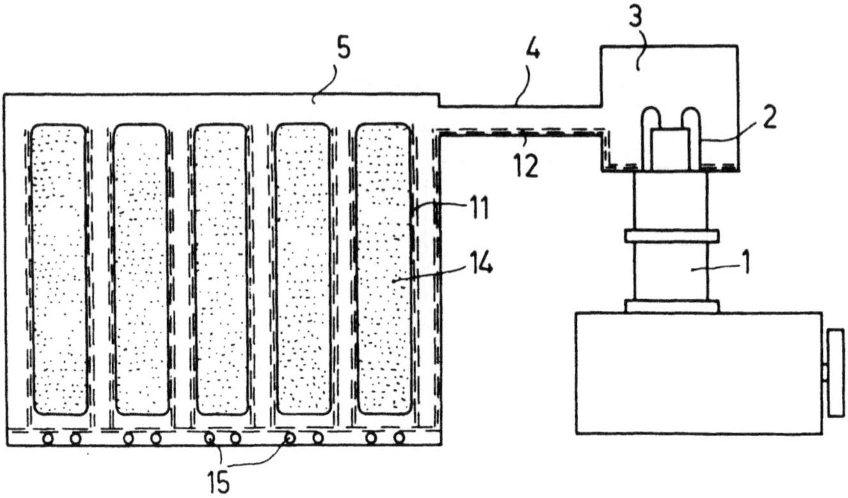

Fig. 335

by changing from the vapour phase into the liquid phase.

The wall parts of the heat-receiving portion, through which thermal energy can be supplied from the heat source to the transporting medium, comprise on their surface contacting this medium of a layer of a porous material having a capillary structure whereby liquid-phase medium is distributed on the walls. See Fig. 335.

Another Patent (197.7) of G.A.A. Asselman, (US Philips Corp.), discloses a heating system for a hot gas machine, characterised in that the heat transfer device comprises at least two closed spaces containing a heat transporting medium and arranged one after the other in the direction of heat transport, the distal ends of these spaces are provided with heat-passing walls through which heat from the heat source can be conducted to the transporting medium, or the latter can give off heat to the heater, the proximal end of the spaces having further heat-passing walls between which a switching element is provided for establishing a thermal contact between the further walls.

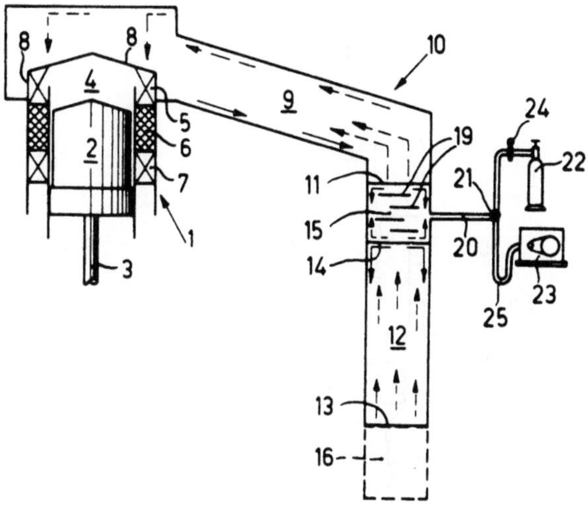

Fig. 336

In this way a machine is obtained in which the heat transfer from the heat source to the heater can be interrupted simply by actuating the switching element. See Fig. 336.

Heat transfer, heat pipes in internal combustion engines (vehicles).

Patent (17.1) of D. Scherenberg et al., (Robert Bosch GmbH.), discloses an ignition chamber for internal combustion engines.

As shown in Fig. 337 the proposed ignition chamber 8 is connected to a main combustion chamber 3 through at least one transfer channel 14. The ignition chamber 8 has a cylindrical wall 23, which is regulated by an annular heat pipe 26 to a nearly constant high temperature level. An ignition device inserted into the ignition chamber 8 forms a spark gap 24 with the cylindrical wall 23 of the ignition chamber by an electrode 22 in the region between the ignition chamber centre and the transfer channels 14. To improve the stability of the electrode 22, it may be provided with heat-dissipating means. The heat pipe 26 may contain a temperature sensor, connected to fuel-air mixture, exhaust gas recirculation or ignition timing regulating means.

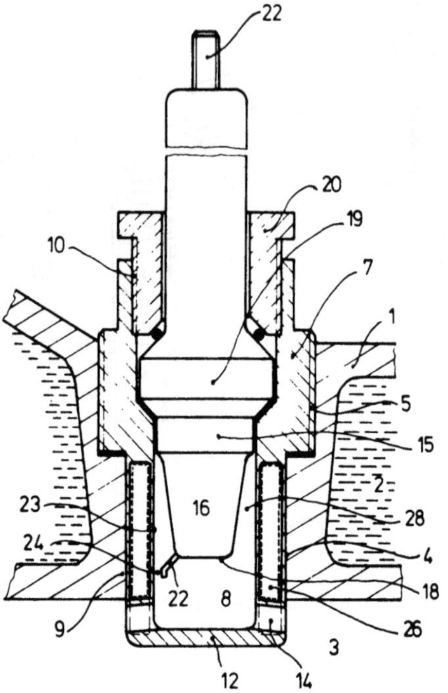

Fig. 337

Another Patent (17.2) of D. Scherenberg et al., (Robert Bosch GmbH.), relates to a method and device for ignition of weak fuel-air mixtures, (see Fig. 338). More particularly, an ignition chamber 8 is connected through a plurality of transfer ports 13, 14 with a main combustion chamber 3. One or more of the transfer ports 14 serve for conveying a part of the fresh mixture which enters the ignition chamber 8 in the form of a jet directly to the region 9 of the ignition chamber 8. Another part of the fresh mixture charge is brought into rotational movement in the ignition chamber by the remaining transfer ports 13 opening substantially tangentially into the ignition chamber 8 so that flow along the chamber wall results the flow being braked by the mixture part flowing back from the chamber region 9. A heat pipe 31 is incorporated in the chamber walls. The ignition chamber 8 may have a frusto-

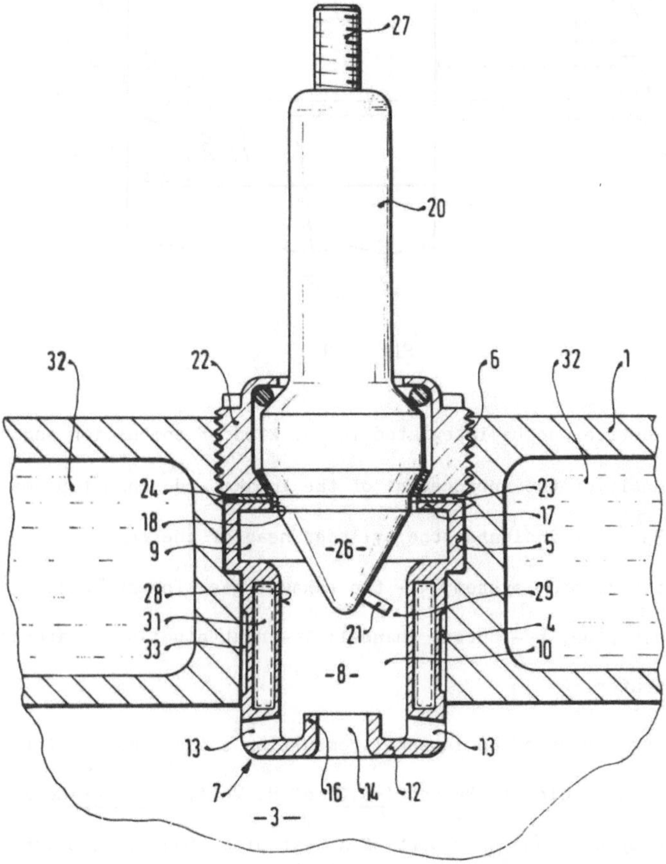

Fig. 338

conical shape and the transfer ports may be variously arranged. The insulator nose 26 may be protected by a sleeve integral with the chamber wall.

In Patent (40.1) of J. Abthoff, et al., (Daimler-Benz AG.), reveals a device for the preheating of a carburator mixture. In contrast to prior art devices, the proposed device has the advantage, that throttling of the sucked-in mixture is completely prevented and

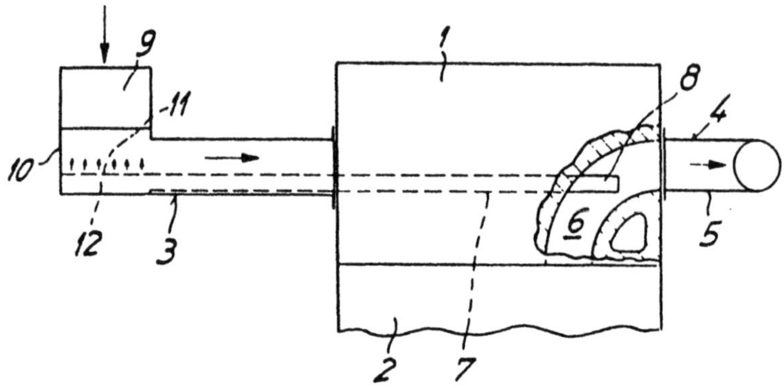

Fig. 339

that the device can be integrated in any kind of combustion engines, independently of the arrangement of the intake and exhaust pipes.

In Fig. 339, 1 - indicates the cylinder head of the engine 2; 3 - the intake pipe system; 4 - the exhaust gas pipe system; 5 - exhaust pipe; 6 - outlet channel; 7 - heat pipe; 8 - evaporator; 9 - carburator.

The cooling device of an internal combustion engine (mainly for vehicles), according to Patent (40.2) of H. Wuif, (Daimler-Benz AG.), is formed by the condenser part of a heat pipe. Such combustion has the advantage that air circulation by means of pumps can be dispensed with, (thus no pumps are needed), and that heat can be removed exactly from the critical zones of the engines, thereby making a forced circulation in the engine cooling jacket superfluous.

In Fig. 340, 1 - indicates a car; 2 - the engine space; 3 - a conventional combustion engine; 4 - cylinder head; 5 - a motor block; 6a, 6b - the evaporator part of a heat pipe system; 7 - a common condensation part thereof.

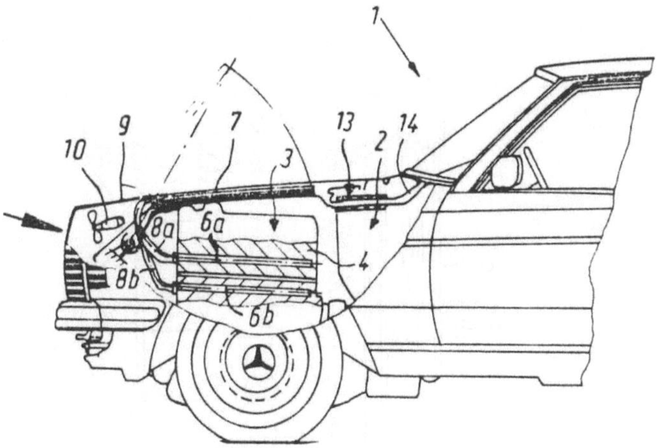

Fig. 340

Fig. 341 illustrates the condensation part of the heat pipe system.

Another Patent (40.12) of H. Wuif, (Daimler-Benz AG.), provides a heat transfer system for vehicles. The system, operating on the heat pipe principle, comprises a heat source 4 connected to a heat exchanger or dissipator 8 by a bundle of capillaries 9. The capillaries permit relative torsional movement of the heat pipe formed by the capillaries about

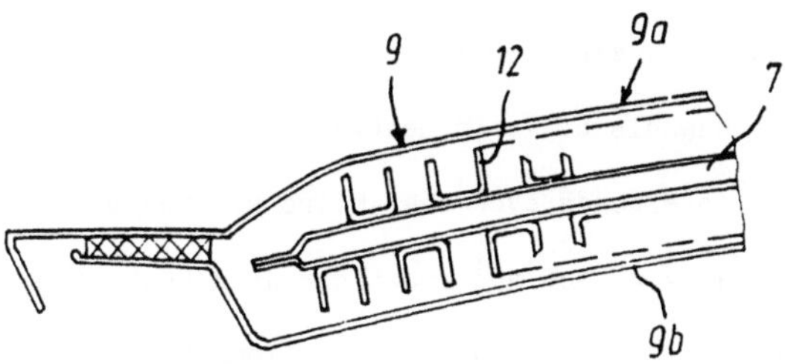

Fig. 341

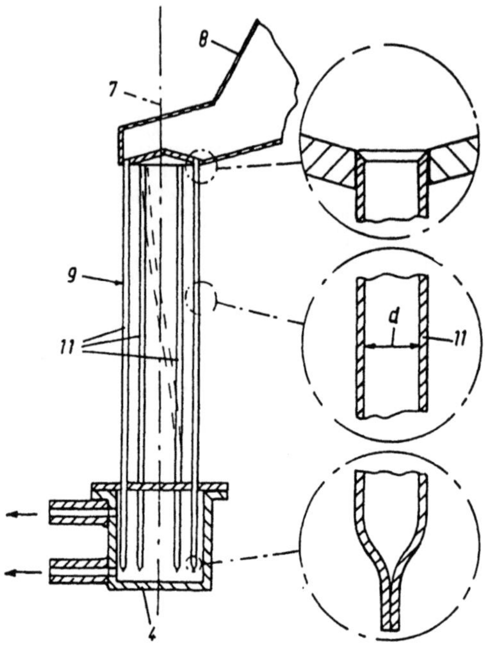

Fig. 342

their common axis 7. In the case of a motor vehicle the heat dissipator 8 may be mounted in a door of the vehicle and the common axis 7 is disposed substantially on the hinge axis of the door, whereby the heat pipe can twist with the movement of the door. See Fig. 342.

Patent (40.10) of R. Andres, et al., (Daimler-Benz), concerns a plane heat body 1 for vehicles, executed in an expanded stratified structure and comprising two channels 6, 7, in heat-conducting contact with one another. The channel 6 extends over the entire surface of the heating body and is hermetically closed and functions as a heat pipe, distributing the heat fed in over a large area. Along the lower edge of the heating body 1 there is arranged a heating channel 7, through which hot water is flowing. Between 6 and 7 there is a non-expanded strip 8. Along this strip 3 the stratified heating body 1 is bent in

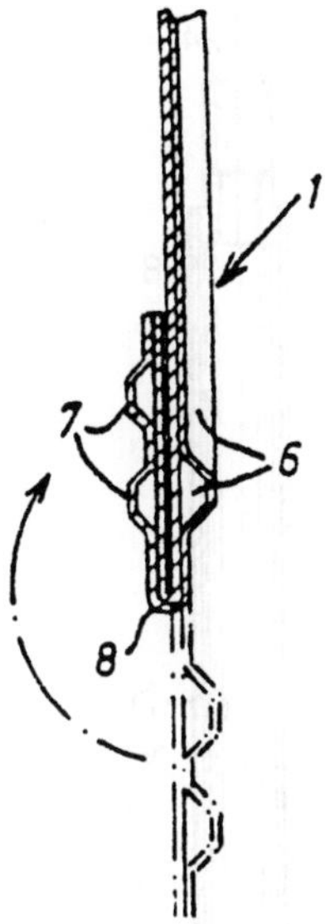

Fig. 343

U-form, in order to realise a superimposed, permanently conductive arrangement of the heat pipe 6 and the heating channel 7. See Fig. 343.

Patent (61.2) of H. Weickmann, (Fiat Veicoli Industriali S.p.A.), refers to a heat pipe, suitable for heat transfer from thermally overloaded areas of internal combustion engines. The invention, revealed in (61.2) aims at increasing the capacity of heat pipes and constructing it in such a form, that it can be fitted in the machine (engine) in a safe, gas and waterproof manner. For this purpose the heat pipe displays a central cylindrical space and a capillary intermediary space at its inner wall, which has an outwardly expanding sector in the proximity of its ends, which is provided with an annular flange, forming a distributor head, limiting the penetration of the heat pipe in an aperture, made in a wall of the engine, permitting a reliable press-fitting of the heat pipe. See Fig. 344.

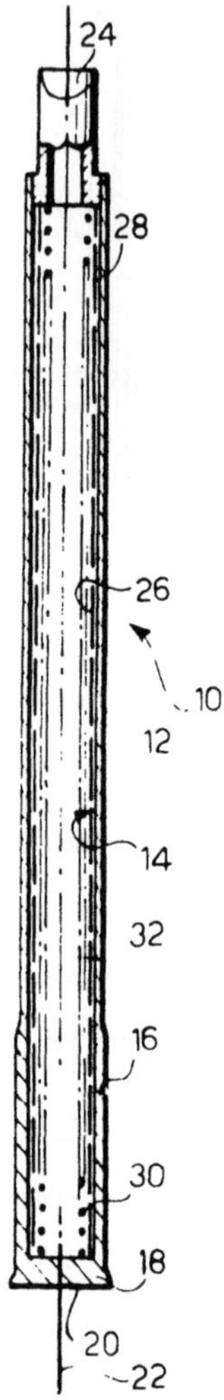

Fig. 344

Patent (144) of Regie Nationale des Usines Renault concerns a preheater for the intake gas of vehicles by means of heat pipes.

As shown in Fig. 345, the heat pipe 1 is provided with a plug 2, in form of a tunnel (or cylinder), the lower part 3 of which is immersed in an exhaust collector 4, while its upper part 5 enters into the intake collector 6. Parts 3 and 5 of the heat pipe 1 are welded together, thereby forming a flange 8 at the level of the joint of

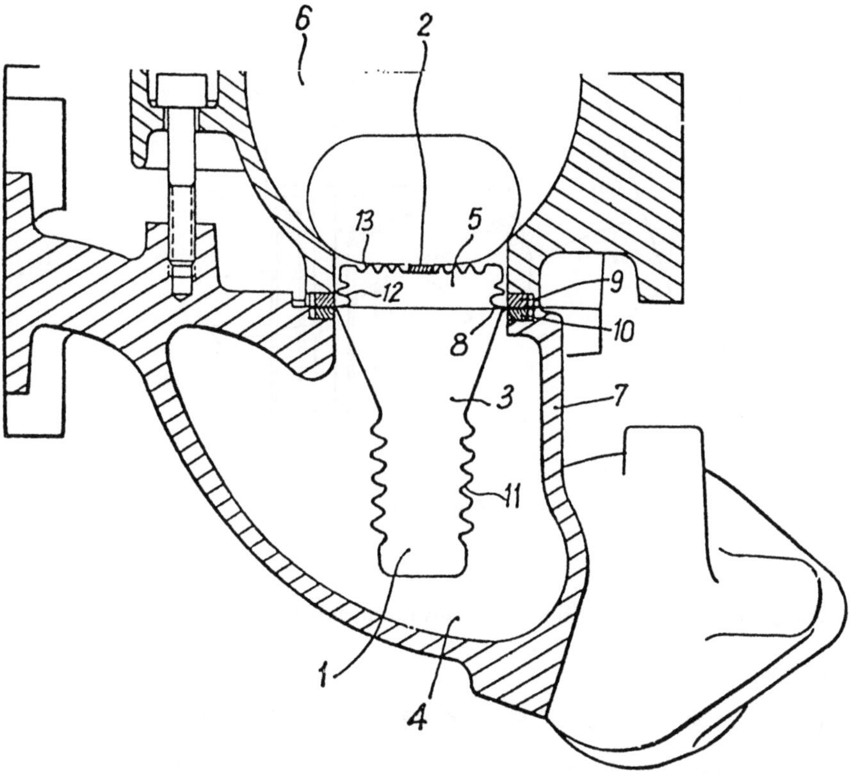

Fig. 345

collectors 4 and 6. Due to the bellows 11, 12 the heat pipe can be expanded, without affecting the impermeability of the arrangement.

Patent (203.1) of K. Emmenthal, (Volkswagenwerk AG), provides a safe heating system for vehicles, by means of a heat pipe, whereby the noxious effect of after-heating through heat radiation from the exhaust pipe, which remains in glowing state for some time, after having stopped the engine, is eliminated.

In Fig. 346, 1 indicates the heat pipe which penetrates at 2 into the exhaust bend 3 of the combustion engine 4. In order to prevent the transfer of radiation heat from the exhaust bend 3 to the sector 2 of the heat pipe 1, after stopping the engine, the invention provides

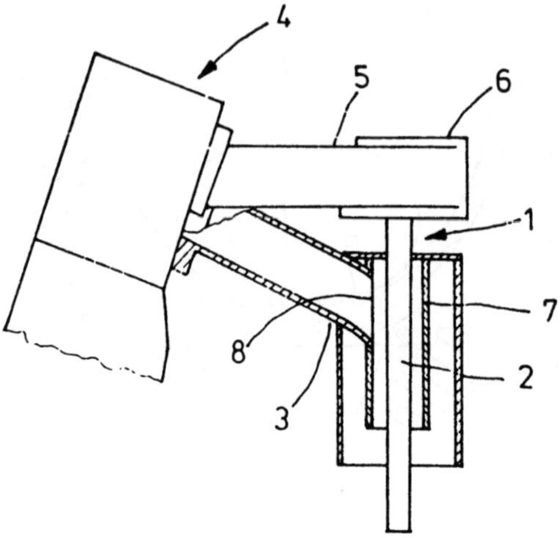

Fig. 346

a pipe 7, which surrounds as a radiation shield the critical sector of the heat pipe.

Another Patent (203.2) of K. Emmenthal, (Volkwagenwerk AG), refers to a similar device, as revealed in (203.1). In order to limit the temperature in the interior of a heat pipe 1, the critical (heat receiving) sector 2 of the heat pipe 1 is partly surrounded by a tube 5, which is removable with regard to sector 2 through a temperature-dependent member 7 in duct 3. Member 7 may contain an expandable material or it can be made in the form of a bellows. See Fig. 347, wherein 8 is an aperture for exhaust gases.

Heat exchangers, heat pipes in turbines, compressors.

Patent (16) of S.R. Bolger relates to a heat exchanger for use in gas turbine engines, and, in particular, to a high efficiency, low bulk heat exchanger utilizing a working fluid to transfer thermal energy from exhaust gases to compressed gas of a turbine engine. The heat exchanger includes a first and second matrix mounted for

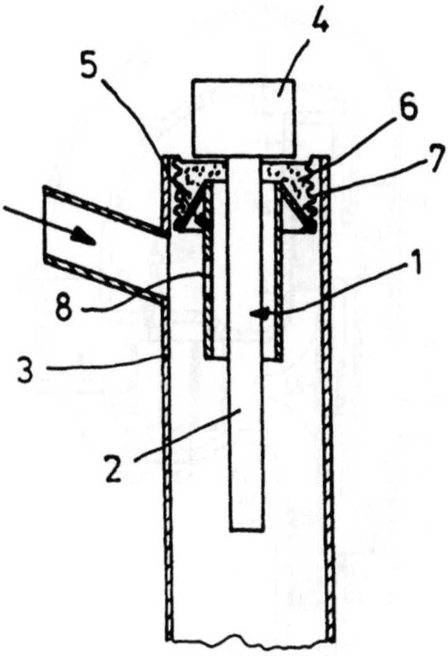

Fig. 347

free rotation about the main shaft of the gas turbine.
The first matrix is in contact with compressed air from the turbine compressor and the second matrix is in contact with the exhaust gases of the turbine. Each matrix includes a plurality of air passageways for flow therethrough of the respective contacting gases. Additionally, each matrix includes at least one and preferably a plurality of fluid chambers positioned perpendicularly to the passageways wherein each chamber is in communication with an associated fluid chamber of the other matrix. Each fluid chamber includes a fluid which is vaporizable in the second matrix to extract thermal energy therefrom and which condenses in the first matrix to release thermal energy thereto to preheat the compressor air.
Fig. 348 illustrates the heat exchanger 20 of gas turbine 10, with an inlet port 11, a drive shaft 12, an impeller 13. 14 indicates the outer casing of the turbine; 16 a diffuser manifold; 17 a compressor manifold; 18 a seal, preventing the escape of compressed air from

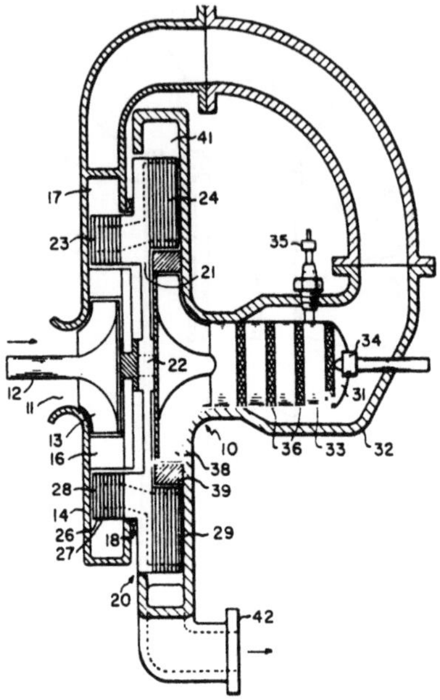

Fig. 348

around the heat exchanger 20; 21 indicates an annular mounting plate, with axial bearing; 23 a first matrix and 24 a second matrix.

A helical expansion condenser, described in Patent (102) of D.A. Kelly, is a special heat exchange unit designed for closed cycle vapour turbine systems. The helical expansion condenser consists of multiple continuous turns or coils of elliptical tubing or built-up metal sheets of elliptical cross-section, which are arranged in a uniform helix. The generally flat elliptical cross-section provides for efficient airflow and heat transfer over the entire surface area of the coils. The airflow over the condenser is provided by vehicular air intake with a large central fan forcing air over forward and aft coils uniformly. Multiple heat pipes are the primary cooling device added to the coils, for a total heat transfer coefficient of 40, approximately. Microfinning of the sheet metal ducts accounts for some of the improved cooling effectiveness, it is claimed.

In Fig. 349, 1 indicates the continuous circular shaped duct, with an elliptical or symmetrical airfoil cross-section 2.

The duct is fabricated from two identical full circle sheet metal sections 1a and 1b, which are lapped-joined endwise to form the continuous circular helical cooling path. The duct coils 3 are uniformly spaced and secured together with the linear spacer fins 4, placed in-line with the normal airflow. A perforated metal sheet 5 is secured to the top of the duct 1, with multiple uniform fins 6, secured to the top of the sheet 5, for heat dissipation and protection from the fan 7. The fan 7 is supported by the shaft 8, which is driven by the turbine expander system 9. The duct entrance connection 10 provides a vapour sealed coupling between the top of the duct and the hot vapour inlet duct 11. A sealed condensate chamber 12 collects the condensate for transfer to another point within the turbine expander system 9. A treaded port 13 within the condensate chamber 12 provides a pipe or duct fluid transfer means.

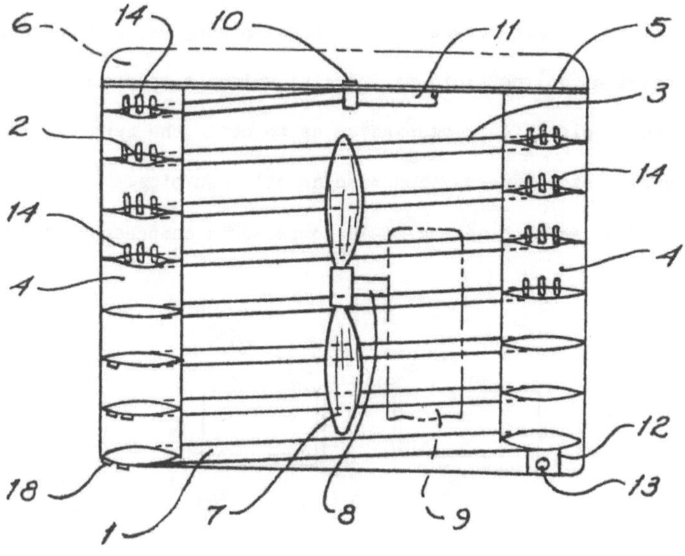

Fig. 349

Multiple short heat pipes 14 are uniformly placed half-way in the duct 1 for heat-sinking. The heat pipes 14 must be canted or angled to the duct wall, and placed in the top wall of the duct to provide gravity feed for the heat pipe working fluid 15. Screen wick 16 returns the vaporized fluid 15 to the hot end of the heat pipe 14, to repeat the transfer cycle. Insulated metallic foil 17 covers the screen wick 16. Multiple cooling thermoelectric modules or thermocells 18 may be added to a portion of the duct 1 as a supplementary cooling means. See Fig. 349.

In Patent (149) of D.W. Artt, (Rolls-Royce Ltd.), an aerofoil blade is provided for a gas turbine engine in the form of a heat pipe, the aerofoil blade having an integral passageway adapted to contain a secondary structure and to thermally insulate the secondary structure from the remainder of the aerofoil blade. The internal passageway is preferably adapted to thermally insulate the secondary structure from the remainder of the aerofoil blade by the provision of a cooling fluid passage between the internal passageway and the secondary structure. The cooling fluid may be air, which is adapted to flow through the cooling fluid passage so as to cool the aerofoil blade. The secondary structure may comprise an oil feed pipe.
In Fig. 350 a gas turbine engine 10 comprises a compressor 11,

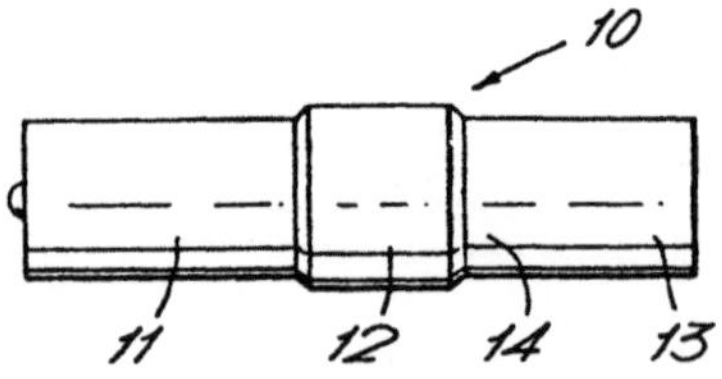

Fig. 350

combustion equipment 12 and a turbine 13. The gas turbine engine 10 operates in the conventional manner, that is, air compressed by the compressor 11 is mixed with fuel and combusted in the combustion equipment 12. The resultant hot gases expand through the turbine 13 to atmosphere thereby driving the turbine 13 which in turn drives the compressor 11. Hot gases from the combustion equipment 12 are directed into the high pressure section 14 of the turbine 13 by an annular array of stationary nozzle guide vanes.

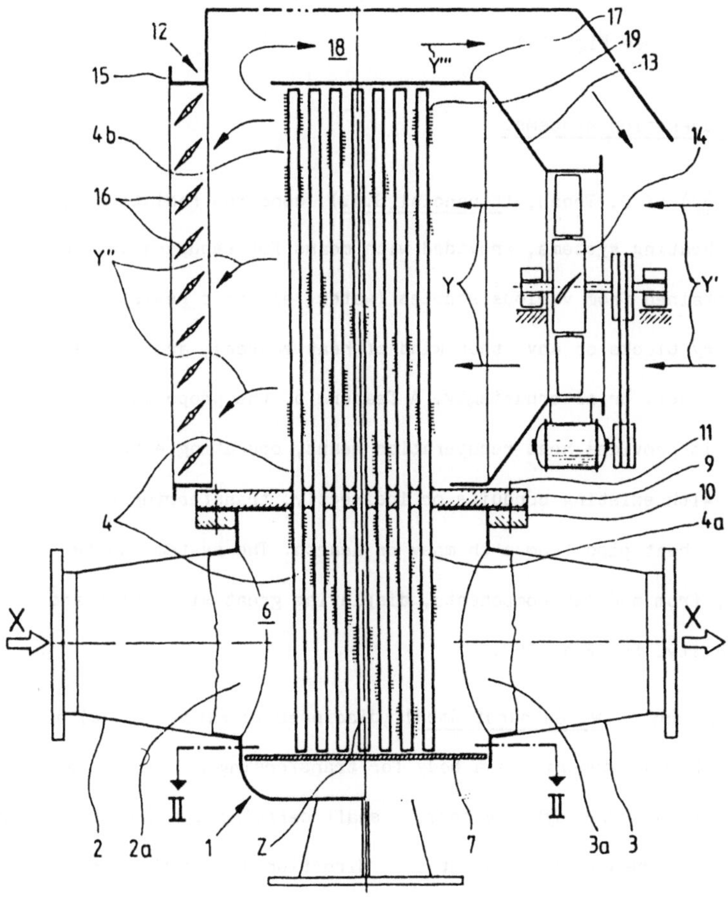

Fig. 351

Patent (74.2) of H. Kassat, (GEA Luftkühlergesellschaft Happel GmbH & Co.), concerns an air-cooled pressure gas cooling device, comprising heat exchanger heat pipes, operating in a static manner, independently of external mechanical or electrical drives, the heat transport being realised by serially arranged evaporation and condensation means. Each heat pipe is partly filled with a working fluid and provided at its inner face with a capillary coating. As working fluid, as a rule, freon is applied. In order to increase the heat capacity of the device the recycling of the condensate can be promoted by utilising the force of gravity. For this purpose the heat pipes are preferably in a vertical arrangement. See Fig. 351.

Heat recuperation systems.

Patent (62) of D. Tresh, (Financial S.A.), concerns radiators for central heating systems, provided with means for recuperating the heat, obtained from various sources (water, electric resistance, refractory blocks or any other kind of heat source), and applied simultaneously or alternatingly. A feature of the proposed system is that it provides heat recuperation means, operating either directly, with existing conduits or indirectly, transferring heat through a heat pipe from a chimney or hearth. The system can be built-up from modular components, displaying great simplicity and adaptability. See Fig. 352.

N. Porzsicsányi, (W. Gebhardt GmbH), discloses in Patent (75.2) a heat transfer device, comprising a body for transferring a fluid; a rotor, moving the fluid through the body, a shaft carrying a plurality of blades, arranged, on the one hand in rotation direction and on the other hand over the axial length of the rotor. The invention features a device for

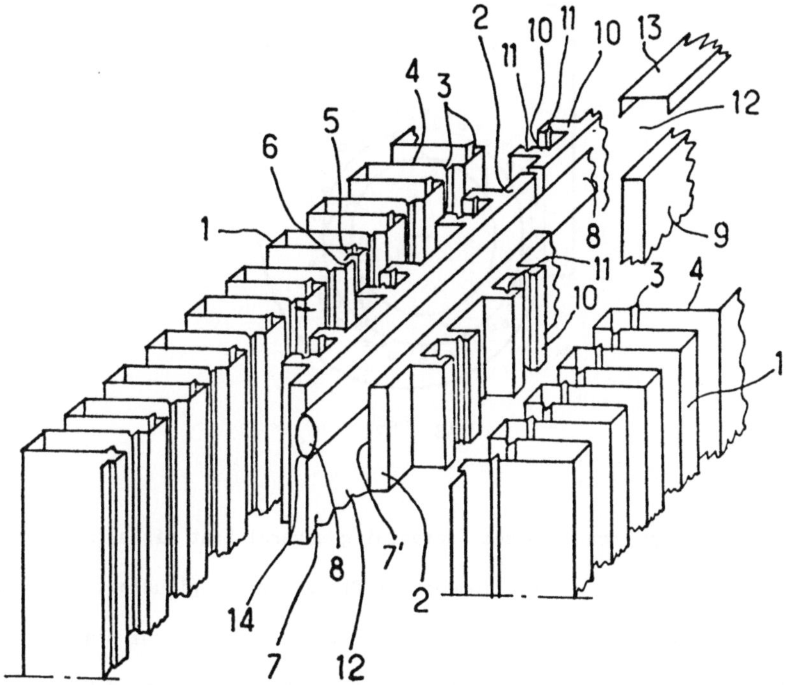

Fig. 352

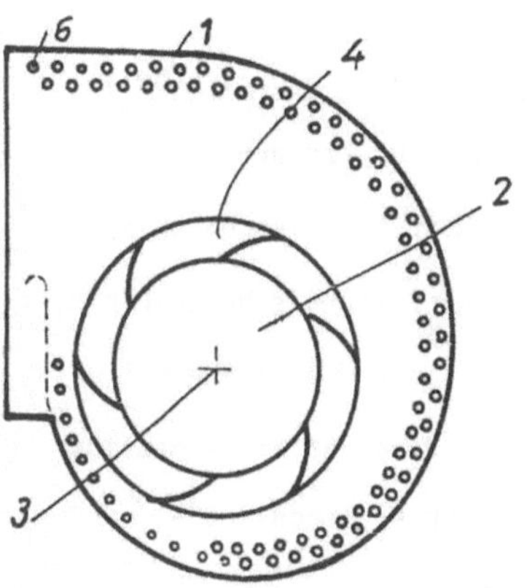

Fig. 353

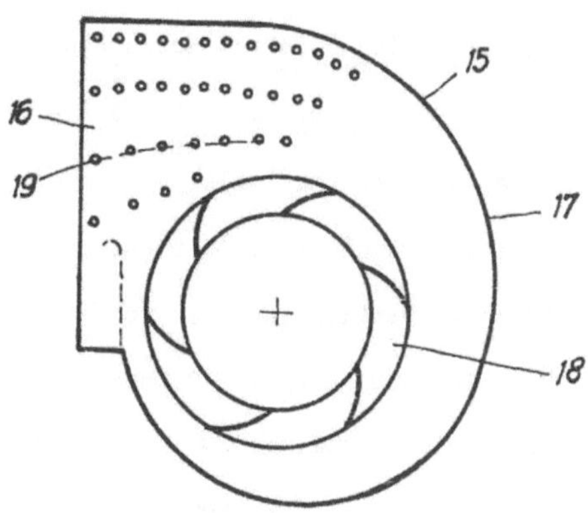

Fig. 354

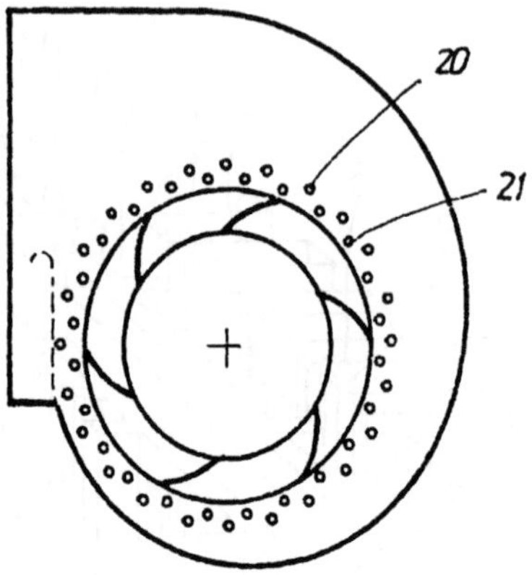

Fig. 355

reducing energy (heat) losses, comprising several heat pipes, arranged crosswise to the flow direction of the fluid, the longitudinal direction thereof being parallel to the longitudinal direction of the blades. Such arrangement eliminates disturbances of the flow in the rotor, furthermore reduces resistance problems.

Figures 353, 354 and 355 illustrate the various modes of arrangement of the heat pipes.

In Patent (82.8) of F. Edelstein, (Grumman Aerospace Corp.), a heat pipe system is provided to remove heat from a light fixture in a room of a building that will transfer the fixture heat to a system in a building which will, in turn, reject or use the fixture's heat in a heating, ventilating, air conditioning system for the building.

See Fig. 356, wherein 10 indicates fluorescent fixtures; 12 - ceiling panel; 18 - ceiling; 20 - emergency water sprinkler heads; 22 - conduits; 36 - heat exchanger; 44 - heat pipes.

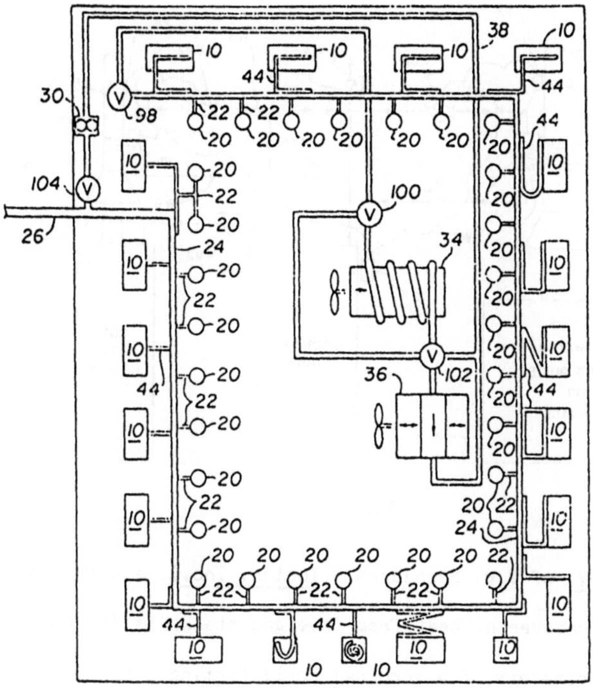

Fig. 356

According to Patent (96) of P.O. Munroe, (International Telephone and Telegraph Corp.), heat pipes are employed to capture energy normally lost in the form of heat in the gases in flue-stacks and to transfer the captured heat to water in a water jacket. The heated water is pumped from the water jacket to a storage tank where it is kept until needed. Control of the system is governed by heat sensors in the flue-stack and in the storage tank which provide control signals to a pump, or pumps, used in controlling the flow of water through the pipes between the water jacket and the storage tank.

See Fig. 357, wherein 2 - a boiler, furnace; 4 - a flue stack; Pa, P2, P3 - heat pipes; 8 - a water jacket; CP1 - a pump; W1, W2 - water conduits.

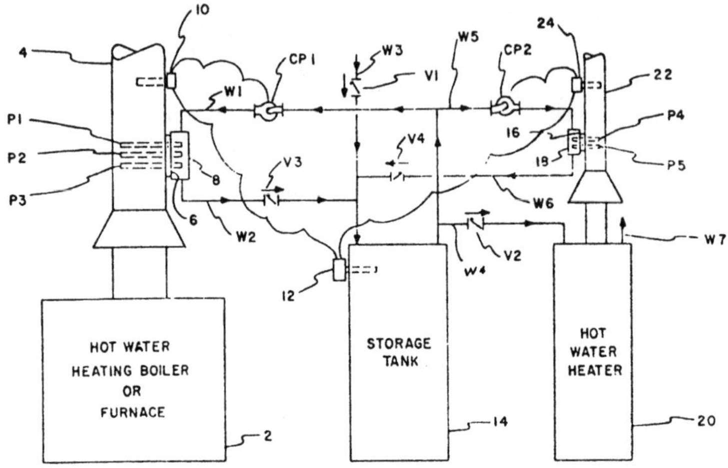

Fig. 357

In Patent (97) of J.W. Seehausen, (Ionics, Inc.), a heat pipe recuperator for recovering heat from flue gas stacks is disclosed. The recuperator consists of a toroidal shell forming a fluid heating chamber having inlet and outlet fluid circulating ports. A plurality of heat pipes are mounted within the chamber and are attached to the inner wall of the shell such that the condensor sides of the pipe reside within the shell and the evaporator sides extend outside the shell into the center of the toroid. The recuperator is positioned in a flue gas stack wherein the hot flue gas stream contacts the heat pipes which transfer heat into the fluid heating chamber. Fluid, gas or liquid, is passed through the chamber resulting in a rise in temperature of the fluid.

The inventor claims various advantages. For example, there is no contamination of the heated fluid since heat transfer takes place in a sealed chamber. There are no moving mechanical parts since heat pipes are self-pumping. The size of the unit is substantially smaller and lighter than existing devices and smaller blowers are needed to transport the fluids to be heated through the system. Thermal expansion

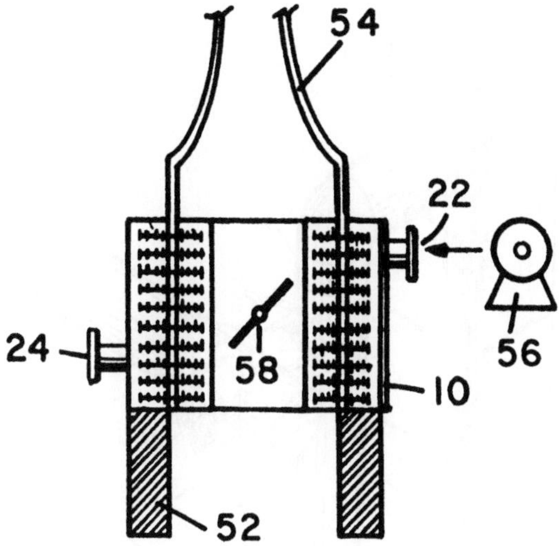

Fig. 358

problems are minimised. The rapid heat transfer provided by heat pipes yields increased efficiency and allows the unit to be used in higher temperature gas streams than would be possible with existing recuperators. See Fig. 358.

Patent (99.2) of Isothermics, Inc. concerns a system for recuperating the residual heat of domestic or industrial heating installations. The system comprises a plurality of heat pipes arranged in the exhaust conduit of the heating installation (oven, combustion chamber). Without affecting in any way the functioning of the exhaust conduit, wherein they are mounted these heat pipes will effect axial heat transfer towards the ends of heat pipes, opposed to those inserted in the conduit. The opposite heat pipe ends, receiving the recuperated heat are enclosed by a second conduit, separated from the first one, by a fluid-impermeable wall. In the second conduit a ventilator is mounted, supplied with air, that is heated by the heat released from the condenser ends of the heat pipes and returning the reheated air into the central heating installation.

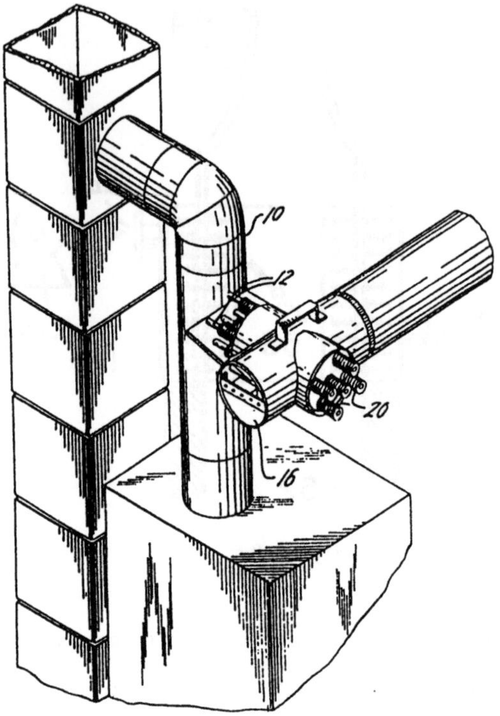

Fig. 359

Fig. 359 illustrates schematically the proposed device, inserted in the exhaust conduit of a combustion chamber.

The heat recuperation system, revealed by Statni Vyzkumny Ustav Pro Stavbu Stroju in Patent (172), comprises a heat exchanger, composed of two sections, containing bundles of heat pipes. The heat pipes in one section are filled with the same medium, consisting of a substance belonging to cooling fluids of low boiling point, cooling fluids of low temperature, fluids of moderate temperature and fluids of high temperature. In the last section of a recuperation heat exchanger stabilising heat pipes are mounted, filled with an inert gas (argon). The evaporator elements of the stabilising heat pipes are arranged in a cold air conduit, while the sectors of the condensation elements, (storage of the inert gas) are disposed outside of the upper structure of the last heat exchanger. The system is capable of operating at temperatures of about 1000°C.

The heat transfer system, disclosed by M. Hage, et al., (Wieland Werke AG.), in Patent (211.2), and consisting of heat exchangers, with heat pipes in bundle arrangement, serves to separate refrigerating fluids from drink water or industrial water, while at the same time indirectly indicating any leakage which may occur in the system, through a reduction of capacity. In order to realise a safe separation of the spaces of refrigerating fluid and drink water, a double wall is provided therebetween. Thus, in case of a leakage (either from the cold fluid or from the water side), one medium will not flow into the other one, but will escape into the atmosphere. See Fig. 360.

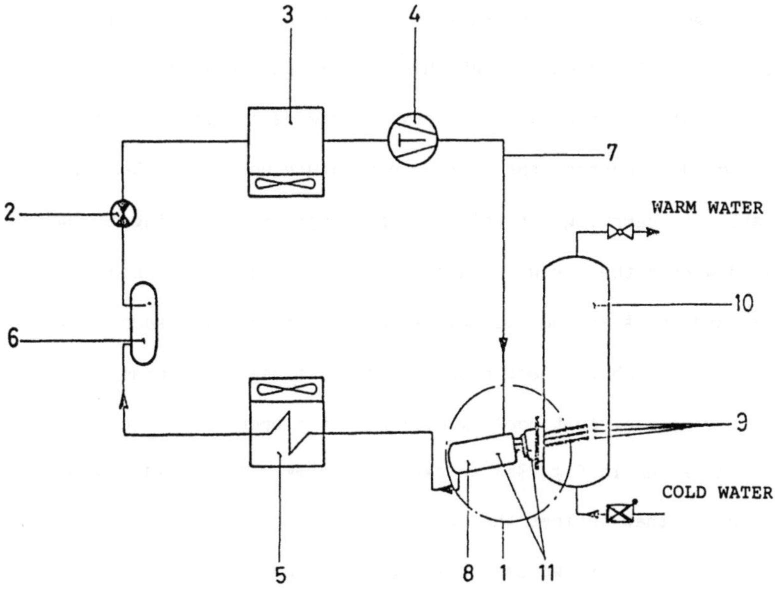

Fig. 360

Molding, extrusion.

According to Patent (94) of J.P. Motte, (Institut de Recherches de la Siderurgie Française (IRSID), the complications caused by aggressive hot slags in the refractory lining of metal smelting equipment can be eliminated by arranging in an upper annular zone of the smelter, lined with refractory material, or in the zone where slag is forming, heat pipes of high thermal conductivity.
The heat pipes comprise a small amount of vaporisable fluid, which evaporates at the hot pipe end and condenses at the cold one and ensure by their operation a satisfactory cooling in the range of the smelting equipment, between the upper and lower levels of slag, thereby reducing the damaging effects thereof.

According to Patent (107) of Kunststoffwerk Gebr. Anger GmbH & Co., the temperature of an extruder screw having an axial bore therein is controlled by providing a closed temperature control system within the screw which is in heat transfer relationship with a second temperature control system through which is circulated a cooling medium from an outside source. The quantity of outside cooling medium may be monitored to control the closed temperature control system. The apparatus for carrying out this process comprises a cooling medium retained within the screw and a heat exchanger extending into the cooling medium. A second cooling medium from an outside source is circulated through the heat exchanger so that heat is transmitted from the closed circuit cooling medium to the open circuit cooling medium. As shown in Fig. 361 an extruder screw 1 is provided with an axial bore 2 that defines the interior space of the screw.
The bore 2 comprises two smooth walled sections of different diameters 2a and 2b. The larger diameter portion 2a has positioned forwardly

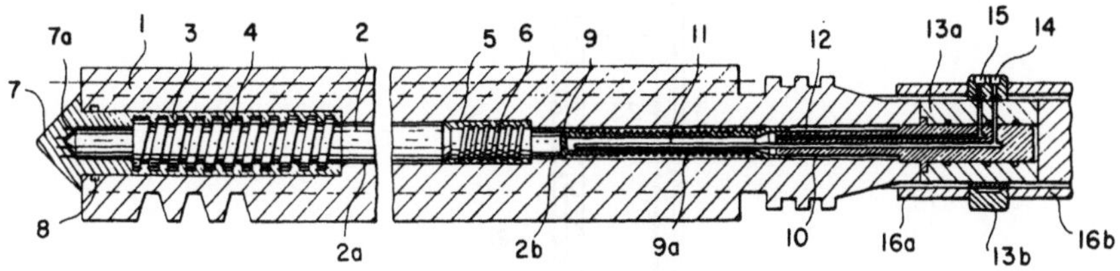

Fig. 361

thereof a large sleeve 3 of a non-ferrous metal, which has good heat conducting properties and comprises an internal helical path 4 or thread to convey the cooling medium from the area of the heat exchanger toward the tip end of the screw. Rearwardly of the smooth wall portion 2a is a smaller sleeve 5 having a helical path 6 for conveying cooling medium and made of the same material as sleeve 3. The inner diameter of the sleeve 3 is substantially equal to the diameter of bore portion 2a.

J.R. Rinderle, (Massachusetts Institute of Technology), discloses in Patent (117) a novel method and device for precision injection molding, providing controlled cooling of a portion of the mold cavity surface during the molding cycle, or several such portions independently, and also, in the preferred embodiment controlled heating of a portion of the mold cavity surface during the molding cycle of several such portions independently, whereby high quality, high precision parts, having close dimensional tolerances, may be produced. In a variant of the invention, a volume-controlled variable conductance heat pipe is disclosed, which novel heat pipe comprises housing means forming a sealed chamber, fluid, such as water or ammonia, within the chamber, wicking means and control means for controlling the thermal conductance of the heat pipe comprising means for controlling the volume of

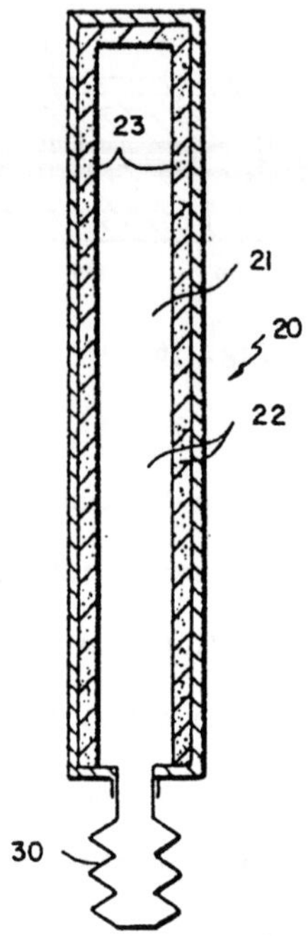

Fig. 362

fluid in the liquid phase in the chamber. Preferably, the novel injection mold comprises a thin mold face supported at least in part by a foundation means which either incorporates, or is itself, a novel volume controlled variable conductance heat pipe.
See Fig. 362.

Soldering element

Patent (197.1) of G.A.A. Asselman, (US Philips Corp.), concerns a soldering device, having a hollow and closed soldering element which contains a heat transporting medium. The soldering element is supported in a holder and may be slidably removed therefrom.
The holder is also provided with an electric element for providing heat to a first heat transmission wall of the soldering element.

The heat transporting medium absorbs thermal energy from the heating element through the first neat transmission wall and changes from the liquid phase into a vapour phase. The vapour supplies the thermal energy to a second heat transmission wall of the soldering element which is at a point of the element forming the soldering place which produces the heat necessary for the soldering operation. The vapour changes back into the liquid phase at the soldering place. A capillary mass of porous material is provided along the inside of the soldering element so that the liquid heat transporting medium can flow back through the porous mass from the second heat transmission wall back to the first heat transmission wall as a result of capillary action. See Fig. 363, wherein: 1 - holder; 2 - grip; 3 - electric heating element; 5 - electric conductor; 6 - soldering element; 7 - first transmission wall; 8 - second transmission wall; 9 - evacuated space; 10 - porous mass.

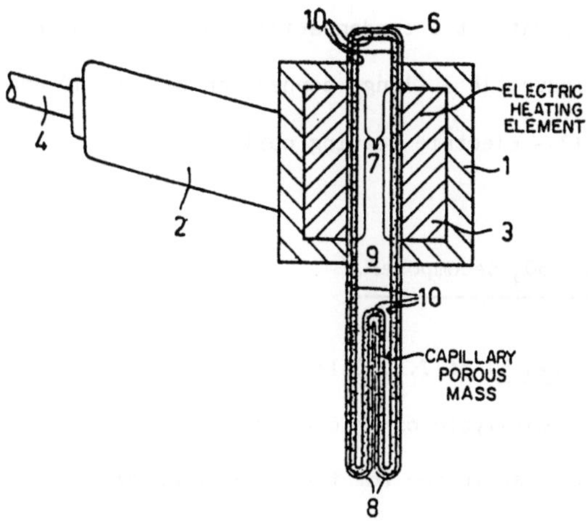

Fig. 363

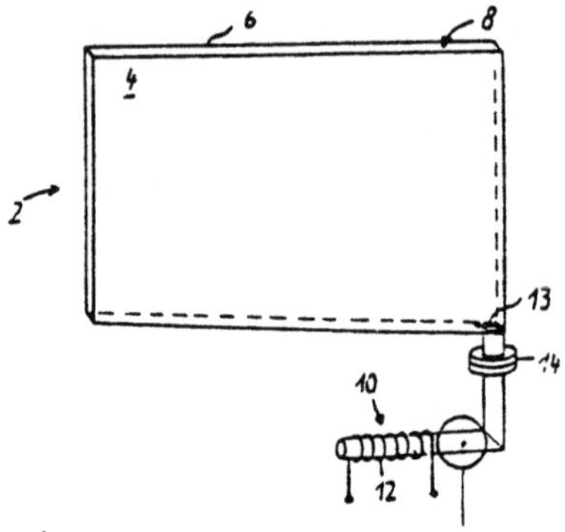

Fig. 364

Composite glass structure.

Patent (47.7) of P. Behrmann, (Dornier System GmbH), refers to composite glass structures with at least two panes, between which a heated space is forming. This space forms the condensation zone of a flat heat pipe, while its evaporator zone is outside of the panes. Between the evaporation zone and the condensation zone a wire tissue is arranged. Fig. 364 shows the transparent glass panes 4, 6, with a heated space 8 therebetween. 8 indicates the condensation zone; 10 - the evaporation zone; 13 - wire tissue; 14 - flange connection between 10 and 2 - the composite glass. 12 - electric heating coil.

Catalytic cartridge SO_3 decomposer.

Patents (193.3 and 195.5) of T.R. Galloway, (The US Department of Energy), disclose a catalytic cartridge (internally heated), which can be used as a SO_3 decomposer for thermochemical hydrogen production. The cartridge has two embodiments, a cross-flow cartridge and an axial flow cartridge. In the cross-flow cartridge, SO_3 gas is flowed through a chamber and incident normally to a catalyst coated tube

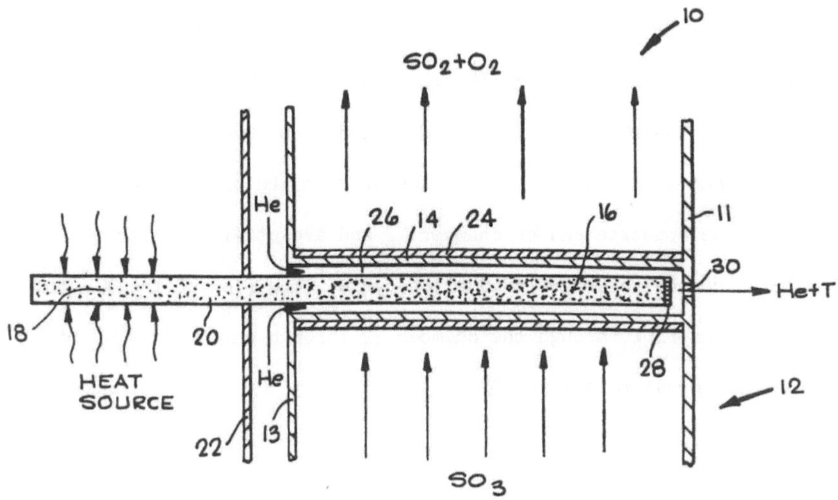

Fig. 365

extending through the chamber, the catalyst coated tube being internally heated. In the axial-flow cartridge, SO_3 gas is flowed through the annular space between concentric inner and outer cylindrical walls, the inner cylindrical wall being coated by a catalyst and being internally heated. The modular cartridge decomposer provides high thermal efficiency, high conversion efficiency, and increased safety.

See Fig. 365, wherein: the tube 14 surrounds the condenser region 16 of a heat pipe 20. The evaporator region 18 of heat pipe 20 is exposed to a heat source such as a fusion reactor or solar collector. The heat pipe transports heat from the heat source through the isolating wall 22 to the catalytic cartridge 10. The tube 14 extending through the chamber 12 is in thermal contact with the condenser region 16 of heat pipe 20 or is separated from the condenser region 16 by a narrow gap 26. A layer 24 of catalyst is coated on the tube 14 and is heated by heat transfer from the heat pipe 20,

being kept at a temperature of about 1070°K. SO_3 gas is flowed in through inlet port 19 at one end of chamber 12 and flowed through the chamber 12 incident to the tube 14 across the heated catalyst layer 24, thereby producing the decomposition reaction of SO_3.
The decomposition products SO_2 + O_2 are removed through outlet port 21 at the opposite end of chamber 12 and are utilized in the thermochemical hydrogen production cycle. The flow of the SO_3 gas is a cross-flow through the chamber 12 through which the tube 14 extends normal to the gas flow.

Cooling of wheels

Patent (23) of E.D. Hemley, (British Railways Board), relates to a wheel having means for cooling the periphery thereof. The invention provides a wheel, including means for cooling the periphery of the wheel, comprising a plurality of closed chambers containing a vapourisable medium, one part of each chamber being in heat conductive contact with the periphery of the wheel, another part of each chamber remote from the one part comprising a heat-emitting condenser part, arranged adjacent the hub of the wheel, the conditions for the medium in each chamber being such that, in operation, vapour is liberated from the medium by heat generated at the periphery which vapour passes to the condenser part, condensing therein to give up its latent heat, so that the latent heat of the medium acts as the cooling agent for the periphery of the wheel.
The invention is applicable to wheels of a traction locomotive.
Fig. 366 shows a metal wheel 3 to which a solid elastomeric tyre 4 is fitted. A heat-conducting chamber 5 formed in a hollow curved metal tube is fitted in a peripheral wheel part 6 adjacent the tyre 4 to

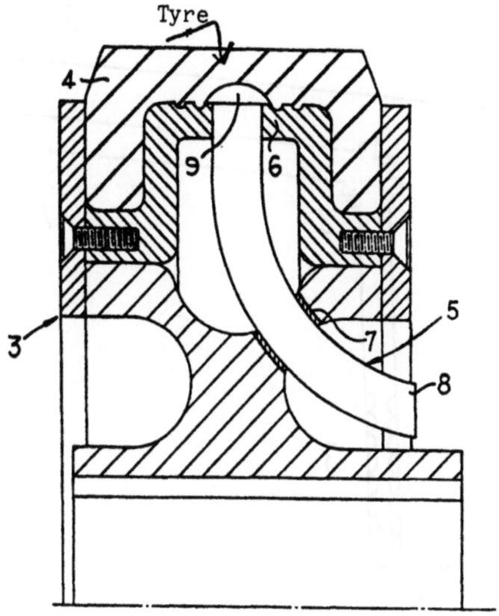

Fig. 366

provide good thermal conductivity between the periphery of the wheel and the chamber. The chamber 5 passes through an orifice in a wall of the wheel where it is supported by packing 7. The end of the chamber 5 remote from the periphery of the wheel forms a heat emitting condenser part 8. The end of the chamber 5 adjacent the tyre 4 is closed by a plug 9. The chamber has 5% of its volume filled with water or other vapourisable liquid and is partially evacuated so that the interior is at less than atmospheric pressure.

Heat pipes used in combination with geothermal systems.

Patent (24) of O. Bschorr concerns the utilisation of the earth's and soil water's heat capacity for preventing the freezing of road covers, landing paths, bridges, water conducts, etc., furthermore, for preventing much too considerable temperature fluctuations (day/night) of, for example, green houses, large spaces and the like.

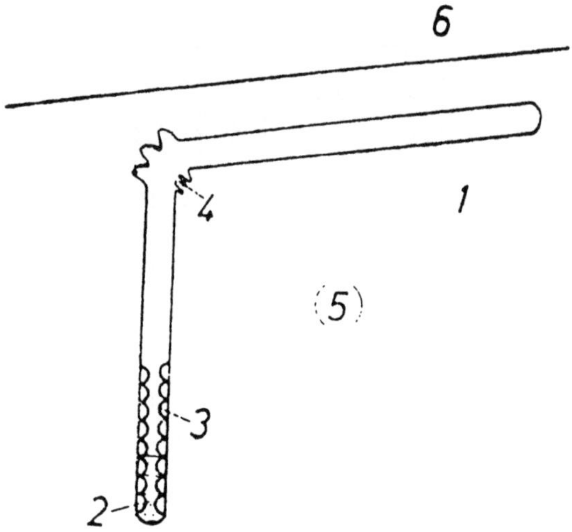

Fig. 367

According to the invention a heat pipe is introduced in a deeper zone of the soil, in combination with a pressure-gradient or temperature-controlled switch (of a bimetal), which activates the heat pipe in the critical temperature range, while inactivating it beyond or under the critical temperature in order to prevent the cooling of the heat reservoir.

Fig. 367 illustrates the basic structure of the proposed pipe system. The object of Patent (48) of W.B. Bienert, (Dynatherm Corp.), is to maintain roadways and other areas free of ice, snow, and slush accumulations in colder weather by utilizing natural earth heat to melt the same, by process that involves depletion of the supply of earth heat, and, in particular, the conservation of earth heat by providing a device effective in warmer weather to augment the depleted supply of heat.

More particularly, the invention provides down-pumping heat pipes to augment natural earth heat when used in association with conventional

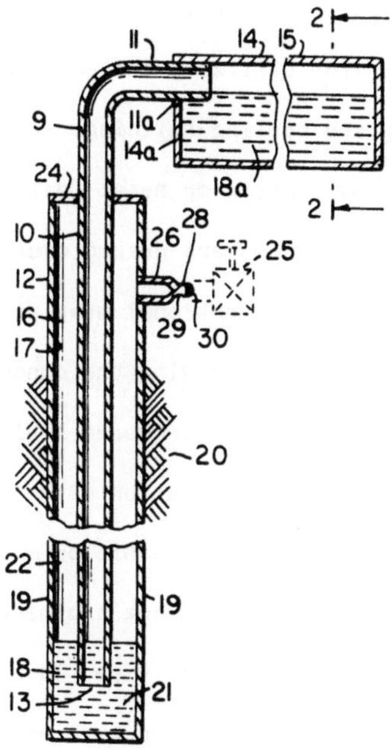

Fig. 368

or up-pumping heat pipes for the purpose of maintaining an area such as a roadway free of ice and snow.
See Fig. 368.

In Patent (187) of B. Tranie, there is provided a geothermal heat pipe-heat pump system, essentially comprising a heat pipe, introduced in a bore hole, the upper end of the heat pipe, which is on ground level, being inserted into a water containing heat exchanger, destined to transferring the geothermal calories, supplied by the heat pipe, to the cold source of a heat pump.
The dimensions of the system depend on local geological conditions.
For example, in case a power of 4 Kw/hr has to be established, a pipe of about 50 m length and 50 mm diameter should be used, since in France the average soil temperature is 14.5^{o}C in a depth of 50 m.

In Patent (198.1) of P.R. Fitzpatrick, (United Technologies Corp.), a method and apparatus for utilizing the earth as a heat source and heat sink for heating and cooling buildings are disclosed. Flexible material, such as hollow plastic or metal foil bags which can be buried in deep, narrow trenches are draped over and clamped to heat transfer fluid conduits or conventional heat pipes, vastly increasing the heat transfer surface area serving the conduits or heat pipes. The heat pipe bags contain a heat transfer fluid and can contain wicks which are in heat transfer communication with the heat transfer fluid conduits or other heat pipes. The plastic or metal of the heat pipe bag similarly encloses the wick forming a closed, sealed system which can also include the conduit or heat pipe. Such conduit-heat pipe bag combination can be used in combination with conventional heat pumps to both heat and cool buildings.

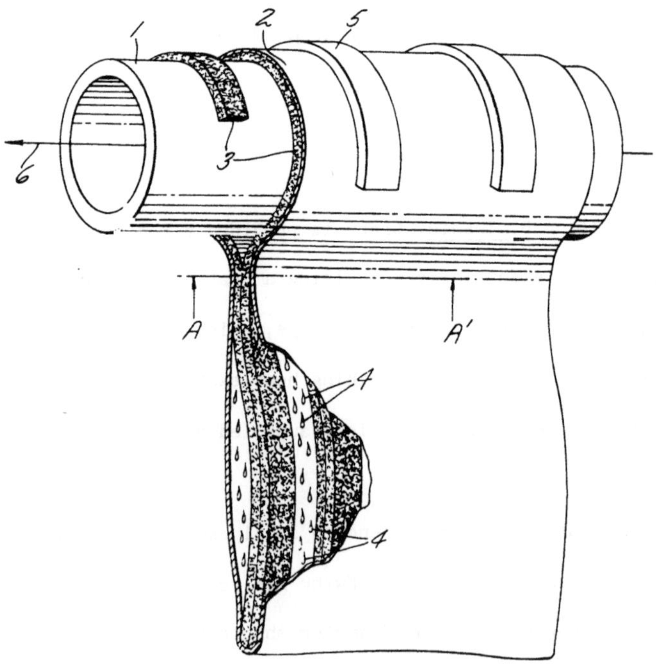

Fig. 369

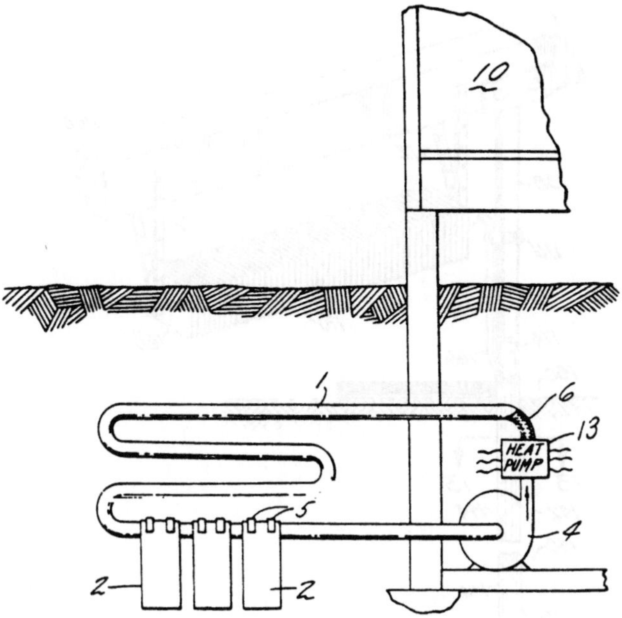

Fig. 370

Fig. 369 illustrates the basic structure of a heat pipe bag with a circulation pipe.

Another Patent (198.2) of P.R. Fitzpatrick refers to a similar system. Fig. 370 illustrates the arrangement with a heat pump included therein.

Support structures, using heat pipes, for arctic or subarctic areas.

Several patents are assigned to McDonnell Douglas Corp., concerning structures, devices for use in permafrost areas or areas having ground layers subject to severe annual freeze-thaw cycles.

Patent (121.1) of E.D. Waters reveals a structural support assembly for use in arctic and subarctic (permafrost) areas including the cooperative combination of a support structure and a heat pipe element installed in generally frozen soil. The heat pipe is of a suitably

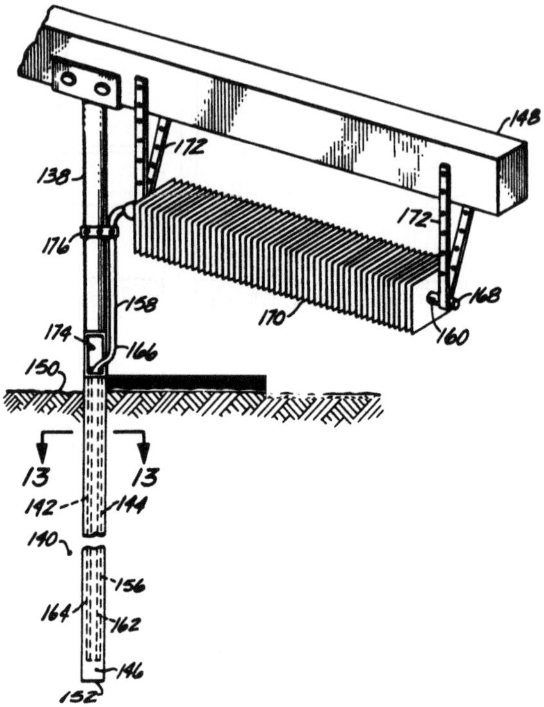

Fig. 371

complementary configuration and/or disposition with respect to the support structure to provide appropriate stabilization of the surrounding frozen soil. In one embodiment, the heat pipe element is disposed externally of the support structure and, in another embodiment, it is disposed internally of such structure. The external embodiment further includes one version employing a linear (straight) heat pipe element and another version employing a helical element. The internal embodiment further includes one version wherein a heat pipe is integrally combined with a support structure and another version wherein a heat pipe is cooperatively installed inside a support structure. See Fig. 371.

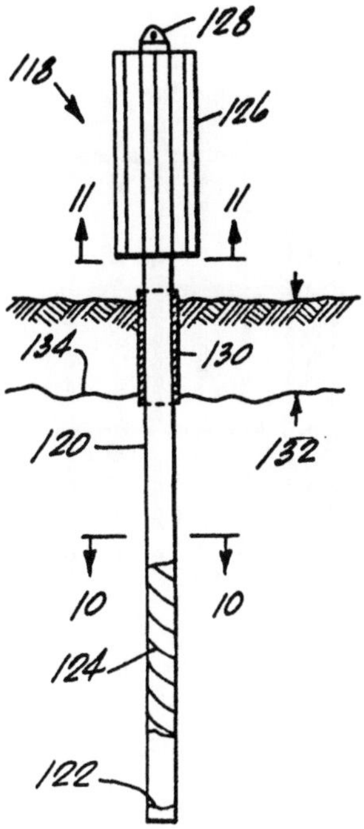

Fig. 372

Another Patent (121.2) of E.D. Waters, (McDonnell Douglas Corp.), also provides a support assembly for arctic and subarctic areas. In an embodiment of (121.2) the heat pipe element is disposed externally of the support structure and, in another embodiment, it is disposed internally of (and integrally combined with) such structure. The external embodiment further includes one version, employing a linear (straight) heat pipe element and another version, employing an angular element. See Fig. 372.

Using heat pipe type heat exchangers in lakes.

L. Backlund, (Forenade Fabriksverken), reveals in Patent (63) a heat pipe type heat exchanger, which can be placed in lakes, wherein the heat exchangers are anchored in a stable manner. More particularly,

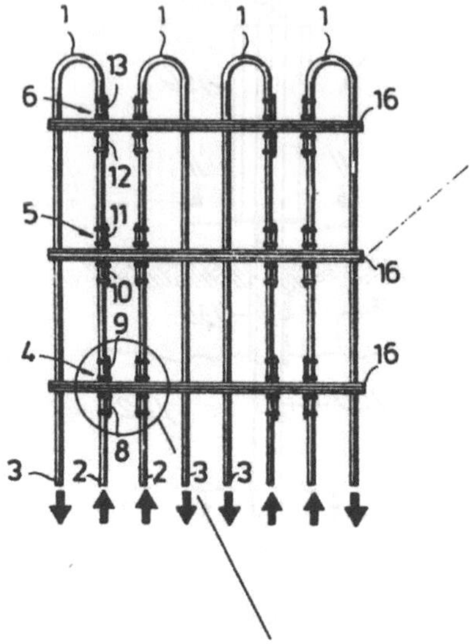

Fig. 373

the coldest parts of a heat exchanger are connected to the end of one or more heat pipes, the other end of which is anchored in the bottom of the lake, so that the heat pipes transfer the heat of the lake bottom to the heat exchanger, as a consequence of which the heat pipes will be solidly frozen into the lake bottom, thereby realising a stable anchoring of the heat exchanger system.

Fig. 373 illustrates the system, wherein 1 indicates the heat pipes of plastic material; 2 - the feed end of the pipes, through which a low freezing point (minus $4^{o}C$) fluid is fed into the pipes; the fluid leaving the pipes at 3, the connection between heat pipes and heat exchangers being realised at 4, 5, 6.

Fabrication process of heat pipe panels.

R.A. Haslett, (Grumman Aerospace Corp.), reveals in Patent (82.10) a process for making a heat pipe panel, comprising the steps of

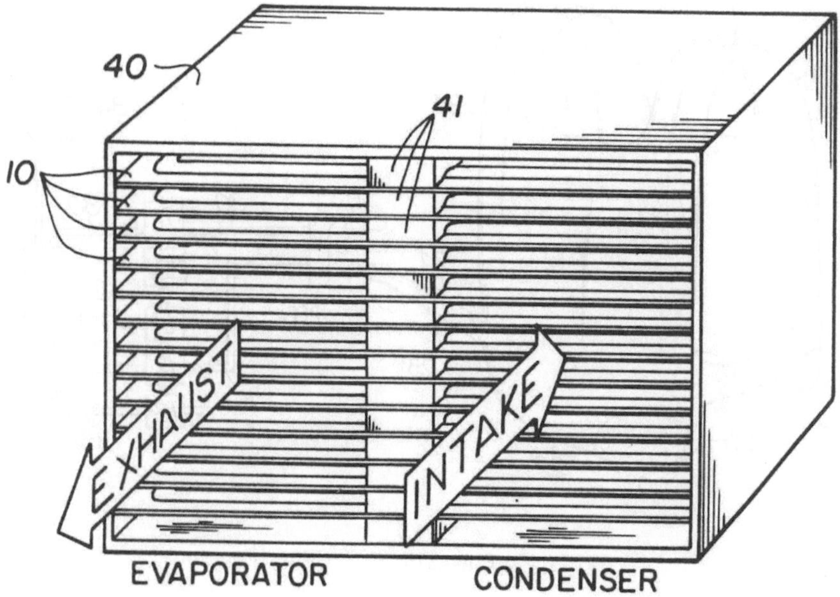

Fig. 374

forming parallel furrows in a sheet metal plate, covering the sheet metal plate with another plate, bonding the plate together around the edges and between the furrows, to form parallel cavities between the plates, charging the cavities with working fluid and sealing the cavities to form individual heat pipes in the panel. The furrows may be filled with wicking material, and capillary grooves may be cut in the furrows before bonding the plates together. The cover plate also may have furrows formed therein. A header furrow may be provided for the fluid charging operations. See Fig. 374.

Thermal conditioning of garments.

Patents (158.4; 158.5) of M.J. Graveney, et al. (assignors to the U.K. Secretary of State for Defence), relate to heat pipes constituted

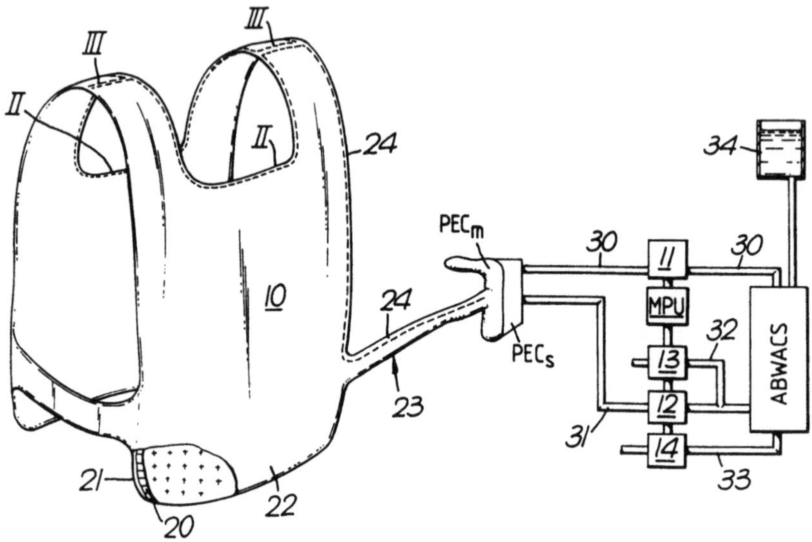

Fig. 375

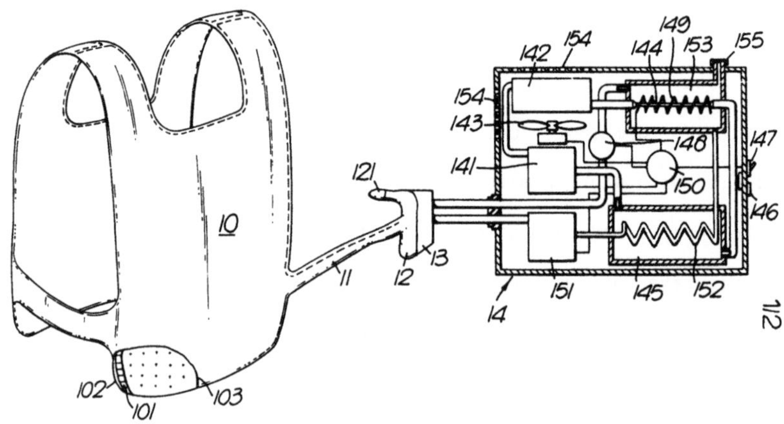

Fig. 376

as thermal conditioning garment assemblies for personnel use in hot environments. It is concerned with the provision of thermal conditioning in circumstances where some power, but not necessarily a lower temperature environment, is available for heat sink purposes.

According to Fig. 375 in (158.4) a heat pipe assembly comprises a heat exchanger of heat pipe form, e.g. a thermal conditioning garment 10, connected via a releasable coupling (PEC) to a heat pump. Liquid from the heat pump is pumped by a pump 11 through the coupling into a tube 24 from which it passes into the garment.
The resulting vapour passes through the coupling, a duct 31 and a compressor 12 into an evaporative condenser (ABWACS).

Patent (158.5) reveals a heat pipe assembly comprising a first heat exchanger of heat pipe form, e.g. a thermal conditioning garment 10, and a thermal reservoir 14, which may include a heat pump 141, 142, 144, 145.
Means are provided for evacuating at least the first exchanger, which is connected to the reservoir by means of a releasable coupling.
See Fig. 376.

LIST OF PATENTEES

Abbreviations used:

PA = patent application; PS = patent specification; AT = Austria; AU = Australia; BE = Belgium; CA = Canada; CH = Switzerland; DE = Western Germany; EU = Europe (Munich); FR = France; GB = Great Britain; IT = Italy; JP = Japan; LU = Luxemburg; NL = The Netherlands; PCT = WIPO; SE = Sweden; US = United States of America.

				Page
121)	McDonnell Douglas Corporation	8)	US-PS 4,351,388	
		9)	US-PS 4,444,249	61
		10)	US-PS 4,461,343	28
122)	Meckler, G.		US-PS 4,027,653	163
123)	Millar, J.H.	1)	GB-PA 2,053,455	164
		2)	GB-PA 2,054,826	165
124)	Motorola Inc.		US-PS 4,377,198	295
125)	Nagron Steel and Aluminium B.V.		EU-PA 0 099 155 = NL-PA 82,02713 + NL-PA 82,02714	166
126)	National Nuclear Corporation Ltd.	1)	GB-PA 2,128,319	4
		2)	GB-PA 2,128,320	4
127)	National Research Development Corp.		EU-PA 0 059 581 = GB-PA 81,06782	66
128)	NCR Corporation		US-PS 3,971,435	50
129)	Nissan Motor Co. Ltd.		DE-PS 2,602,933 = JP-PA 50-11628/75	62 + 111
130)	Nolan Design Limited, John		GB-PS 1,583,857	6
131)	O'Hanlon, E.J.		US-PS 4,089,366	167
132)	Otdel Fiziko-Tekhnicheskikh Problem Energetiki Uralskogo Nauchnogo Tsentra Akademii Nauk SSSR	1)	US-PS 4,467,861	5
		2)	DE-PS 3,301,998 = FR-PA 2,540,613	6 + 113
133)	Pa Management Consultants Limited		GB-PS 1,503,903	169
134)	Passarelli, F.J.		US-PS 4,335,706	170
135)	Perkin-Elmer Corporation, The		US-PS 3,892,273	39
136)	Philips' Gloeilampenfabrieken, N.V.	1)	DE-PS 2,517,080 = US-PS 4,033,130 = NL-PA 74,05947	171
		2)	GB-PA 2,013,870 = DE-PA 2,804,411	172
		3)	GB-PA 2,024,408 = NL-PA 78,07174	173
		4)	GB-PA 2,053,453 = NL-PA 79,05057	173

				Page
136)	Philips' Gloeilampenfabrieken, N.V.	5)	GB-PA 2,053,456 = NL-PA 79,05158	174
		6)	GB-PA 2,107,035 = NL-PA 81,04565	272
		7)	FR-PS 2,300,220 = NL-PA 75,01273	223
		8)	FR-PS 2,349,806 = DE-PA 2,618,651	175
		9)	FR-PS 2,370,251 = NL-PA 76,12360	37
		10)	FR-PS 2,438,244 = NL-PA 78,10047	274
		11)	FR-PS 2,434,347 = NL-PA 78,08774	177
		12)	EU-PA 0 000 217 = NL-PA 77,06880 + NL-PA 77,14306	274
		13)	EU-PA 0 013 777 = NL-PA 79,00621	177
		14)	EU-PA 0 015 017 = NL-PA 79,01030	178
		15)	EU-PA 0 054 319 = NL-PA 80,06716	178
		16)	EU-PA 0 054 320 = NL-PA 80,06717	179
		17)	EU-PA 0 061 222 = NL-PA 81,01442	180
		18)	EU-PA 0 066 315 = NL-PA 81,02468	178
		19)	EU-PA 0 070 580 = NL-PA 81,02619	173
		20)	EU-PA 0 109 716 = NL-PA 82,04482	181
137)	Philips Patentverwaltung GmbH		DE-PS 3,214,886	
138)	Phillips, M.E.		US-PS 4,076,016	182
139)	Pöhlmann, E.	1)	DE-PS 2,851,867	183
		2)	DE-PS 3,120,520	184

MIX
Papier aus verantwortungsvollen Quellen
Paper from responsible sources
FSC® C105338

If you have any concerns about our products,
you can contact us on
ProductSafety@springernature.com

In case Publisher is established outside the EU,
the EU authorized representative is:
Springer Nature Customer Service Center GmbH
Europaplatz 3, 69115 Heidelberg, Germany

Printed by Libri Plureos GmbH
in Hamburg, Germany